(*continued on back*)

Detectors for Capillary Chromatography

CHEMICAL ANALYSIS

A SERIES OF MONOGRAPHS ON
ANALYTICAL CHEMISTRY AND ITS APPLICATIONS

Editor
J. D. WINEFORDNER
Editor Emeritus: **I. M. KOLTHOFF**

VOLUME 121

A WILEY-INTERSCIENCE PUBLICATION

JOHN WILEY & SONS, INC.

New York / **Chichester** / **Brisbane** / **Toronto** / **Singapore**

Detectors for Capillary Chromatography

Edited by

HERBERT H. HILL

Department of Chemistry
Washington State University
Pullman, Washington

and

DENNIS G. McMINN

Department of Chemistry
Gonzaga University
Spokane, Washington

A WILEY-INTERSCIENCE PUBLICATION

JOHN WILEY & SONS, INC.

New York / Chichester / Brisbane / Toronto / Singapore

Copyright © 1992 by John Wiley & Sons, Inc.

Library of Congress Cataloging in Publication Data:

Detectors for capillary chromatography / edited by H. H. Hill and D. G.
McMinn.
   p. cm. — (Chemical analysis ; v. 121)
"A Wiley-Interscience publication."
Includes bibliographical references and index.
ISBN 0-471-50645-1 (cloth)
1. Gas chromatography. 2. Detectors. I. Hill, H. H., (Herbert
H.) II. McMinn, Dennis Gordon, 1944– . III. Series.
[DNLM: 1. Chromatography, Gas—methods. QD 79.C4 D479]
QD79.C45D45 1992
543'. 0896—dc20
DNLM/DLC   92-29
for Library of Congress   CIP

Printed in United States of America

10 9 8 7 6 5 4 3 2 1

CONTRIBUTORS

Fikry F. Andrawes, American Cyanamid Company, Stamford, Connecticut

Hiromi Arimoto, Analytical Application Department, Shimadzu Corporation, Nakagyo-ku Kyoto, Japan

Darryl J. Bornhop, Citation Medical Corp., Reno, Nevada

Raymond E. Clement, Ontario Ministry of Environment LSB, Rexdale, Ontario, Canada

Jos de Wit, Tennessee Eastman Co., Kingsport, Tennessee

John N. Driscoll, HNU Systems Inc., Newton, Massachusetts

Toshihiro Fujii, Division of Chemistry and Physics, National Institute for Environmental Studies, Tsukuba, Ibaraki, Japan

Eric P. Grimsrud, Department of Chemistry, Montana State University, Bozeman, Montana

Donald F. Gurka, Environmental Monitoring Systems Laboratory, Office of Research & Development, U.S.E.P.A., Las Vegas, Nevada

Randall C. Hall, Randy Hall & Associates, Cloudcroft, New Mexico

Herbert H. Hill, Department of Chemistry, Washington State University, Pullman, Washington

Richard S. Hutte, Sievers Research Inc., Boulder, Colorado

James W. Jorgenson, Department of Chemistry, University of North Carolina, Chapel Hill, North Carolina

Dennis G. McMinn, Department of Chemistry, Gonzaga University, Spokane, Washington

Paul L. Patterson, Detector Engineering and Technology, Walnut Creek, California

Roswitha S. Ramsey, Oak Ridge National Laboratory, Oak Ridge, Tennessee

John D. Ray, Sievers Research Inc., Boulder, Colorado

Eric J. Reiner, Ontario Ministry of Environment LSB, Rexdale, Ontario, Canada

Bruce E. Richter, Lee Scientific Division of Dionex Corporation, Salt Lake City, Utah

Peter C. Uden, Department of Chemistry, University of Massachusetts at Amherst, Amherst, Massachusetts

PREFACE

Numerous articles and books have been written detailing the benefits that capillary chromatography offers in the area of separation and analysis. This book does not attempt to reproduce that information. Rather, it focuses on the detection methods commonly used with capillary chromatography and describes the unique features of each detector which are required for successful interfacing with capillary columns.

Over the past few years, fused silica column technology has advanced to the point where tremendous separation power is available. However, without the ability to "see" the separated components, this high resolution would be wasted. It is true that the column is the heart of the chromatographic process, but detectors are the "eyes" that make the separation visible. A number of methods for "looking" at the results of chromatographic separations have been developed, and each provides a different "view." To understand the information gained in high resolution chromatography, one must be able to interpret correctly the view provided by the detector. We hope this book will provide some insights into how that can be accomplished.

The authors of the chapters in this volume are all experts in their field. In many cases, they are the researchers who originally developed the detector, or they have made major contributions to understanding its operation. With twenty different contributors, some overlap is inevitable. We view this as a positive feature, and in most cases have not attempted to alter the presentations within each chapter. We wish to express our appreciation to these authors for their patience and help in putting this book together.

HERBERT H. HILL
DENNIS G. MCMINN

Pullman, Washington
Spokane, Washington
June 1992

CONTENTS

Detectors for Capillary
Chromatography

DETECTION IN CAPILLARY CHROMATOGRAPHY

HERBERT H. HILL

Department of Chemistry
Washington State University
Pullman, Washington

DENNIS G. McMINN

Department of Chemistry
Gonzaga University
Spokane, Washington

This book marks the thirty-fifth anniversary of two major developments in chromatography that have served as the basis for modern high-resolution chemical separation methods. During the years 1957 and 1958 reports first appeared describing the potential of capillary columns for increasing chromatographic efficiency. Coincidentally, detection methods were first described that had sufficient sensitivity and response times to track concentration profiles of components eluting from capillary columns.

Using gas chromatography, Golay (1, 2) demonstrated that small-diameter open tubular columns in place of those packed with small particles significantly reduced eddy and multiple path band broadening during the separation process. While Golay was describing the concept of capillary chromatography, several other researchers were laying the groundwork for methods to detect the highly resolved components. McWilliam and Dewar (3) along with Harley et al. (4) demonstrated the use of flame ionization (Chapter 2) for the detection of organic compounds. In addition, Lovelock (5) introduced a sensitive ionization method that was to become the predecessor to both the helium ionization detector (Chapter 3) and the electron capture detector (Chapter 5). This combination of high separation efficiency with rapid and sensitive on line

Detectors for Capillary Chromatography, edited by Herbert H. Hill and Dennis G. McMinn.
Chemical Analysis Series, Vol. 121.
ISBN 0-471-50645-1 © 1992 John Wiley & Sons, Inc.

detection has, over the last 35 years, revolutionized analytical methods for the determination of components in complex mixtures.

1.   DETECTOR REQUIREMENTS

The evolution of high-resolution chromatography has brought with it stringent requirements for rapid, low-volume, sensitive, and selective methods of detection. The rapid elution of narrow chromatographic zones requires a fast response time by the detector to accurately track the concentration profile of the sample as it elutes from the column and a low detector cell volume to ensure that the resolution gained from the separation is not lost. In addition, reduced sample capacities of capillary columns limit the quantities of sample that can be introduced into the chromatograph and thus increase the need for high detector sensitivity. High-resolution separation from capillary columns also means that more complex samples can be separated, thus necessitating the use of detectors that respond selectively to compounds of specific interest.

1.1.   Sensitivity

Detectors for capillary chromatography must have high sensitivity. Here, the term *sensitivity* refers to the change in detector response as a function of the change in the amount or concentration of the analyte. That is,

$$S = dR/dC$$

or

$$S = dR/dQ$$

where S is sensitivity, R is detector response, C is concentration of the analyte in the detector, and Q is the total quantity of the analyte in the detector. Some detectors are responsive to concentration whereas others respond to quantity. Detector sensitivity is best measured as the slope of the calibration graph, a plot of detector response vs. analyte concentration or analyte quantity.

The range over which the detector sensitivity is constant is called the linear dynamic range, and the entire range over which response varies with concentration or quantity is called the dynamic range of the detector. The upper limit of the dynamic range is determined when detector sensitivity falls to zero and the detector is said to be saturated. The lower limit of the dynamic range occurs at the detection limit.

For most detection methods used in capillary chromatography, the saturation point of the detection method usually occurs above the saturation point of the capillary column. Thus, for the combined chromatography–detector system, the upper limit of the operating dynamic range is determined by chromatographic conditions whereas the lower limit of the operating dynamic range is determined by detector conditions.

This lower limit of detection is a function not only of the detector sensitivity but of detector noise. Detector noise is defined as the standard deviation of the detector response when no sample is present and is referred to as the root-mean-square noise (N_{rms}). The detection limit is defined as the quantity or concentration required to produce a response which is three times the detector noise. Detection limit (D_L), noise (N_{rms}) and sensitivity (S) have the simple relation

$$D_L = 3N_{rms}/S$$

1.2. Selectivity

Because high-resolution chromatography is capable of separating hundreds of compounds from single mixtures, it is sometimes practical to detect only those compounds that are of interest for a particular analysis. Thus detectors have been developed that respond selectively or specifically to certain classes of compounds. For example, compounds containing certain elements such as Si, N, P, S, and Cl can be detected selectively in the presence of other species not containing these elements. Molecular mass, ionic mobility, optical emission, and optical absorbance are also used as the basis for selective detection.

The selectivity of a given compound over a potentially interfering compound can be measured by the ratio of the detector sensitivities. Selectivity (SEL) is reported in terms of relative molar response (RMR) or as relative weight response (RWR):

$$SEL = S_1/S_2$$

where S_1 is the detector sensitivity of the compound of interest and S_2 is the detector sensitivity of the potentially interfering compound. Because selectivity is a unitless number, it is best to define it as either RMR or RWR. When selectivity is greater than three orders of magnitude for most potentially interfering compounds, it is sometimes referred to as specificity and the detector is said to be specific for that compound or class of compounds.

1.3. Post–Column Band Broadening

A primary requirement of capillary chromatographic detectors is that post–column band broadening be minimized in order to retain the integrity of the chromatographic separation. Several dispersive processes can contribute to post–column band broadening: (1) longitudinal diffusion; (2) parabolic velocity profile dispersion; (3) stagnant mobile phase pools; and (4) detector volume.

Careful attention must be paid to interfacing detectors to capillary columns if these dispersive processes are to be minimized or eliminated. Short and low-volume transfer lines from the column to the detector aid in reducing the effects of longitudinal diffusion and velocity profile dispersion. Minimization of stagnant mobile phase areas can be accomplished by designing the interface such that the entire transfer volume is continuously swept by the mobile phase. With many detectors, these transfer broadening processes can be nearly eliminated by inserting the column directly into the detector. When direct insertion is possible, the major contribution to post–column band broadening becomes the volume of the detection zone.

The generally accepted rule is that the volume of the detector should not be greater than 5% of the peak volume. This requirement can be severe, since typical peak volumes are on the order of 10^{-6} L for gas chromatography and 10^{-9} L for liquid chromatography. A 5% increase in the peak volume due to the detection volume will result in a 10% decrease in separation efficiency.

1.4. Response time

Significant distortions of peak shape can also occur if response times of the detector are too slow. Response times required for the detector in which the peak is distorted by no more than 1% can be determined from the following relation:

$$R_t = t_r/N^{1/2}$$

where R_t is the detector response time, t_r is the retention time of the chromatographic peak, and N is the number of theoretical plates of the column. Thus, for a peak eluting in 10 min with an efficiency of 100,000 plates, a minimum detector response time of less than 2 s is required. Modern chromatographic detectors fall well within these requirements.

2. DETECTOR CLASSIFICATIONS

A complex array of detection methods has been developed that are both sensitive to and selective for compounds eluting in narrow bands from capillary chromatographs. The following chapters provide an introduction to many of the most common of these capillary chromatography detection methods.

These detectors can be classified in a variety of ways. They can be grouped according to the type of chromatographic application, response mechanism, or degree of selectivity.

The overall order of this book presents the detectors according to their chromatographic application: Chapters 2–13 focus on detection methods used in gas chromatography; Chapters 14 and 15 provide reviews of detection methods used in supercritical fluid chromatography and liquid chromatography, respectively.

The detectors used for capillary gas chromatography are grouped according to both selectivity and mechanism of response. Nonselective detection methods are presented first (Chapters 2–4), followed by selective and specific detection methods (Chapters 5–10). Chapters 11–13 discuss the use of dispersive detection methods for both quantitative and qualitative analysis.

Gas chromatographic detectors are also arranged according to mechanism of response. Ionization detectors are by far the most common and are presented in Chapters 2–8. Optical detection methods are presented in Chapters 9–11. Ion dispersive detection methods are presented in Chapters 12 and 13.

While this book has been organized in a manner such that reading the chapters in order will provide a logical progression (gas to supercritical fluid to liquid chromatographic detectors; or nonselective to more selective to dispersive detection methods; or ionization to optical detection), each chapter is meant to stand on its own so that the book can be read selectively or in any order best suiting the requirements of the reader.

REFERENCES

1. M. J. E. Golay, *Anal. Chem.* **29**, 928 (1957).
2. M. J. E. Golay, in *Gas Chromatography 1958* (D. M. Desty, ed.), p. 36. Butterworth, London, 1958.
3. I. G. McWilliam and R. A. Dewar, *Nature (London)* **181**, 760 (1958).
4. J. Harley, W. Nel, and V. Pretorius, *Nature (London)* **181**, 177 (1958).
5. J. E. Lovelock, *J. Chromatogr.* **1**, 35 (1958).

CHAPTER

2

THE FLAME IONIZATION DETECTOR

DENNIS G. McMINN

Department of Chemistry
Gonzaga University
Spokane, Washington

HERBERT H. HILL

Department of Chemistry
Washington State University
Pullman, Washington

The flame ionization detector (FID) is undoubtedly the most common detector available for use in the analysis of trace levels of organic compounds. It was one of the early detectors used with packed column gas chromatography and has been easily adapted for use with capillary columns. As a universal detector for carbon-containing compounds, it responds with high sensitivity and is relatively forgiving in the sense that modest changes in flow, pressure, or temperature have only a small effect on the response characteristics. When the FID is properly installed, there is a stable baseline, especially when used with modern low-bleed columns. Because of its sensitivity and linearity, the FID continues to be used for the majority of routine applications involving capillary gas chromatography and new uses continue to be developed. As discussed elsewhere in this volume, it has more recently been applied to supercritical fluid chromatography (SFC) and liquid chromatography.

Many suppliers of commercial detectors provide kits for converting packed column chromatographs to accept capillary columns and to allow the introduction of makeup gas (1). Nonetheless, there are more stringent requirements on the detector when capillary columns are used because of the reduced sample size. In addition, a faster response time is required because of improved resolution and shorter retention times. Use of modern transistor based

Detectors for Capillary Chromatography, edited by Herbert H. Hill and Dennis G. McMinn.
Chemical Analysis Series, Vol. 121.
ISBN 0-471-50645-1   © 1992 John Wiley & Sons, Inc.

electrometers has allowed display of the low (picoamperes to nanoamperes) current produced in the flame.

1. BACKGROUND

Developed by McWilliam and Dewar (2) in Australia and almost simultaneously by Harley et al. (3) in South Africa, the FID utilizes a mixture of hydrogen, nitrogen, and the column effluent. This fuel-rich mixture is combusted in air at the exit of a flame-jet, producing ions that are collected at an electrode to produce an increase in current proportional to the amount of carbon in the flame. Since the magnitude of the ion current is independent of the sign of the polarization voltage (4), instruments that collect positive ions or negative ions are available commercially. As normally configured, the FID is insensitive to inorganic gases and to compounds such as CO, CO_2, and CS_2.

2. MECHANISM OF RESPONSE

Over the past several years, much attention has been paid to the mechanism responsible for the ion current in the FID. Many of the proposed steps are based on experimental results that come from mass spectral analysis of premixed flames. The generally accepted mechanism (5) involves a series of pyrolytic reactions that begin at the tip of the flame jet. Since the oxygen is provided by diffusion from the side of the flame, it is proposed that materials eluting from the column undergo degradation reactions in the hydrogen-rich region of the flame to yield a group of single carbon species. This suggestion is supported by the experimental observations that the response to various hydrocarbons is equal for each gram-atom of carbon combusted (the "equal per carbon response") and that response is linear with concentration over many orders of magnitude (6). A series of reactions with hydrogen producing an equilibrium mixture of single carbon species is likely:

$$CH_3 + H = CH_2 + H_2$$
$$CH_2 + H = CH + H_2$$
$$CH + H = C + H_2$$

As these radicals move through the flame into a zone containing oxygen, the following chemi-ionization reaction occurs:

$$CH + O = CHO^+ + e$$

The CHO^+ ions are unstable and react rapidly with water produced in the flame to generate hydroxonium ions which are the primary positive charge carrying species.

$$CHO^+ + H_2O = H_3O^+ + CO$$

Mass spectroscopic measurements have supported the chemi-ionization reaction by revealing the presence of H_3O^+ above the flame and the absence of CHO^+ (7). The current is carried by clusters of hydrated protons $(H_2O)_n H^+$. Mobility measurements have given an apparent value of $n = 2.2$ (7). One such charged species is produced for approximately every 100,000 carbon atoms introduced, which enables the flame to act as a carbon counter. In the upper part of the flame, several other ions have been observed. None are considered important as charge carriers.

Less well understood is the process by which the organic materials are broken down into single carbon species. Sternberg et al. (8) have suggested that partial pyrolysis near the outer boundary of the oxygen-free precombustion zone followed by Rice–Herzheld cracking processes (9) initiated by back–diffusion of hydrogen from the main reaction zone could produce the required species. Blades (10) questioned the Rice–Herzfeld portion of this suggestion because it requires alkene products that are not consistent with the requirement of single carbon products. Experiments using the hydrogen atmosphere flame ionization detector (HAFID; see Section 6, below) similarly do not support the Rice–Herzfeld postulate (11). Peeters et al. (12) have proposed oxidative degradation. Nicholson (13) used a computer simulation of proposed thermal reactions with alkanes, ethers, and alcohols and determined that the calculated yield of methane was consistent with the relative ionization yield. He suggested that these additives give methane as a major product, with negligible contributions from other species and that the H-atom cracking mechanism was relatively unimportant. He further suggested that methane decomposes without producing CH_3. However, in a later paper, Blades (14) determined an effective carbon number of 0.5 for trioxane and argued that because its normal decomposition product is formaldehyde (which gives a trivial response in the FID) some other process must be operating. At this point it seems that the safest conclusion is that of Blades: "the H-atom induced decomposition is a reasonable candidate for this process" (14), but the Rice–Herzfeld scheme is not likely a part of the overall process.

Less attention has been paid to the negative ions found in the flame, although their initial formation is presumed to be by attachment of electrons (formed during the chemi-ionization reaction) to molecules or radicals in the flame. A plausible mechanism would include formation of OH^- and O_2^-

followed by clustering and charge exchange reactions with CO_2 to give CO_3^-, HCO_3^-, and HCO_4^-, which are apparently not hydrated (15, 16).

3. OPTIMIZATION

Installing a capillary column into an FID is a relatively simple matter given the inherent straightness of the fused silica. Most workers score and discard the last centimeter or so of the end of the capillary column in order to ensure that ferrule particles do not remain. This also serves to provide a smooth surface, thus minimizing sample adsorption, which otherwise would result in anomalous peak shapes. One should be aware that the position of the column in the detector is important, as shown in Figure 2.1 (17), which demonstrates the improvement in peak shape that comes from having the capillary column extend to within a few millimeters of the FID flame. A position much below the flame tip allows the eluting materials to contact metal surfaces, with accompanying loss of peak integrity. On the other hand, if the fused silica column extends into the flame, the polyimide coating will decompose, resulting in excess noise or spiking (17).

The design of the jet is important so that the flows do not convolute and cause a turbulent flame. If the flame is too small, diffusion limited combustion

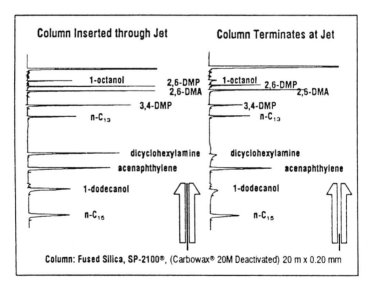

Figure 2.1. The effect of capillary column positioning on the quality of the chromatographic data. Reproduced with permission from K. J. Hyver, ed., *High Resolution Gas Chromatography*, 3rd ed. Copyright 1989, Hewlett-Packard Company.

is retarded and the flame cannot be kept above the ignition temperature. If flows are too low, the flame burns out before being fed from the jet; if too high, tongues of flame are produced that result in an unwanted turbulence (18). In general, use of narrow-bore jet tips gives a more sensitive response with capillary columns, but such tips are somewhat prone to plugging or contamination, especially if silylated derivatives are being chromatographed. Excessive peak tailing observed when halogenated solvents are used has been shown to be due to traces of involatile or polymeric materials in the flame tip, and performance can be restored by sonication of the tip (19). The detector housing typically is heated to prevent condensation of water, although the exact temperature is not too important as long as it is above the maximum operating temperature of the oven. The housing is constructed so as to allow the heat and gases produced to be removed effectively.

Although one of the major advantages of the FID is the fact that it gives satisfactory performance even if not truly optimized, attention should be paid to optimization and calibration since variations in detector design may require different flow rates of hydrogen and air. In addition, this allows one to compare sensitivities of different detectors (20) and judge performance on a day-to-day basis.

In most applications, addition of a makeup gas is recommended in order to optimize the detector (see Figures 2.2 and 2.3) and to sweep the detector volume, thus minimizing band spreading. In addition, at the low flow rates encountered with capillary columns, the response is sensitive to flow rate

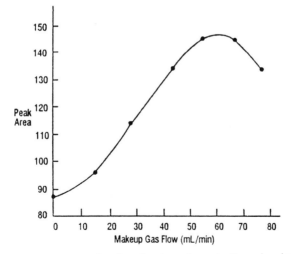

Figure 2.2. Detector response as a function of makeup flow rate. Reproduced with permission from K. J. Hyver, ed., *High Resolution Gas Chromatography*, 3rd ed. Copyright 1989, Hewlett-Packard Company.

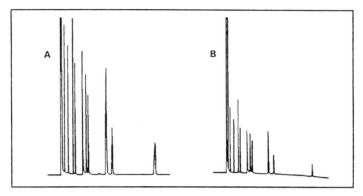

Figure 2.3. Detector response as a function of total carrier gas flow: (A) He makeup at 48 mL/min; (B) no makeup gas. Reproduced with permission from K. J. Hyver, ed., *High Resolution Gas Chromatography*, 3rd ed. Copyright 1989, Hewlett-Packard Company.

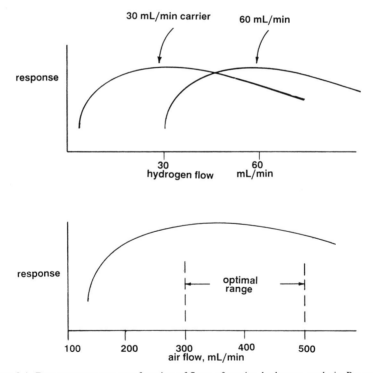

Figure 2.4. Detector response as a function of flows of carrier, hydrogen, and air. Reproduced with permission from *Detectors for Gas Chromatography: A Practical Primer*. Copyright 1987, Hewlett-Packard Company.

changes, as in the case of a temperature programmed chromatographic run using a pressure regulated carrier gas supply (21). Figure 2.4 depicts the interdependency of carrier (plus make up) flow and combustion gas flows. Nitrogen is a better choice of makeup gas than is helium (see Figure 2.5). Early experiments in characterizing the FID showed that there is a maximum in sensitivity at a particular ratio of carrier plus makeup gas to hydrogen flow (22) and that this ratio depends on the particular makeup gas. The controlling feature appears to be that maximum ionization comes from flames having approximately the same temperature, which was shown to be on the order of 2100 K (23) and corresponds to a composition index, I_d, of 0.21. For a N_2-diluted flame with a stoichiometric H_2/O_2, I_d is given by

$$I_d = F(H_2)/\{F(N_2) + 3.5F(H_2)\}$$

where F is the flow rate of the respective gas.

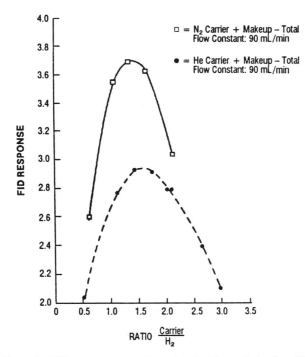

Figure 2.5. Gas ratio–FID response curves. Reproduced with permission from K. J. Hyver, ed., *High Resolution Gas Chromatography*, 3rd ed. Copyright 1989, Hewlett-Packard Company.

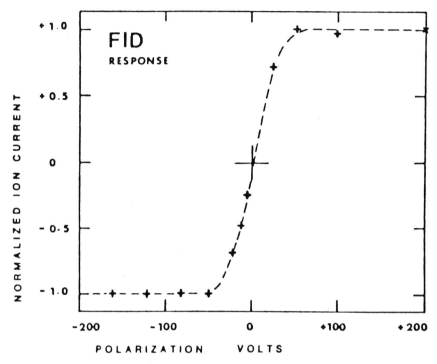

Figure 2.6. FID response current at positive and negative polarization voltages. Reproduced with permission from P. L. Patterson, *J. Chromatogr. Sci.* **24**, 487 (1986). Copyright 1986, Preston Publishers, a division of Preston Industries..

Another desirable feature of the FID is the wide linear range that is available, most often quoted as about 10^7. It has been demonstrated that in some cases the linear range of the detector can be increased by operating at higher hydrogen flows (24). When properly configured, and with the use of high purity flame and carrier gases, the background noise is on the order of a few picoamperes, due mainly to ionization of organic impurities.

As shown in Figure 2.6, the voltage difference between the jet tip and the collecting electrode can be quite low and still collect the ions effectively. Most collectors are cylindrical and placed around the top of the flame itself so that ions are collected near the base of the collector where the electric field is greatest. The practical result of this is a linear response to varying sample sizes, at least at low concentrations. However as the amount of material in the sample gets larger, the flame characteristics change from a pure hydrogen/air diffusion flame to a larger, more hydrocarbon-like flame. This increase in flame size means that ion production takes place higher in the collector, where the electric field is weaker and the flame is cooler. These ion collection and formation processes thus become less efficient, resulting in a nonlinear response.

4. QUANTIFICATION AND SENSITIVITY

The FID is a mass flow sensitive detector in that the area of a given peak is proportional to the mass of single carbon fragments. Flame response factors are well behaved, with normal hydrocarbons giving a response of > 0.015 coulombs/g(C) in a typical detector. However, heteroatoms affect the response to varying degrees, which has led to introduction of the term *effective carbon number* (ECN). The ECN is estimated by using an adjusted value for a particular grouping of atoms that is roughly transferable from one molecule to another. Contributions to the ECN are shown in Table 2.1 (8, 25). Thus one can estimate the relative response for any compound being chromatographed (26). The accuracy of this estimation is questionable when halogens are present. Similar caution should be applied to perfluoroalkanes and phthalate esters (27).

Recently, Yieru et al. have suggested that a new calculation method be employed, utilizing the concept of "relative carbon weight response factors"

Table 2.1. Effective Carbon Number of Various Groups

Group	Carbon Number
Aliphatic C	1.0
Aromatic C	1.0
Olefinic C	0.95
Acetylenic C	1.30
Carbonyl C	0
Carboxyl C	0
Nitrile C	0.3
Ether O	-1.0
Primary alcohol O or amine N	-0.6
Secondary alcohol O or amine N	-0.75
Tertiary alcohol O or amine N	-0.25
Trimethyl silyl derivatives of alcohols	3.69–3.78
Trimethyl silyl derivatives of carboxylic acids	3.0
Trimethyl silyl derivatives of oximes	3.3

(28). In this method, the weight percent of the compound under consideration $(W_i\%)$ is given by

$$W_i\% = \frac{A_i M_i / C_i}{\Sigma A_i M_i / C_i} \times 100$$

where A_i is the peak area of the component, M_i is the molecular weight of the component, and C_i is the total atomic weight of carbon in the compound.

A typical FID can respond to approximately 20 pg of each component eluting from a high-resolution capillary column (29), although detection limits as low as 1 pg/s have been claimed (30). As already mentioned, sensitivity, defined as peak area (in coulombs) divided by the mass of the material (in grams), is on the order of 0.015 coulombs/g(C).

5. LIMITATIONS

The features that make the FID so useful also lead to its limitations. Although only a small percentage of the carbon-containing compound is ionized, the rest is destroyed in the flame, thus precluding subsequent analysis by another detector. With the few exceptions noted earlier, it is nonselective, so that small quantities of a desired component in a larger sample matrix may be hard to determine and quantify.

An additional limitation is imposed by the column itself. For small-diameter analytical columns coated with a thin stationary film, the sample capacity is quite small, often only about an order of magnitude higher than the minimum detectable quantity of the detector. Thus the upper end of the effective linear range of capillary GC–FID is limited by the column capacity.

6. MODIFICATIONS TO IMPROVE SELECTIVITY

Much of the emphasis of modern chromatography is on the development of selective detectors, and the FID has not been overlooked. The nitrogen phosphorus detector, discussed in a later chapter, is the most obvious example of such a modification. Most investigations focus on flame characteristics that result in selectivity over hydrocarbons. The use of carbon monoxide, ammonia, formamide, formic acid, and carbon disulfide as flame-forming agents in the FID has been discussed (31). Although their main utility seems to be that they can also be used as the carrier gas and thus influence the chromatography, some detection selectivity is introduced.

With CO, sensitivity to hydrocarbons was essentially the same as in a typical FID, but response to chlorinated hydrocarbons was about doubled. Many other features of this detector were similar to an FID, but response to positive ions was greater than that to negative ions. Similar results were obtained by using ammonia as a carrier gas and fuel, although oxygen-enriched air was required. This also provided chromatographic advantages, especially for amines. Since much of this work was done with packed columns, it remains to be seen if extension to capillary columns is straightforward.

A similar comment can be made with regard to the use of FIDs modified to produce response to inorganic gases. In general, these have methane or some other hydrocarbon supplied to the flame, resulting in a high background current. When certain inorganic gases are eluted, there is a decrease in the current and hence a negative response (32). Doping with methane was also employed in an attempt to detect silicon-containing compounds as a depletion of ionization (33). Response was non-linear and subject to several complications.

Of more interest perhaps when capillary chromatography is considered is the hydrogen atmosphere flame ionization detector (HAFID) (34). In this detector the hydrogen and air flows are interchanged to provide an air diffusion flame burning in hydrogen. In addition the collecting electrode is positioned several centimeters above the flame. When both metal atoms and silicon atoms are present, the response characteristics change markedly. Thus, two versions are available: the HAFID in which the carrier gas is doped with silane so that the detector is sensitive to metal-containing compounds being eluted (35), and the HAFID–Si in which the carrier gas is doped with a metal and response is to silylated compounds (36).

Depending on parameters chosen, response in the HAFID can be either positive (i.e., the analyte produces an increase in signal) or negative (i.e., presence of the analyte reduces the signal, resulting in an inverted peak). The controlling factor seems to be the amount of silane present in the flame. This inverted response has potential analytical utility. Experiments to date have utilized only the configuration wherein the collecting electrode is negatively biased.

The HAFID responds to a variety of metals, although the sensitivity and selectivity is slightly different for each. While most of the sensitivity data is for packed columns, the HAFID has been connected to a capillary SFC (37). Of more potential interest is the fact that a sheathed-flow modification of the HAFID has been developed (38) and applied to capillary GC. By redesign of the detector body, an order of magnitude reduction of hydrogen flow was achieved without compromising significantly the sensitivity and selectivity. Figure 2.7 shows both negative and positive mode responses to tin compounds in a spiked fish sample compared to a chromatogram using a normal FID.

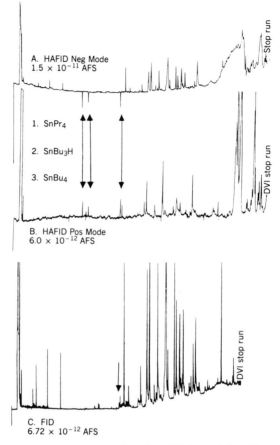

Figure 2.7. Chromatograms of an extract from fish: (A) sheathed-flow HAFID negative mode response; (B) sheathed-flow HAFID positive mode response; (C) FID response. (AFS = amperes full scale.) Reproduced with permission from M. M. Gallagher, D. G. McMinn, and H. H. Hill, *J. Chromatogr.* **518**, 297 (1990). Copyright 1990, Elsevier Science Publishers, Physical Sciences & Engineering Division.

The HAFID–Si has been used with both packed and capillary columns. Its potential is for the analysis of silylated derivatives after chromatography without the need to derivatize with electron capture active substituents. Figure 2.8 shows a chromatogram of a derivatized air sample where components were detected using a HAFID–Si (39). Response is a function of the number of silicon atoms in the compound.

While the mechanism responsible for the selective response of the HAFID

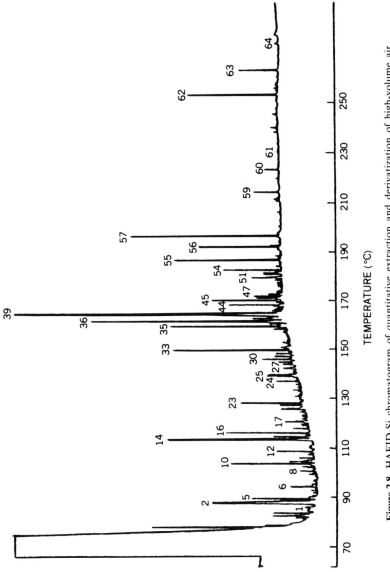

Figure 2.8. HAFID-Si chromatogram of quantitative extraction and derivatization of high-volume air sample. Reproduced with permission from A. R. Gholson, R. St. Louis, and H. H. Hill, Jr., *J. Chromatogr.* **408**, 329 (1987). Copyright 1987, Elsevier Science Publishers, Physical Sciences & Engineering Division.

has not been worked out, some clues are beginning to emerge. Mass spectral studies on the HAFID flame (40) have revealed that the positive ions differ markedly from those found in a normal hydrogen/air diffusion flame. In particular, hydrates of SiO or SiO_2 were observed, as well as hydrated ammonia species. No metal-containing species were observed, at least up to m/e 300. However, when organometallics were present in the flame, higher hydrates of silicon oxide were present. Presumably, the less mobile silicon-containing species are collected more efficiently at the distant electrode than are the lighter clusters of hydronium ions. This speculation was tested by placing a bias voltage on the walls of the detector to produce a different electric field (41). Enhancement of response was observed when the bias voltage was placed in a narrow collar roughly corresponding to the height of the collecting electrode.

To date, no specific role of the metal has been delineated, nor is there an explanation for the inverted (negative) response.

REFERENCES

1. J. E. Boyle, P. H. Silvis, W. F. Fatula, L. M. Sidisky, and R. J. Bartram, *Am. Lab.* **17**, 54 (1985).

2. I. G. McWilliam and R. A. Dewar, *Nature (London)* **181**, 760 (1958).

3. J. Harley, W. Nell, and V. Pretorius, *Nature (London)* **181**, 177 (1958).

4. P. L. Patterson, *J. Chromatogr. Sci.* **24**, 487 (1986).

5. H. H. Hill, Jr. and M. A. Baim, *Trends Anal. Chem.* **1**, 206 (1982).

6. A. J. C. Nicholson, *J. Chem. Soc., Faraday Trans. 1* **78**, 2183 (1982).

7. H. C. Bolton, J. Grant, I. G. McWilliam, A. J. C. Nicholson, and D. L. Swingler, *Proc. R. Soc. London* **A360**, 265 (1978).

8. J. C. Sternberg, W. S. Gallaway, and T. L. Jones, *Gas Chromatography*, pp. 231–267. Academic Press, New York, 1962.

9. F. O. Rice and K. F. Herzfeld, *J. Am. Chem. Soc.* **56**, 284 (1934).

10. A. T. Blades, *J. Chromatogr. Sci.* **11**, 251 (1973).

11. J. H. Wagner, C. H. Lillie, M. D. Dupuis, and H. H. Hill, Jr., *Anal. Chem.* **52**, 1614 (1980).

12. J. Peeters, J. F. Lambert, P. Hertoghe, and A. Van Tiggelen, *Symp. (Int.) Combust.* [*Proc.*] **13**, 321 (1971).

13. A. J. C. Nicholson, *J. Chem. Soc., Faraday Trans. 1* **79**, 2183 (1982).

14. A. T. Blades, *J. Chromatogr. Sci.* **22**, 120 (1984).

15. A. N. Hayhurst and D. B. Kittelson, *Combust. Flame* **31**, 37 (1978).

16. T. McAllister, A. J. C. Nicholson, and D. L. Swingler, *Int. J. Mass Spectrom. Ion Phys.* **27**, 43 (1978).

17. K. J. Hyver, in *High Resolution Gas Chromatography* (K. J. Hyver, ed.), 3rd ed., Chapter 4. Hewlett-Packard Co., Avondale, Pennsylvania, 1989.

18. M. J. O'Brien, in *Modern Practice of Gas Chromatography* (R. L. Grob, ed.), 2nd ed., p. 242. Wiley, New York, 1985.

19. B. V. Burger and P. J. Pretorius, *J. Chromatogr. Sci.* **25**, 118 (1987).

20. M. M. Thomason, W. Bertsch, P. Apps, and V. Pretorius, *HRC & CC, J. High Resolut. Chromatogr. Chromatogr. Commun.* **5**, 690 (1982).

21. K. Grob, Jr., *HRC & CC, J. High Resolut. Chromatogr. Chromatogr. Commun.* **3**, 286 (1980).

22. I. G. McWilliam, *J. Chromatogr.* **6**, 110 (1961).

23. B. A. Schaefer, *Combust. Flame* **54**, 71 (1983).

24. D. E. Albertyn, C. B. Bannon, J. D. Craske, N. T. Hai, K. L. O'Rourke, and C. Szonyi, *J. Chromatogr.* **247**, 47 (1982).

25. J. T. Scanlon and D. E. Willis, *J. Chromatogr. Sci.* **23**, 333 (1985).

26. H. Y. Tong and F. W. Karasek, *Anal. Chem.* **56**, 2129 (1984).

27. G. Perkins, Jr., R. E. Laramy, and L. D. Lively, *Anal. Chem.* **35**, 360 (1963).

28. H. Yieru, O. Qingyu, and Y. Weile, *Anal. Chem.* **62**, 2063 (1990).

29. J. V. Hinshaw, *LC-GC* **7**, 305 (1989).

30. R. K. Simon, Jr., *J. Chromatogr. Sci.* **23**, 313 (1985).

31. V. G. Berezkin, *CRC Crit. Rev. Anal. Chem.* **20**, 291 (1989).

32. V. A. Schaefer and D. M. Douglas, *J. Chromatogr. Sci.* **9**, 612 (1971).

33. B. Lengyel, G. Garzo, D. Fritz, and F. Till, *J. Chromatogr.* **24**, 8 (1966).

34. H. H. Hill, Jr. and W. A. Aue, *J. Chromatogr.* **140**, 1 (1977).

35. D. R. Hansen, T. J. Gilfoil, and H. H. Hill, Jr., *Anal. Chem.* **53**, 857 (1981).

36. M. A. Osman and H. H. Hill, Jr., *Anal. Chem.* **54**, 1425 (1982).

37. M. A. Morrissey and H. H. Hill, Jr., *HRC & CC, J. High Resolut. Chromatogr. Chromatogr. Commun.* **11**, 375 (1988).

38. M. M. Gallagher, D. G. McMinn, and H. H. Hill, Jr., *J. Chromatogr.* **518**, 297 (1990).

39. A. R. Gholson, R. St. Louis, and H. H. Hill, Jr., *J. Chromatogr.* **408**, 329 (1987).

40. C. H. Lillie, D. G. McMinn, and H. H. Hill, Jr., *Intl. J. Mass Spectrom. Ion Proc.* **103**, 219 (1991).

41. M. M. Gallagher and H. H. Hill, Jr., *J. High Resolut. Chromatogr.* **13**, 694 (1990).

CHAPTER

3

THE MODERN HELIUM IONIZATION DETECTOR

ROSWITHA S. RAMSEY

Oak Ridge National Laboratory
Oak Ridge, Tennessee

FIKRY F. ANDRAWES

American Cyanamid Company
Stamford, Connecticut

This volume is devoted to detectors appropriate for capillary separations, but within this class the helium ionization detector (HID) is unusual. Only recently has its potential for the sensitive analysis of higher-molecular-weight compounds or the analysis of complex mixtures been explored. Because of its unique ability to detect ultra-trace quantities of the permanent gases, most of the work has centered on packed column determinations of these species. Since this continues to be an important application, especially for those industries (such as the semiconductor industry) where high-purity gases are an essential requirement, the analysis of lower-molecular-weight gases will not be overlooked. Rather, the advances made in separating these species on micro packed or wide-bore capillary columns will be discussed. The theoretical basis of the HID, design, operating conditions, and their effect on response will be examined. The detector will also be evaluated for liquid sample analysis. This should provide some guidelines in the use of this frequently overlooked but highly sensitive detector for capillary applications.

The HID is currently one of the most sensitive detectors available for gas chromatography. It is nonselective, that is, capable of responding to all volatile species ranging from the permanent gases to complex organic molecules. Despite these advantages, the detector has enjoyed a somewhat limited usage. The factors responsible for this include the stringent requirements for high-sensitivity operation, variations in response for selected species coupled

Detectors for Capillary Chromatography, edited by Herbert H. Hill and Dennis G. McMinn. Chemical Analysis Series, Vol. 121.
ISBN 0-471-50645-1 © 1992 John Wiley & Sons, Inc.

with a lack of understanding of the conditions that generate these responses, and the belief that the detector is primarily suited for the analysis of gases which can be separated on low-bleed adsorption columns. While the criteria for obtaining the maximum sensitivity from the detector are stringent, they are not insurmountable. A well-conditioned column, ultra-high-purity helium carrier gas, and the elimination of atmospheric leaks are the basic requisites. These will be discussed in the following sections. Response variations (i.e., changes in peak shape or polarity) are obtained only with a few gases (neon, hydrogen, argon, oxygen, and nitrogen) or only when the detector is overloaded. In the latter case an M-shaped signal is obtained (1). Methods for controlling these variations have recently been described (2, 3) and will also be addressed below. Although the physical basis for these phenomena are not understood, it can be argued that the mechanisms of most chromatographic detectors are also not fully understood at this time.

Finally we will review the recent work on the HID showing that the detector has been used with a variety of chromatographic columns for diverse applications and is therefore not limited in range. We are hopeful that this chapter will help to dispel some of the misconceptions about the HID.

1. DETECTION MECHANISM

The mechanism by which the HID operates is generally regarded to be based upon the Penning effect, which is the transfer of the excitation from metastable helium (He*) to other atoms or molecules. Since He* has an ionization potential that is higher than all other species with the exception of neon (i.e., 19.8 eV for He vs. 21 eV for Ne), it is capable of ionizing all other compounds. A response is, in fact, also obtained for Ne, although it is ordinarily observed as a negative signal or peak. This negative response is most likely due to a "vacancy" effect. As neon enters the detector the species responsible for the background current are diluted, resulting in a negative signal. Under certain conditions a positive response may also be obtained (2).

The reactions that may occur in the detector are as follows (4, 5):

$$B + He \rightarrow He^* + B' \tag{1}$$

$$B + He \rightarrow He^+ + e + B' \tag{2}$$

$$e + He \rightarrow He^* + e' \tag{3}$$

$$e + He \rightarrow He^+ + e + e' \tag{4}$$

$$He + He^* \rightarrow 2\,He \tag{5}$$

$$He + He^* \rightarrow He_2^+ + e \tag{6}$$

$$He^* + S \rightarrow He + S^+ + e \tag{7}$$

$$He^* + S \rightarrow HeS^+ + e \tag{8}$$

$$B + S \rightarrow S^+ + e + B' \tag{9}$$

$$B + S \rightarrow S^* + B' \tag{10}$$

$$e + S \rightarrow S^+ + e + e' \tag{11}$$

$$e + S \rightarrow S^* + e' \tag{12}$$

where B and B' are the primary electrons and slowed-down primary electrons (see Section 3); e and e' are the secondary electrons and slowed-down secondary electrons, respectively; and S is any analyte except for neon. The metastable helium is predominately generated as a result of collision with high-energy electrons (Equation 1). Some helium is also ionized (Equations 2, 4, and 6), which contributes to the background current along with any primary or secondary electrons that may be collected. The secondary electrons formed as a result of the ionization of the carrier gas (Equations 2 and 6) or analyte (Equations 7, 8, and 9) may take further part in the ionization process (Equations 4 and 11) if they have sufficient energy. This occurs at high applied potentials as discussed below. Equations 6 and 8 are referred to as associative ionizations, and Equation 5 as a quenching interaction (6).

Although Equations 7, 8, 9, and 11 express the ionization of the analyte, these reactions do not represent the detector's response (4). As the sample enters the detector and increases in concentration, the metastable atoms and consequently the secondary electrons will decrease in concentration. Only the direct ionization of the analyte by the primary electrons (Equation 9) will always be directly proportional to the sample concentration (4). The sum of all the reactions, many of which are competing, will comprise the response.

According to the mechanism outlined above, the detector should provide a positive signal for all molecules except neon. It has been well documented, however, that when the detector is operated under conditions where there is essentially no contamination (i.e., with ultra-high-purity helium carrier gas and a clean chromatographic system), the response for hydrogen, argon, nitrogen, and oxygen is negative (1, 2, 7). It has also been reported recently that tetrafluoromethane responds negatively (8). All other species examined produce positive signals, independent of the operating conditions. At this time there is no satisfactory explanation for the negative peaks. Note, however, that all compounds that react in this manner have relatively high ionization potentials.

Hurst and Klots have stated that the origins of the Penning phenomenon are "complex, subtle, and subject to reinterpretation" (9). Indeed, the mechanism of the HID is still not fully understood. The response for most species, however, can be adequately explained by the Penning effect. It accounts for the high sensitivity of the detector and provides a plausible explanation for the increase in ionization observed when most compounds are introduced into the detector. A detailed, systematic study, however, will be necessary to identify the actual species generated and to elucidate the complete mechanism.

2. DETECTOR GEOMETRY

The first HIDs were essentially argon ionization detectors, as originally described by Lovelock (10), used with helium carrier gas. The two detectors are, in fact, identical in principle, differing only in the energy of the metastable species produced. Replacing the argon carrier gas with helium allows the determination of species with higher ionization potentials (up to 19.8 eV vs. 11.8 eV for argon). Because of the extended range of applications of the HID, the metastable detectors are now used almost exclusively with helium carrier gas.

These early detectors had asymmetric or displaced coaxial geometries (i.e., an anode slightly removed and placed above a cylindrical cathode) (5). To minimize atmospheric leaks they were also often enclosed in glass envelopes. This asymmetric geometry produces a nonhomogeneous field (11). If the electrode spacing is maximized, positive space charges and some of their adverse effects (i.e., anomalous responses) may be minimized. This is because the electric field is concentrated near the anode and is weaker near the cathode. The volume of this type of detector, however, is relatively large (about 1 mL). In fact, the detector is often referred to as a "macro cell". The plane parallel or the symmetrical coaxial configuration is most frequently employed today. These designs are depicted in Figures 3.1 and 3.2. The electrodes are closely spaced, which minimizes the internal volume. Typical "dead" volumes of commercially available detectors range from 100 to 200 μL, which preserve column efficiency and make the detectors suitable for use with conventional capillary columns (i.e., 0.25 mm i.d. or larger). In addition to these configurations a miniaturized triaxial detector has been described that consists of a two electrode array contained within a cylindrical radioactive source (12). This design results in a compact detector with an 80 μL internal volume and a weight of only 1 g. It was explicitly developed for on-board space exploration.

Radioactive beta emitters ordinarily provide the primary electrons that

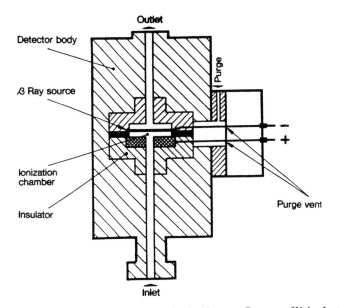

Figure 3.1. Plane parallel design of the helium ionization detector. Courtesy of Valco Instruments Co., Inc., Houston, TX.

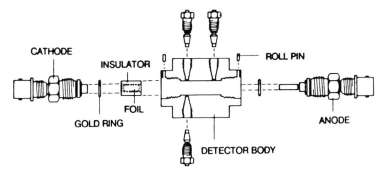

Figure 3.2. Coaxial cylindrical configuration of the helium ionization detector. Courtesy of Haake Buchler Instruments, Inc., Carlo Erba Product Group, Saddle Brook, NJ. Copyright, Fisons Instruments.

promote the helium atoms to the metastable state. In the early detectors ^{90}Sr and ^{226}Ra (an alpha emitter) were used (4). These two sources emit high-energy particles that increase the noise level in the detector. The alpha source also emits gamma radiation that could be harmful unless the cell is properly shielded. Today beta-emitting tritium sources in the form of Ti^3H_2 and Sc^3H_3 are most frequently employed, with activities ranging from 250 mCi to 1 Ci. In general, higher activity yields greater sensitivity. Scandium foils have the advantage of being stable at higher temperatures as compared to titanium foils (i.e., 325 °C vs. 225 °C). This may be especially advantageous if the detector is used to analyse high-boiling compounds since contamination is usually reduced by heating the detector cell at higher temperatures. Discharge ionization has also been used to promote helium to the metastable state (13), although no commercial devices based on this principle are currently available.

3. MEASURING DETECTOR CURRENT

Charged particles created by ionization in the detector volume are collected by applying a bias potential across two electrodes in the cell. A d.c. electrometer is connected to one of the electrodes, commonly held at ground potential, and a bias potential is applied to the opposite electrode. This induces a current flow due to the drift of the particles in an electric field. Typical background currents are in the nanoampere range but are critically dependent upon the applied potential, carrier gas purity, atmospheric leakage, column "bleed," or other contamination. When a sample is introduced, the current becomes a function of the total ionization rate (sample plus background) and is modified by recombination reactions or other loss of charged particles in the detector.

The dependence of the current on the applied electric field is shown in Figure 3.3. The range between 0 and 20 V is called the *collection region*, where the current is linearly dependent upon the applied voltage. Above 20 V to about 200 V the response remains independent of the field strength. This area is referred to as the *saturation region*, where there is an equilibrium between collected and generated ions. As the detector bias is increased above the saturation region, the response increases exponentially. This is also the region where the detector is most frequently operated. It affords the highest sensitivity but also reduced stability and increased noise in correlation with the voltage. The increased sensitivity is derived from the action of secondary electrons that are accelerated by the field, thereby obtaining sufficient energy to contribute to the overall ionization.

The response of the HID in the saturation region has been examined by

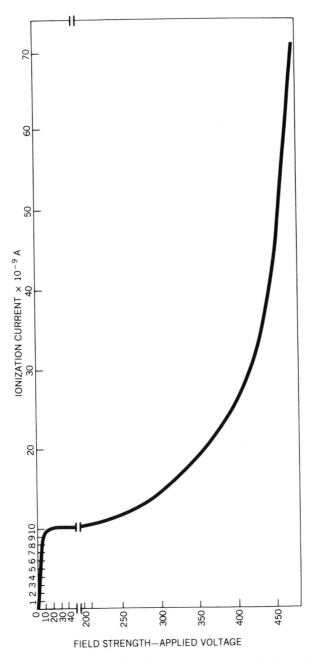

Figure 3.3. The volt–ampere curve of the helium ionization detector. Reprinted with permission from F. F. Andrawes, R. S. Brazell, and E. K. Gibson, Jr., *Anal. Chem.* **52**, 891 (1980). Copyright 1980, American Chemical Society.

Andrawes et al. (14). Excellent sensitivity was obtained for both the permanent gases and for various organic compounds that were tested, primarily because of the reduced noise levels in this region. Response surface methodology has also been used to determine the conditions that provide the maximum signal (15). The greatest sensitivity for the noble gases using a plane parallel detector was obtained at 550 V. Increases above this value resulted in high-frequency noise and instability.

In addition to the operation of the HID with a continuous d.c. voltage, designs that allow the detector to operate in a fixed-frequency or variable-frequency (constant-current) pulsed mode have recently been described (16, 17). The concepts involved in pulsing the detector are similar to those of the pulsed electron-capture detector (ECD), but there are major design and response differences due to the very different mechanisms involved. The pulser for the HID must be capable of operating at a bias potential of several hundred volts at high frequencies as compared to the ECD, which is nominally operated at 50 V or lower. This has necessitated the development of new electronic circuitry (16, 18). The pulser that has been described is capable of operating at frequencies from d.c. to greater than 300 kHz over a 0–500 V range. The rise and fall times of the pulse output are in the order of 50 ns, and the minimum pulse width is 125 ns. The maximum operating frequency is limited only by the power dissipation in the output devices and by the available power supply. With an appropriate power supply and thermal design of the output stage, the pulser is capable of operating in excess of 1 MHz. High-speed electronics have also been developed to collect the current signal. In the variable frequency mode, the detector is combined with a feedback control loop that pulses the bias voltage at a frequency designed to maintain a selected constant current through the detector as the analyte concentration varies.

The effects of pulsed operation have been reduced background current levels and, in the case of fixed-frequency operation, reduced noise levels as compared to d.c. operation. The greatest response is obtained at high frequencies, high duty cycles, and at high voltages. The signal-to-noise ratios for the fixed-frequency mode were reported to be at least comparable, if not slightly superior, to those for the d.c. mode. Since background current levels are lower, the stability of the detector is greater. It was also determined that operation at certain frequency and duty cycle combinations would automatically invert the signal polarity for the gases that characteristically provide a negative response with the HID. This effect is similar to adding a dopant to the carrier gas (2) and extends the upper end of the dynamic range. Overall pulsing of the detector allows the chromatographer considerable control over the output response. For this reason it may also serve as an important analytical tool for studying the species that are actually generated in the detector, which may lead to an understanding of the negative signals and the inversion process.

4.   EFFECT OF EXPERIMENTAL CONDITIONS ON DETECTOR RESPONSE

The purity of the carrier gas at the detector cell, carrier gas flow rate, detector temperature, and applied potential are all important variables that are easily adjusted and that affect the response of the detector. The latter two variables have already been addressed. In this section we will examine how the two former factors affect the sensitivity and the overall response of the detector.

4.1.   Carrier Gas Purity

Carrier gas purity plays an important role in determining the magnitude of the detector response. High impurity levels will increase the noise level, increase the lower detection limit, and narrow the linear dynamic range. Although helium is commercially available at a minimum purity level of 99.9999%, the purity at the detector cell may be significantly lower than at the immediate outlet of the gas cylinder. Contaminants due to atmospheric leakage may be introduced at any connection between the cylinder and the detector. Poorly designed detectors will also allow leakage through the cell itself (19). The more modern detectors are operated in a helium atmosphere to eliminate this problem.

A method for evaluating the purity of the carrier gas at the detector cell has been reported (20). The method is based upon the response to hydrogen, argon, oxygen, and nitrogen on a molecular sieve column. The polarity of the signals given by these gases is negative if the carrier gas at the detector cell is highly pure. If a negative response is obtained for these gases at the low ppb level, then one can be assured that the helium in the cylinder is unadulterated and that no significant leaks or contamination in the lines transporting the gas are present. Under these conditions careful replacement of the packed molecular sieve column with a capillary column should also yield excellent sensitivity for both organic or inorganic compounds.

While it has been reported that the addition of some impurities to the carrier gas may be useful in the analysis of the permanent gases mentioned above, any impurity will decrease the sensitivity of the detector (20). The role of carrier gas purity is illustrated in Figure 3.4. When one uses high-purity carrier gas, the response is high for all compounds. For the permanent gases that respond with a negative signal, the intensity of this negative signal is likewise enhanced under these conditions. Note also that these negative

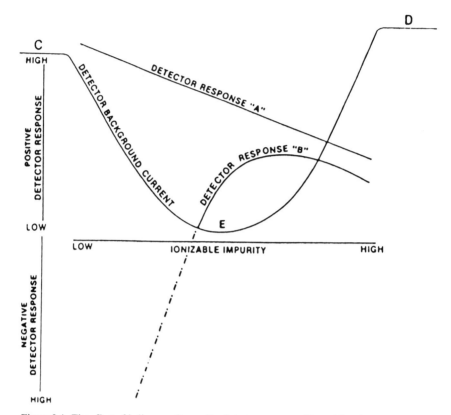

Figure 3.4. The effect of helium purity on the detector response. Dotted line indicates negative response. Reprinted with permission from F. F. Andrawes, *J. Chromatogr.* **290**, 65 (1984). Copyright 1984, Elsevier Science Publishers, Physical Sciences & Engineering Division.

signals can be calibrated and legitimately quantified (7). Any increase in the impurity level, however, will result in an overall reduction in response, for both positive and negative-going peaks, as illustrated by curves A and B in Figure 3.4, respectively. The background current will also drop to some minimum value before it increases and levels off. At the minimum background current the negative peaks will undergo an inversion in polarity. At point E the response for all species is positive. In the region between points E and D, the magnitude of the background current may also be similar to that obtained in the region between points C and E. The response, however, will be significantly lower. This explains why it is not possible to solely use background current values as an indicator of the cleanliness of the system and consequently of the degree of sensitivity that may be expected from the detector. The test described above that examines the polarity of the signals of a gaseous sample is easy to perform and will provide the assurance that the detector is operating at optimum conditions for high-sensitivity analysis.

4.2.   Effect of Carrier Gas Flow Rate

The HID can be classified as both a mass flow-dependent and a concentration-dependent detector. However, the effect of flow rate on the response depends on the detector design and the carrier gas purity at the detector cell. If the detector is not fully sealed to the atmosphere, air diffusion will generate a high background current (14, 19). Increasing the flow rate in this case will purge the detector, lower the background current, and increase the detector response to a certain level, after which the response will decline. If the detector is fully sealed to the atmosphere, any increase in flow rate will cause a decrease in the response and background current (2, 19). Figure 3.5 shows the effects of flow rate on response for both a well-sealed detector and for one with some ambient air leakage.

Most of the reported work with capillary columns has used makeup gas,

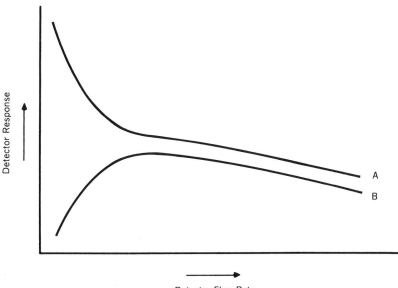

Figure 3.5. The effect of flow rate on detector response: (A) in a well-sealed detector; (B) in a detector with air leakage. Reprinted with permission from F. F. Andrawes, *J. Chromatogr.* **290,** 65 (1984). Copyright 1984, Elsevier Science Publishers, Physical Sciences & Engineering Division.

introduced with "T" connector at the base of the detector cell, to minimize peak broadening (14, 17, 20). A total flow of 30–50 mL/min is frequently used. The design of lower volume detectors could eliminate the need for makeup, however, resulting in even greater sensitivity. A recent report shows that the HID can be operated at a flow rate of 5 mL/min and still provide optimum response (19). However, at low flow rates the detector should be well sealed to the atmosphere.

5. SUITABILITY OF THE HID FOR CAPILLARY GAS CHROMATOGRAPHY

Most analyses with the HID in the past have been performed on gaseous samples separated on packed adsorption columns. It was generally regarded that the bleed generated from liquid stationary phases was sufficiently great so as to prohibit the use of these phases with the detector (5, 21). Although it is possible to control carrier gas purity and atmospheric leakage, column bleed is largely an inherent property of the chromatographic phase and of the column used. Prior to 1980 most WCOT (wall coated open tubular) columns exhibited high bleed, but in recent years this has improved significantly owing to new developments in column technology. A variety of stationary phases immobilized on FSOT (fused silica open tubular) columns are now available commercially. They have excellent resolving power and exhibit very low bleed even at high temperatures. Other stationary phases that are appropriate for use with the HID are molten salts and liquid crystals, both of which have low bleed properties.

It is commonly understood that heating even the most stable of chromatographic columns will generate "bleed" that can be easily detected with the sensitive HID. In a recent report (20) six columns were evaluated for use with the detector: Carbosieve S, Carbopack C containing 0.2% Carbowax 1500, Porapak Q, Super Q, Chromosorb 101, and a fused silica column containing immobilized Carbowax 20 M. The results are summarized in Figure 3.6. Of the six columns tested, the Carbowax 20 M capillary produced the lowest background current. The bleed for this column at 180 °C is lower than the bleed for all the packed columns at 20 °C. From these data it also appears that background level has some correlation with the surface area available for separations in the analytical column. This may be because higher surface areas per column length provide more opportunities for desorption of impurities. Carbosieve S, Porapak Q, Super Q, and Chromosorb 101 have surface areas of 1000, 840, 840, and 35 m^2/g, respectively. The Carbopack column with a surface area of 12 m^2/g generated higher bleed due to the presence of the

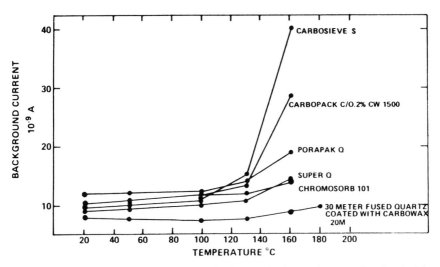

Figure 3.6. Effect of column temperature on the detector background current. Reprinted with permission from F. F. Andrawes, *J. Chromatogr.* **290**, 65 (1984). Copyright 1984, Elsevier Science Publishers, Physical Sciences & Engineering Division.

Carbowax liquid phase. The surface area of the open tubular column is of course much smaller ($0.0118 \, m^2$).

The amount of column bleed or the magnitude of the background current also depends upon the cleanliness of the packing material and the upper temperature limit. Note the difference between Porapak Q and Super Q (which is essentially purified Porapak Q from which monomers and polymerization catalysts have been removed). It also appears from Figure 3.6 that the capillary column can be used quite adequately and without significant loss in sensitivity at least up to 180 °C. This column can of course be operated at much higher temperatures (225 °C). The real limiting factor is the thermal stability of the radioactive source, as discussed in Section 3.

Owing to the extreme sensitivity of the HID, some attention must be directed toward understanding its baseline behavior. Figure 3.7 shows the background obtained from a heating and cooling cycle of a capillary column that was being conditioned. While the column is heated there is an increase in background followed by a decrease. The initial increase is due to bleed from the stationary phase and elution of trace impurities trapped on the front end of the column during a previous cooling period. The decrease is primarily due to the drop in column flow rate at the higher temperatures. As the column is cooled, the cycle is reversed with the background decreasing and then

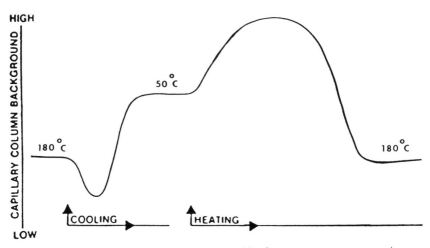

Figure 3.7. Changes in background current resulting from temperature programming.

increasing. The overall background will drop in accordance with the degree of conditioning.

6. INJECTION SYSTEM

If the injection system is not properly designed it may introduce air and moisture to the analytical column. Continuous leakage through the inlet system will increase the background current and decrease the detector response. For these reasons, direct injection through a septum is generally not recommended. A modified direct injector that minimizes atmospheric leakage has been described (14) and is shown in Figure 3.8. Essentially it contains a small compartment that is continuously purged with helium so as to reduce the amount of air introduced when a syringe pierces the septum closest to the column. Sample splitters have also been used with high spilt ratios to minimize atmospheric leakage (20). Capillary on-column injection temporarily interrupts column flow and may produce baseline disturbances, making it difficult to analyze early eluting components. This technique has, however, been used for some applications without any difficulty (17).

Gas sampling valves have conventionally been used for the analysis of gases as well as light organic vapors (14, 22). The problems that may be encountered in this case involve irreversible adsorption on the valve body and ghosting of more highly concentrated species. These problems may be eliminated by heating the valve. Note, however, that heated valves off-gas traces of water.

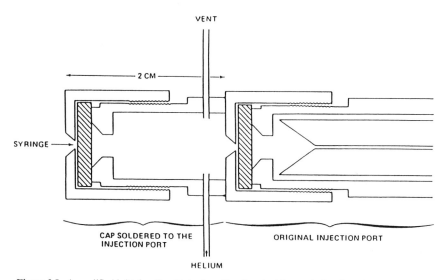

Figure 3.8. A modified inlet for direct injection. Reprinted with permission from F. F. Andrawes, R. S. Brazell, and E. K. Gibson, Jr., *Anal. Chem.* **52**, 891 (1980). Copyright 1980, American Chemical Sociery.

This is only a problem if water is to be analyzed at trace levels (23). A sample introduction system based on non moving parts such as the one introduced by Dean (24) should be useful for some HID applications, although this remains to be determined.

7. APPLICATIONS

The use of the HID in trace analysis has recently been reviewed (25), and only the work specifically conducted on micro packed, PLOT (porous-layer open tubular), or WCOT columns will be discussed here.

7.1. The Analysis of Liquid Samples

As stated earlier, the HID has been used primarily to analyze gaseous samples. Several reports, however, have shown that the detector can also be applied to the analysis of organic compounds that are volatile only at elevated temperatures. The important factor here is to minimize contributions to the background current from sample introduction methods and from column bleed so as to preserve the sensitivity of the detector.

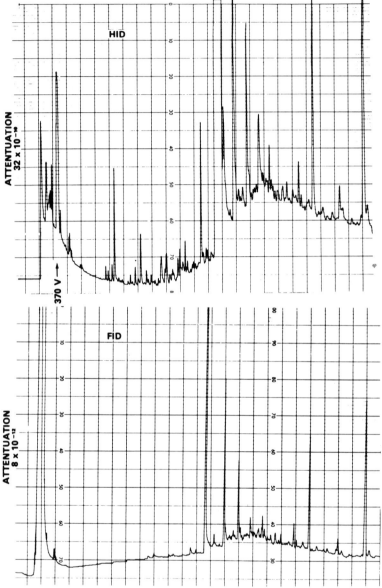

Figure 3.9. Particulate matter extract of sidestream tobacco smoke analyzed on the FID and HID. Column 25 m × 0.32 mm I.D., coated with 5% phenylmethylsilicone. Column conditions: initial temperature, 60 °C for 2 min; then programmed 4 °C/min to 180 °C, with an additional hold for 16 min. Reprinted with permission from R. S. Ramsey and R. A. Todd, *J. Chromatogr.* **399**, 139 (1987). Copyright 1987, Elsevier Science Publishers, Physical Sciences & Engineering Division.

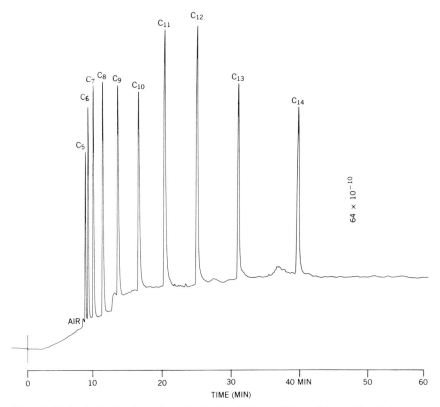

Figure 3.10. Analysis of hydrocarbons C_5 to C_{14}. Column: 100 m × 0.5 mm i.d. stainless steel coated with Witconol LA-23. Column conditions: initial temperature, 90 °C for 2 min; then programmed at 10 °C/min to 160 °C. Reprinted with permission from F. F. Andrawes, R. S. Brazell, and E. K. Gibson, Jr., *Anal. Chem.* **52**, 891 (1980). Copyright 1980, American Chemical Society.

A temperature-programmed analysis (up to 180 °C) of a complex sample separated on a FSOT column is shown in Figure 3.9. The chromatogram was obtained under pulsed conditions for the HID, using an on-column injection and adding makeup gas of 25 mL/min to the detector. It shows a moderate increase in background enabling the determination of compounds eluting at higher temperatures. A comparison with an FID analysis is also shown. Based on signal-to-noise ratios, the HID has been found to be about 50 times more sensitive than the FID for compounds that provide a good response by flame ionization (17). For example, a detection limit of 2 pg has been reported for

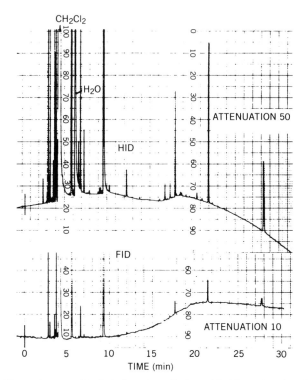

Figure 3.11. Chromatogram of technical grade methylene chloride analyzed on the FID and HID. Column: 50 m × 0.25 mm i.d. fused silica coated with Carbowax 20M. Column conditions: initial temperature, 50 °C for 2 min; then programmed at 5 °C/min to 180 °C. Split injection: 385:1. Reprinted with permission from F. F. Andrawes, *J. Chromatogr.* **390**, 65 (1984). Copyright 1984, Elsevier Science Publishers, Physical Sciences & Engineering Division.

dodecane (17) and less than 80 pg for benzene (14) on the HID. The latter determination was made at 200 V in the saturation region of the detector. An increase in sensitivity may be obtained in this case by operating at higher voltages. The results of a temperature-programmed analysis is also shown in Figure 3.10 for some hydrocarbons (C_5 through C_{15}) separated on a WCOT column. The sample was introduced by direct injection through the specially designed port illustrated in Figure 3.8. Figure 3.11 shows the results of an analysis of technical grade methylene chloride in a temperature-programmed run and shows the impurities present in the sample. It also compares the response to the FID.

Because the HID is truly non selective, it is especially important for determining species that are difficult to analyze with other chromatographic detectors (for example, inorganic gases, perfluorocarbons, water, formic acid, and formaldehyde). Andrawes has reported on the determination of formaldehyde in aqueous solution (20). Other chromatographic methods for this compound require sample preconcentration or derivatization for trace level analysis. Figure 3.12 shows the response to 18.5 ppm CH_2O in H_2O, separated on a capillary column coated with Carbowax 20M. The sample was introduced by

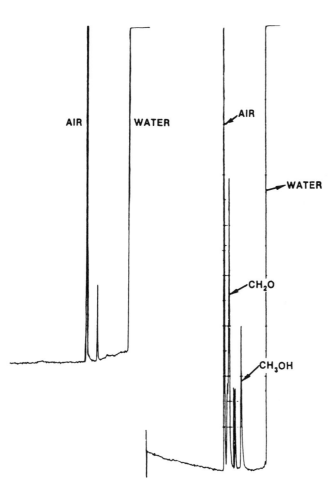

Figure 3.12. Chromatogram of 18.5 ppm formaldehyde in aqueous solution. Column: 50 m × 0.25 mm i.d. fused silica coated with Carbowax 20M. Column conditions: 70 °C isothermal. Split injection: 110:1. The figure to the left is a blank. Reprinted with permission from F. F. Andrawes, *J. Chromatogr.* **290**, 65 (1984). Copyright 1984, Elsevier Science Publishers, Physical Sciences & Engineering Division.

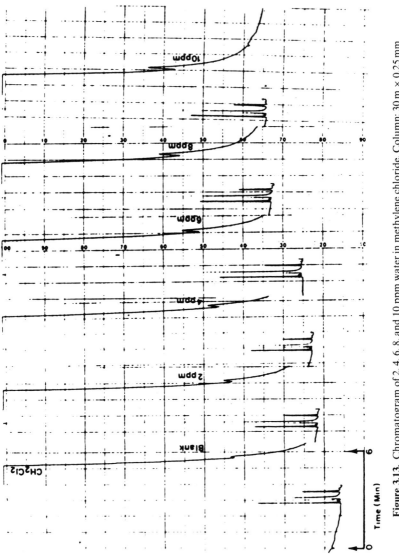

Figure 3.13. Chromatogram of 2, 4, 6, 8, and 10 ppm water in methylene chloride. Column: 30 m × 0.25 mm i.d. fused silica coated with Carbowax 20M. Column temperature: 50 °C isothermal. Split injection: 385:1 ratio. Reprinted with permission from F. F. Andrawes, *J. Chromatogr.* **290**, 65 (1984). Copyright 1984, Elsevier Science Publishers, Physical Sciences & Engineering Division.

using a split injection technique with a 385:1 ratio. Based on the magnitude of the response, it was reported that less than 1 ppm of the compound can be determined easily.

One of the most difficult analyses to perform is that for trace levels of water. The HID is the only chromatographic detector that is capable of ppm-to-ppb determinations of this compound. The problems involved in this analysis, however, extend beyond detection. Water is ubiquitous and therefore difficult to eliminate in a chromatographic system at the sub-ppm range. It is polar and adsorbs on most surfaces. Also it is liberated from internal surfaces of sample injection valves as a result of switching of the rotor, or from syringes when an injection is made (23). Finally, it is extremely challenging to generate reliable standards of this compound at low levels. The analysis of water vapor in gaseous samples using the HID and a packed column has been reported (23). For the determination of water in liquid samples, a Carbowax 20M capillary column has been used with a split injection technique (20). Less than 2 ppm water in methylene chloride could be determined. The response was linear from the detection limit (about 1 ppm) to 700 ppm. It was not possible to analyze lower concentrations because the solvent in which the standards were prepared contained trace amounts of water (i.e., slightly less than 1 ppm). Figure 3.13 shows the response from 2 to 10 ppm. To perform this type of analysis it is especially important to minimize the background or blank from syringe injections and to minimize exposure of standards and sample to the atmosphere (20). The latter may be accomplished by working in a dry glove box. It is also important to dry the solvent that will be used in preparation of the standards. Figure 3.14 shows the analysis of high-purity toluene (analyzed reagent grade, Baker, Phillipsburg, NJ) after it was dried with well-conditioned molecular sieves. Trace amounts of water are still detectable in the solvent. Note also in this figure the number of low-boiling impurities that are present in the sample. The three peaks eluting after toluene are the three xylene isomers (ortho, meta, and para).

7.2. The Analysis of Gaseous Samples

Gaseous samples are often analyzed with packed columns that accommodate relatively large sample sizes (0.5–1.0 mL or greater). Owing to the high sensitivity of the HID, however, 100 μL or less are often sufficient to achieve ppb detection. With such small sample quantities, micropacked, PLOT, or wide-bore capillary columns can be used. Figure 3.15 shows the analysis of ppm levels of neon, hydrogen, argon + oxygen, nitrogen, methane, and carbon monoxide on a molecular sieve 13X micropacked column (2 m × 1.1 mm i.d.). The analysis is completed in 2 min, and the peaks are well separated, with the exception of argon and oxygen. These species may be resolved on a longer

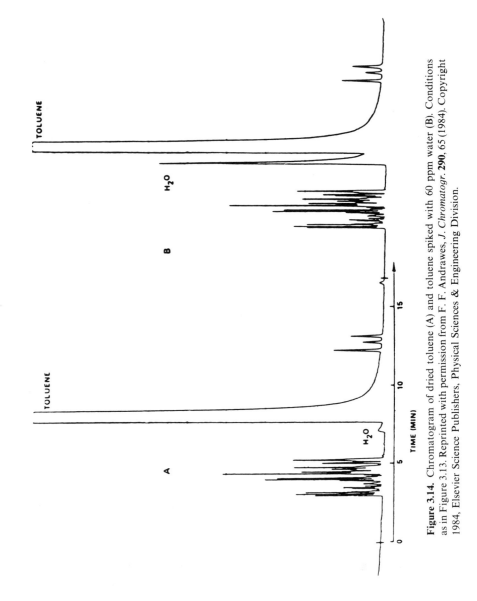

Figure 3.14. Chromatogram of dried toluene (A) and toluene spiked with 60 ppm water (B). Conditions as in Figure 3.13. Reprinted with permission from F. F. Andrawes, *J. Chromatogr.* **290**, 65 (1984). Copyright 1984, Elsevier Science Publishers, Physical Sciences & Engineering Division.

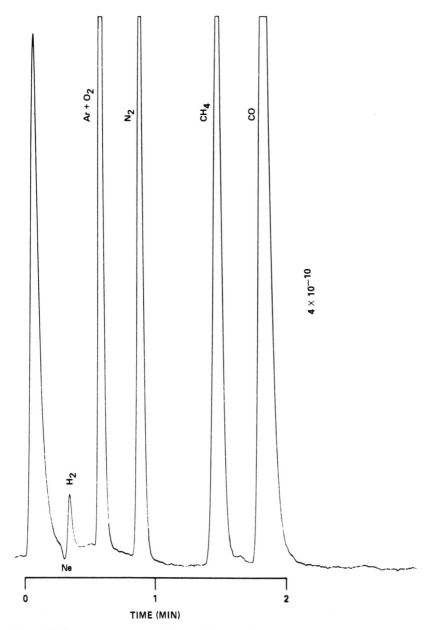

Figure 3.15. Detector response to 11 ppm hydrogen, 24 ppm argon, 9 ppm oxygen, 29 ppm nitrogen, 11 ppm methane, and 23 ppm carbon monoxide. Column: 2 m × 1.1 mm i.d. packed with molecular sieve 13X, 60/80mesh. Carrier gas flow rate, 14 mL/min; column temperature, 40 °C isothermal; sample size, 100 μL. Reprinted with permission from F. F. Andrawes, R. S. Brazell, and E. K. Gibson, Jr., *Anal. Chem.* **52**, 891 (1980). Copyright 1980, American Chemical Society.

molecular sieve column. The detection limits for the gases shown in Figure 3.15 are reported to be in the ppb range (14). The molecular sieve column used was conditioned at 150 °C; higher temperatures will result in increased retention and broadening of the carbon monoxide peak. Figure 3.16 shows the analysis of air on the same column. In this case both hydrogen and methane (0.5 and 2 ppm, respectively) are easily detected in the presence of large quantities of nitrogen and oxygen, illustrating that trace components can be determined in a matrix other than helium.

PLOT fused silica columns coated with molecular sieve 5A have been used to analyze carbon monoxide in air. When a split ratio of 10:1 was used, 2–3 ppm of carbon monoxide were determined. It seems that even with such a small split ratio, the sensitivity of the detector remained high. Apparently back-diffusion through the splitter was not significant. On the same column a standard mixture of six gases was separated as shown in Figure 3.17. The PLOT columns coated with porous polymers such as Porapak Q have been

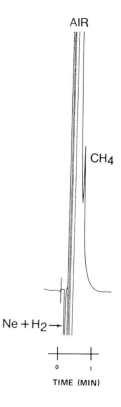

Figure 3.16. Chromatogram showing hydrogen and methane in air. Column: 2 m × 1.1 mm i.d. packed with molecular sieve 13X, 60/80 mesh. Carrier gas flow rate, 14 mL/min; column temperature, 60 °C isothermal; sample size, 100 μL.

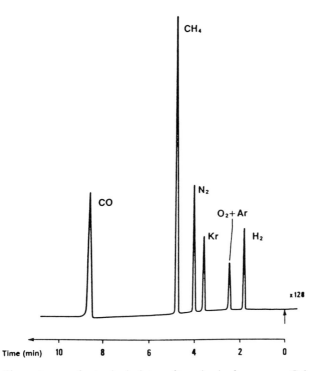

Figure 3.17. Chromatogram of a standard mixture of trace levels of seven gases. Column: 25 m × 0.32 mm i.d. PLOT column coated with molecular sieve 5A. Column temperature, 35 °C iso-thermal; split ratio 10:1; sample size 1.5 mL. Reprinted with permission from G. R. Verga, *HRC & CC, J. High Resolut. Chromatogr. Chromatogr. Commun.* **8**, 456 (1985). Copyright 1985, Alfred Huethig Publishers.

used to analyze of a variety of gases and vapors with the flame ionization and thermal conductivity detectors (27). The potential of these columns for the analysis of trace water, light alcohols, formaldehyde, and light hydrocarbons with the HID should be explored. It should be possible to detect these compounds in the ppb range.

Wide-bore capillary columns (0.5–0.75 mm i.d.) have also been used with the HID. The analyses of 1-butene, 1-pentene, and 1-hexene have been reported at the 40 ppb level on a 100 m × 0.5 mm i.d. column coated with a moderately polar stationary phase (14). Ethane, propane, isobutane, and butane have been separated on a 50 m × 0.5 mm i.d. column coated with PS-255 using a split injection (27).

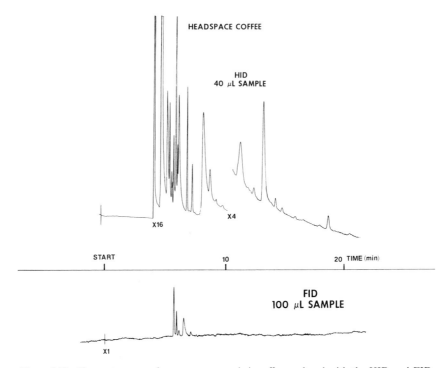

Figure 3.18. Chromatograms of aroma compounds is coffee analyzed with the HID and FID. Column: 100 m × 0.5 mm i.d. stainless steel coated with Witconol LA-23. Column temperature: 60 °C isothermal. Sample introduced via a gas sampling valve. Reprinted with permission from F. F. Andrawes and E. K. Gibson, *HRC & CC, J. High Resolut. Chromatogr. Chromatogr. Commun.* **5**, 265 (1982). Copyright 1982, Alfred Huethig Publishers.

The compounds in the headspace of a ground coffee sample have been analyzed on a 100 m × 0.5 mm i.d. stainless steel column coated with Witconol LA-23, a slightly polar phase (22). Figure 3.18 shows the chromatograms obtained on both the FID and HID. More than 20 compounds were detected on the HID before a large water peak was eluted, interrupting the analysis. It is obvious that the HID provided a much greater response than the FID. The sample was introduced with a gas sampling valve. The more inert wide-bore fused silica columns and on-column injection techniques that are available today should improve this type of analysis. It is well known, for example, that some compounds are adsorbed on stainless steel, resulting in peak tailing or in some cases complete loss. The HID can be useful for the analysis of aroma compounds found in headspace samples. The sensitivity of the detector enables the analysis to be performed directly without preconcentration steps that often change the nature or the composition of the sample. Since water will be the most abundant compound present, it will mask part of the chromatogram. The column could, however, be programmed to elute higher boiling or more polar species after the water has eluted. It may also be possible to obtain information on the compounds that are masked by water by analyzing the sample on a column of different polarity that shifts the retention of water.

REFERENCES

1. E. Bros and J. Lasa, *Chromatographia* **13**, 567 (1980).
2. F. F. Andrawes and E. K. Gibson, Jr., *Anal. Chem.* **50**, 1146 (1978).
3. F. F. Andrawes, T. B. Byers, and E. K. Gibson, Jr., *Anal. Chem.* **53**, 1544 (1981).
4. S. Lukac and J. Sevcik, *Chromatographia* **5**, 258 (1972).
5. J. Sevcik, *Detectors in Gas Chromatography*, p. 131. Elsevier, New York, 1976.
6. F. W. Lampe, *Ion-Mol. React.* **2**, 601 (1972).
7. F. F. Andrawes and E. K. Gibson, Jr., *Anal. Chem.* **52**, 846 (1980).
8. F. F. Andrawes. E. K. Gibson, Jr., and D. A. Bafus, *Anal. Chem.* **52**, 1377 (1980).
9. G. S. Hurst and C. E. Klots, in *Advances in Radiation Chemistry* (M. Burton and J. L. Magee, eds.), p. 1. Wiley (Interscience), New York, 1976.
10. J. E. Lovelock, *J. Chromatogr.* **1**, 35 (1958).
11. S. Lukas and J. Sevcik, *Chromatographia* **5**, 311 (1972).
12. F. H. Woeller, D. J. Kojiro, and G. C. Carle, *Anal. Chem.* **56**, 860 (1984).
13. M. Yamane, *J. Chromatogr.* **14**, 355 (1964).
14. F. F. Andrawes, R. S. Brazell, and E. K. Gibson, Jr., *Anal. Chem.* **52**, 891 (1980).
15. P. Deng and F. F. Andrawes, *J. Chromatogr.* **349**, 415 (1985).
16. R. S. Brazell and R. A. Todd, *J. Chromatogr.* **302**, 257 (1984).
17. R. S. Ramsey and R. A. Todd, *J. Chromatogr.* **399**, 139 (1987).
18. R. A. Todd, *Pulsed Helium Ionization Detector Electronics System for Gas Chromatography*, Ph.D. Dissertation, University of Tennessee, Knoxville, 1988.
19. F. F. Andrawes and P. Deng, *J. Chromatogr.* **349**, 405 (1985).
20. F. F. Andrawes, *J. Chromatogr.* **290**, 65 (1984).
21. J. D. David, *Gas Chromatographic Detectors*, p. 165. Wiley (Interscience), New York, 1965.
22. F. F. Andrawes and E. K. Gibson, *HRC & CC, J. High Resolut. Chromatogr. Chromatogr. Commun.* **5**, 265 (1982).
23. F. F. Andrawes, *Anal. Chem.* **55**, 1869 (1983).
24. D. R. Dean, *J. Chromatogr.* **289**, 43 (1984).
25. F. F. Andrawes and S. Greenhouse, *J. Chromatogr. Sci.* **26**, 153 (1988).
26. J. de Zeeuw, R. C. M. de Nijs, and L. T. Henrick, *J. Chromatogr. Sci.* **25**, 71 (1987).
27. G. R. Verga, *HRC & CC, J. High Resolut. Chromatogr. Chromatogr. Commun.* **8**, 456 (1985).

CHAPTER

4

FAR-UV IONIZATION (PHOTOIONIZATION) AND ABSORBANCE DETECTORS

JOHN N. DRISCOLL

HNU Systems Inc.
Newton, Massachusetts

High-resolution chromatography has become more important during the past decade as a result of the development of fused silica capillary columns that change the average column resolution in gas chromatography (GC) from several thousand to more than 20,000 plates. The photoionization detector (PID) has emerged over the past decade as a versatile detector for a variety of organic compounds and has applications in capillary GC, high-performance liquid chromatography (HPLC), and supercritical fluid chromatography (SFC). Application of the PID to the latter techniques is possible because of the high ionization potential of many of the solvents used as the mobile phases.

The far-ultraviolet absorbance detector (FUVAD) is of a more recent vintage (about 1987) but has shown some promise for the detection of low-molecular-weight organic compounds, and low- or sub-ppm levels of inorganic materials such as water and oxygen after separation by GC. No applications of this detector are apparent in HPLC or SFC since most solvents used as mobile phases strongly absorb in the far-UV range, thus eliminating the possibility of trace level detection.

The PID and FUVAD are concentration sensitive detectors with a response that varies inversely with flow rate of the carrier gas:

$$C = 1/F$$

where C is the solute concentration and F is the carrier gas flow rate. This concentration sensitivity is particularly important for high resolution GC,

Detectors for Capillary Chromatography, edited by Herbert H. Hill and Dennis G. McMinn.
Chemical Analysis Series, Vol. 121.
ISBN 0-471-50645-1 © 1992 John Wiley & Sons, Inc.

HPLC, and SFC, where the lower flow rates can increase the sensitivity of such detectors.

Background information, analytical applications and sources for the far-UV radiation are described in detail in the following sections.

1. PHOTOIONIZATION DETECTION

1.1. Background

The use of far-UV ionization (photoionization) detection for GC was first described in the late 1950s along with flame ionization (FID), electron capture (ECD), mass spectrometer, and cross-section detection (1). During the 1960s, there was considerable research on the PID (2–4) but the FID gained popularity so quickly that all commercial gas chromatograph manufacturers offered the FID along with the thermal conductivity detector (TCD). In the meantime, the PID continued to be a sensitive, novel detector but required a vacuum (1–10 Torr) for operation, was easily fouled by column bleed, was unstable, and required a skilled operator. By the later 1960s, a decade after its discovery, the PID had been virtually abandoned. By the mid-1970s, however, the PID was reborn when several researchers (5, 6) found that separating the lamp discharge from the ion chamber improved the stability of the PID, and the detector was simplified. Driscoll and Spaziani (7) described additional improvements to the PID that resulted in an improved range ($> 10^7$ instead of 10^4), improved stability, and lower background. It also minimized the problem of column bleed that plagued earlier PIDs. After a slow start in the 1960s, the PID was finally on its way in the later 1970s to becoming an important tool in the arsenal of the analytical chemist. Davenport and Adlard have reviewed

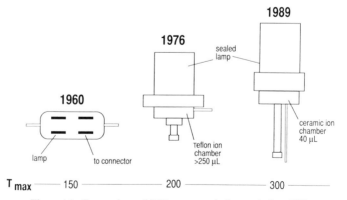

Figure 4.1. Comparison of PID geometry before and after 1970.

the PID for its application with respect to analysis of alkenes and aromatic compounds; they found that it has a high sensitivity and selectivity to these materials (8). Some suggestion for an improved detector for analysis via capillary columns was also included.

1.2. Theory of Operation

In 1976, the first commercial PID was described by HNU Systems, Inc., Newton, Massachusetts (9). Schematic drawings of that detector and a recent (third-generation) PID are shown in Figure 4.1. The process of photonionization starts with the absorption of a photon (hv) by a molecule R. If the energy of the photon is more than or equal to the ionization potential of species R and the carrier gas is given by C, then ionization and the other processes shows here occur:

Direct

$$R + hv \longrightarrow R^+ + e^- \qquad (k_1)$$

Indirect

$$C + hv \longrightarrow C^* \qquad (k_2)$$

$$C^* + R \longrightarrow R^+ + e^- + C \qquad (k_3)$$

$$R + hv \longrightarrow R^* \qquad (k_4)$$

$$R^* \longrightarrow R_1^+ + e^- \qquad (k_5)$$

Quenching

$$R^+ + e^- \longrightarrow R \qquad (k_6)$$

$$e^- + C \longrightarrow C^- \qquad (k_7)$$

$$R^+ + C^- \longrightarrow R + C \qquad (k_8)$$

The number of ions formed is equal to the summation of the direct and indirect process minus the quenching process:

$$(R_{sum}^+) = [k_1 + k_3 + k_5 - (k_6 + k_8)]$$

Note that in k_5, ions may also result from a photochemical rearrangement process leading to the formation of R_1. This process may account for the photoionization signal observed for the 10.2-eV lamp with methylene chloride (ionization potential $= 11.3$ eV). Here, the rearrangement and resultant loss of Cl_2 may form an ethylenic species with a lower ionization potential. The energy of a 10.2-eV lamp, which corresponds to more than 230 kcal/mol, is greater than the 80 kcal/mol required to break a typical carbon–carbon bond. One can calculate the energy (E) in eV or wavelength (nm) from the following equation:

$$W(nm) = 1234.5/E(eV)$$

Thus, a 10.2-eV lamp has an energy of 121 nm.

The photoionization yield (PI_{yield}) is the number of ions produced per photon absorbed:

$$PI_{yield} = 100(R^+_{sum})/(\text{number of photons absorbed})$$

From this equation, the number of ions produced (R^+_{sum}) is proportional to the product of the absorption coefficient and the intensity of the discharge lamp.

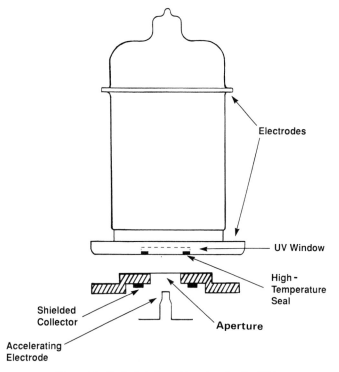

Figure 4.2. Exploded view of ion chamber for PID.

A typical PID has two functional parts: an excitation source and an ionization chamber. An expanded view of the lamp and ionization chamber is shown in Figure 4.2. A potential of 100–200 V is applied to the accelerating electrode to push the ions formed by UV ionization (above) to the collection electrode where the current (proportional to concentration) is measured. Note that the collection electrode is shielded to reduce the background (9) and the configuration of the ion chamber is axial. It can be shown that the most effective design of an ionization chamber is one with a coaxial configuration where the field is given by

$$E = V/2.3r \log(a/b)$$

Here V is the applied voltage between the collector of radius a and the accelerating electrode of radius b, while E is the electric field at any point in distance r from the center of the accelerating electrode. Thus, the field increases rapidly as $r \rightarrow b$. If the collection electrode is in the UV beam, the background will increase by one or two orders of magnitude. On the other hand, placing the accelerating electrode in the UV beam does not increase the background at all.

The HNU ion chamber is novel because the collection electrode is shielded from the UV beam. This design resulted in the lowest background possible with the widest dynamic range. These two features provided a PID with a low background current, high sensitivity, and an improvement of 2–3 orders of magnitude in linear range (7, 9) with a dynamic range of 10^7.

1.3. Excitation Sources

1.3.1. Discharge

The excitation source may be a discharge lamp excited by direct current (1–2 kV), a radio frequency (75–125 kHz), a microwave (2450 MHz), or a laser. For the sealed discharge lamp, the lamp is filled with a nonreactive gas at low pressure and produces an emission line or lines in the far-UV region. A typical emission spectrum for a d.c. excited discharge is shown in Figure 4.3. Note that the 11.7 eV has two emission lines at 104.4 and 106.6 nm, the two argon resonance lines.

Alkali or alkaline earth fluorides are used most commonly for the lamp windows since few other materials transmit in this region. The short-wavelength cutoffs for these windows are lithium fluoride, 105 nm; magnesium fluoride, 112 nm; calcium fluoride, 122 nm; strontium fluoride, 128 nm. The 10.2-eV lamp has a high temperature seal that will withstand temperatures > 300 °C for short periods of time and temperatures of 275 °C for proionged

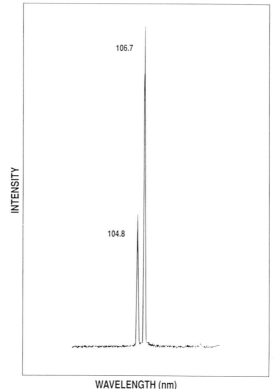

Figure 4.3. Emission spectrum of an 11.7-eV lamp.

periods. This lamp has a lifetime of > 5000 h. The lifetimes of the 9.5- and 8.3-eV lamps are similar. The 11.7-eV lamp, however, has a lithium fluoride window that has a problem with solarization (color center formation) that limits its useful lifetime to a few hundred hours. A recently improved lamp, shown in Figure 4.4, has a nickel cathode to bleach the color centers (formed by absorption of the short-wavelength photons and trapping of electrons just below the conduction band of lithium fluoride) and extends the lifetime to 500–1000 h. The trapped electrons are released by irradiation with UV (about 5–7 eV) from the Ni lines in the lamp. Another problem with this lamp is the shift in short-wavelength cutoff to a longer wavelength as the temperature of the window increases. Since the argon emission lines (104.4 and 106.6 nm) are close to the cutoff at ambient temperatures, heating the window above 125 °C reduces the transmission of lithium fluoride and therefore the photon flux. At

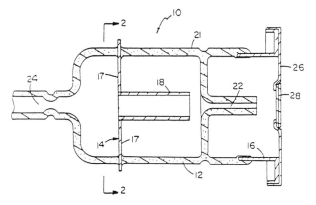

Figure 4.4. Construction of an improved 11.7-eV lamp: #18, metal cathode; #28, lithium fluoride window; #14 and #26, discharge electrodes.

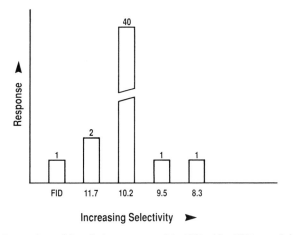

Figure 4.5. Comparison of the relative response of the FID with a PID containing various lamps (energies).

the same time, the solarization of the window increases as a result of the increased absorption of photons by lithium fluoride.

The most popular PID lamp is the 10.2-eV model, which has the highest photon flux and therefore the greatest sensitivity. This lamp uses a magnesium fluoride window. Figure 4.5 illustrates the variation in sensitivity relative to lamp energy. In comparing the two chromatograms in Figure 4.6, the difference in sensitivity (attenuation) of benzene should be noted (10). The 10.2-eV

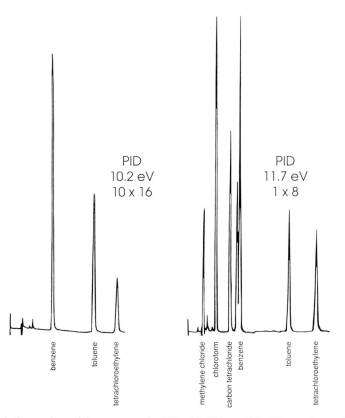

Figure 4.6. Comparison of the response of a PID with 10.2- or 11.7-eV lamps for a mixture of aromatic and chlorinated hydrocarbons.

lamp is more selective than the 11.7-eV lamp, as it does not respond to the low-molecular-weight chloroalkanes. The peak heights are approximately the same, but the attenuation differs by a factor of 20. There are two reasons for the difference in intensity of these lamps: (1) the transmission of lithium fluoride (in the 11.7-eV lamp) is only 20% that of magnesium fluoride (in the 10.2-eV lamp) at the appropriate resonance line output, and (2) the absolute energy output of the 10.2-eV lamp is much higher than that of the 11.7-eV lamp.

1.3.2.   Lasers

The UV discharge lamps just described differ from lasers in several respects. They are of low intensity (about 10^{12} photons/s) and are noncoherent. Since

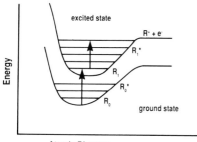

Figure 4.7. Morse curve of potential energy vs interatomic distance for the ground state (R_0) and the first electronically excited state (R_1).

the sensitivity of the PID is proportional to the intensity of the light source, lasers will have an inherently higher sensitivity. The development of high-peak-power and high-repetition-rate tunable UV lasers has made multiphoton ionization (MPI) feasible (11). In this process, the first photon excites the molecule from the ground state to the first excited electronic state and the absorption of the second photon produces ionization. Typical Morse curves for photoionization and two-photon ionization are shown in Figure 4.7. In the two-photon ionization process, the sum of the energy of the two photons must be greater than the ionization potential of the molecule and the two photons can have either identical or different frequencies. Since many large molecules have ionization potentials from 7 to 11 eV (176 to 112 nm), two-photon ionization can be initiated with near-UV laser sources. Typical laser sources can be excimer pumped frequency doubled dye lasers.

Two-photon ionization can produce ionization yields of several percent compared with $10^{-5}\%$ for conventional photoionization processes (above). Provided the repetition rate (duty cycle) is sufficiently high, the "potential" sensitivity is higher than that obtained with a typical discharge lamp. Although Klimet and Wessel (12) showed that laser photoionization is "potentially" more sensitive than ionization with a discharge lamp, their data is not conclusive since the low-picogram detection limits reported can be achieved using a conventional UV discharge lamp in the PID. Ogawa et al. (13) also report detection limits of 6 pg for pyrene using laser two-photon ionization.

Two major problems with a laser replacing a discharge lamp in a PID are the cost ($50,000 vs. $300) and the size (several cubic meters vs several cubic centimeters). Should these problems be overcome, the possibility of selectivity determining isomers or classes of compounds by two-photon ionization will become a reality.

Table 4.1. Ionization Potentials of Selected Molecules

Atoms and simple molecules	IP
H₂	15.426
N₂	15.580
O₂	12.075
CO	14.01
NO	9.25
F₂	15.7
Cl₂	11.48
Br₂	10.55
I₂	9.28
HF	15.77
HCl	12.74
HBr	11.62
HI	10.38
SO₂	12.34
CO₂	13.79
COS	11.18
CS₂	10.08
N₂O	12.90
NO₂	9.78

Aliphatic Alcohols, Ethers,	IP
Ethyl chloride	10.98
1,2-Dichloroethane	11.12
1-Chloropropane	10.82
Methyl bromide	10.53
Dibromomethane	10.49
CH₂BrCl	10.77
CHBr₂Cl	10.59
Ethyl bromide	10.29
1,1-Dibromoethane	10.19
1-Bromopropane	10.18
Methyl iodide	9.54
CFCl₃ (Freon 11)	11.77
CF₂Cl₂ (Freon 12)	12.31
CF₃Cl (Freon 13)	12.91
CHClF₂ (Freon 22)	12.45
CFBr₃	10.67
CF₂Br₂	11.07
CH₃CF₂Cl (Genetron 101)	11.98
CFCl₂CF₂Cl	11.99
CF₃CCl₃ (Freon 113)	11.78

Some Derivatives of Olefins	IP
Vinyl chloride	9.995
Trichloroethylene	9.45
Tetrachloroethylene	9.32
Vinyl bromide	9.80
3-Chloropropene	10.04
Acrolein	10.10
Crotonaldehyde	9.73
Vinyl acetate	9.19

	IP
Acrolein	10.10
Crotonaldehyde	9.73
Benzaldehyde	9.53
Acetone	9.69
Methyl ethyl ketone	9.53
Cyclohexanone	9.14

Aliphatic Acids and Esters	IP
Formic acid	11.05
Acetic acid	10.37
Propionic acid	10.24
Ethyl acetate	10.11

Aliphatic Amines and Amides	IP
Methyl amine	8.97
Ethyl amine	8.86
n-Propyl amine	8.78

Heterocyclic Molecules	IP
Pyrrole	8.20
Pyridine	9.32
Furan	8.89
Tetrahydrofuran	9.45
Thiophene	8.860

Aromatic Compounds	IP

	IP
H_2O	12.59
H_2S	10.46
NH_3	10.15

Paraffins and Cycloparaffins	IP
Methane	12.98
Ethane	11.65
Propane	11.07
n-Butane	10.63
n-Pentane	10.35
n-Hexane	10.18
n-Heptane	10.08
Cyclohexane	9.88

Alkyl Halides	IP
Methyl chloride	11.28
Dichloromethane	11.35
Trichloromethane	11.42
Tetrachloromethane	11.47

Thiols, and Sulfides	IP
Methyl alcohol	10.85
Ethyl alcohol	10.48
n-Propyl alcohol	10.20
Dimethyl ether	10.00
Methanethiol	9.440
Ethanethiol	9.285
1-Propanethiol	9.195
Dimethyl sulfide	8.685
Ethyl methyl sulfide	8.55
Diethyl sulfide	8.430

Aliphatic Aldehydes and Ketones	IP
Formaldehyde	10.87
Acetaldehyde	10.21
Propionaldehyde	9.98

	IP
Formamide	10.25
Acetamide	9.77
Nitromethane	11.08
Nitroethane	10.88

Other Aliphatic Molecules with a N Atom	IP
HCN	13.91
Acetonitrile	12.22
Propionitrile	11.84

Olefins, Cyclo-olefins, Acetylenes	IP
Ethylene	10.515
Propylene	9.73,
1-Butene	9.58
trans-2-Butene	9.13
1-Hexene	9.46
1,3-Butadiene	9.07
Acetylene	11.41
1-Butyne	10.18

	IP
Benzene	9.245
Toluene	8.82
Ethylbenzene	8.76
Biphenyl	8.27
Phenol	8.50
Naphthalene	8.12
Styrene	8.47
o-Xylene	8.56
p-Xylene	8.445
Mesitylene	8.40
Aniline	7.70
Fluorobenzene	9.195
Chlorobenzene	9.07
Bromobenzene	8.98
Iodobenzene	8.73

Miscellaneous Molecules	IP
Ethylene oxide	10.565
Propylene oxide	10.22
p-Dioxane	9.13
Methyl disulfide	8.46
Ethyl disulfide	8.27
Phosgene	11.77

1.4.   Gas Chromatography Applications

The 11.7-eV lamp will respond to many low-molecular-weight compounds that have more tightly bound electrons and hence higher ionization potentials. Typical applications include the detection of low- or sub-ppm (nanogram) levels of formaldehyde (14), which cannot be measured directly by any other detector at these levels. The 11.7-eV lamp is also useful for the detection of low-molecular-weight compounds (< 150 amu) such as CCl_4, $CHCl_3$, C_2H_6, and C_2H_2 that have IPs > 10.5 eV. A comparison of the response for some typical hazardous waste components using a 10.2- and a 11.7-eV lamp is shown in Figure 4.6. Chloroalkanes are not detected with the 10.2-eV lamp. The 11.7-eV lamp PID is similar in response to the FID for many hydrocarbons (except for methane, which does not respond), and it also responds to inorganic compounds such as Cl_2, PH_3, and I_2. The FID only responds to carbon-containing compounds. A list of ionization potentials (IPs) is given in Table 4.1.

The major difference between the 10.2- and 9.5-eV lamps is the absolute intensity of the lines. There are certain applications where the 9.5-eV lamp is preferable to the 10.2-eV lamp. These applications include aromatics in an aliphatic matrix such as pentane, mercaptans in the presence of H_2S, and amines in the presence of ammonia (15). With the 8.3-eV lamp there is a considerable increase in selectivity compared to the 10.2-eV lamp. A benzene or toluene solvent produces no response on a PID (8.3 eV). The 8.3-eV lamp was selected for the determination of polyaromatic hydrocarbons (16) because of the increased selectivity. Typical detection limits were low- to sub-nanogram levels. Chromatograms of complex mixtures using these three lamps and the FID are shown in Figures 4.8 and 4.9. The pattern for the 10.2-eV PID lamp is similar to that of the FID, as noted previously (17), whereas the chromatograms are simpler for the 8.3-eV lamp. In a previous publication (10), we reported a lower "apparent" sensitivity for biphenyl (IP = 8.27 eV) used as a normalization compound. In later work, we found that if anthracene (IP = 7.5 eV) is used for normalization, a 10-fold higher response is obtained for the 8.3-eV lamp, which is attributable to the increased efficiency of anthracene.

Langhorst (18) determined the sensitivities for nearly two hundred compounds for a PID with a 10.2 eV lamp. She found that the PID was a carbon counter (on a molar basis); that the sensitivity for alkanes < alkenes < aromatics; that sensitivities for cyclic > monocyclic and branched > nonbranched; and that for substituted benzenes, ring activators increased the sensitivity whereas ring deactivators decreased the sensitivity. Figure 4.10 depicts a typical curve for a homologous series of hydrocarbons compared to benzene. A comparison of the sensitivity (normalized to benzene) for a variety

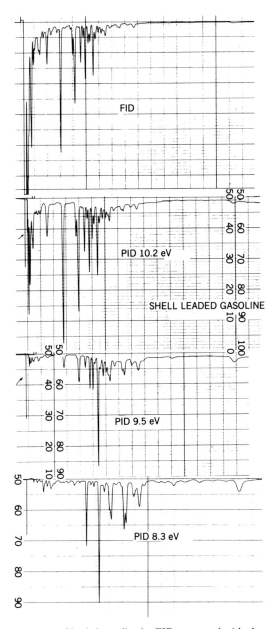

Figure 4.8. Chromatogram of leaded gasoline by FID compared with chromatograms by PID (10.2, 9.5, and 8.3 eV).

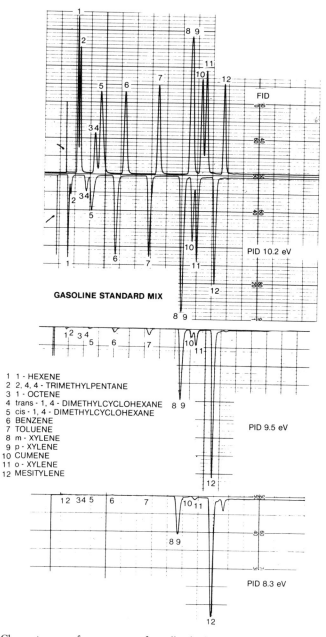

1 1 - HEXENE
2 2, 4, 4 - TRIMETHYLPENTANE
3 1 - OCTENE
4 trans - 1, 4 - DIMETHYLCYCLOHEXANE
5 cis - 1, 4 - DIMETHYLCYCLOHEXANE
6 BENZENE
7 TOLUENE
8 m - XYLENE
9 p - XYLENE
10 CUMENE
11 o - XYLENE
12 MESITYLENE

Figure 4.9. Chromatogram of components of gasoline by FID compared with chromatograms by PID (10.2, 9.5, and 8.3 eV).

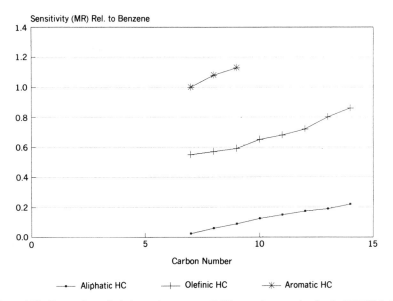

Figure 4.10. Comparison of relative molar response (MR) vs. carbon number for the PID (10.2 eV).

of substituted benzenes is shown in Figure 4.11. The GC–PID response for substituted benzenes changes slightly with electronegative substituents, yet the response for the same compounds using LC–PID changes dramatically (see Section 1.5, below).

In 1985 (19), a review of the first decade following the commercial introduction of the PID showed that only 15% of the applications involved capillary GC. During the past few years, this has changed significantly—with 30% of the references using capillary columns. The primary reasons for this change involve wider use of capillary columns and the development of an improved PID (10, 20). A review of capillary column applications with the PID was done by Driscoll (21). Several papers on the analysis of aromatics in water were later published (22, 23).

The PID described in 1986 was a third-generation detector with many improved features (21). We investigated two different cell volumes (175 and 40 μL) with similar geometries and found that the latter cell operated optimally at low flow rates (< 1–15 mL/min) whereas the larger cell provided better results at the higher flow rates required for packed columns (10). The lower volume cell displayed a threefold lower sensitivity at 30 mL/min compared to the 175-μL cell. The data are shown in Figure 4.12.

How do the PIDs differ? As a result of the axial ion chamber design described earlier, the only way to reduce the dead volume is to reduce the

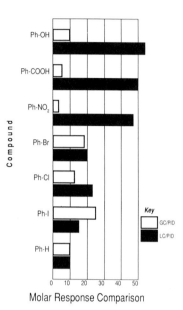

Figure 4.11. Comparison of the molar response for GC–PID with LC–PID.

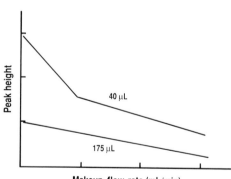

Figure 4.12. Comparison of peak height (sensitivity) vs. flow rate for the 40- and 175-μL PID cell volumes.

aperture size. This also reduces the light source flux and therefore the sensitivity of the PID. If we can reduce or eliminate the makeup flow needed, we can regain some of the lost sensitivity. We find that if we use a total flow rate of 2–3 mL/min for the carrier gas on a capillary column (no makeup), the low dead volume detector (40 μL) will be about three to five times more sensitive

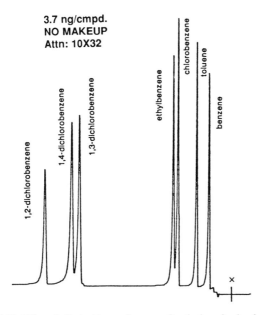

**3.7 ng/cmpd.
NO MAKEUP
Attn: 10X32**

Figure 4.13. Effect of eliminating makeup gas for the low dead volume PID.

than with the 175-μL cell as a result of the sharper peaks. Peak height vs. flow rate curves for the 175- and 40-μL detectors are shown in Figure 4.12, as well as the sensitivity vs. flow rate for each detector. It is obvious from the curves in Figure 4.12 that the 175-μL cell is more suitable when packed and capillary columns are to be used, since makeup gas offsets the dead volume at low flow rates. The 40-μL cell is most useful for capillary (0.05–0.75 mm) applications. A comparison of two chromatograms with and without makeup gas using the 40-μL PID is shown in Figure 4.13. The addition of 1.5 mL/min of makeup gas eliminated the tailing and achieved baseline resolution for all the components. With the standard 175-μL cell, it took 20–30 mL/min of makeup gas to achieve the same result. Other capillary chromatograms obtained for the PID are shown in Figures 4.14 and 4.15 and compared with the FID using 0.32-mm capillary columns. Note the large difference in sensitivity for the solvents between the two detectors.

Development of "fast GC" using a PID has been underway at the University of Michigan (24) for some time. By using a cryogenic inlet system, which was heated rapidly to inject the sample onto the capillary column (0.32 mm), the chromatogram in Figure 4.16 was obtained. This shows fast and excellent (baseline) separation of six components within 4. Some additional work on a fast PID for GC at the University of Eindhoven, Netherlands,

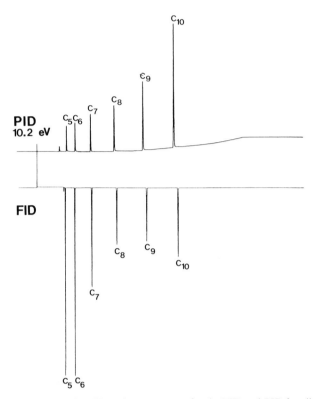

Figure 4.14. Comparison of capillary chromatograms for the FID and PID for alkanes (Nordi-bond SE54, 25 M × 0.32 mm; temp. prog. 50–180 °C at 10 °C/min). Courtesy of HNU-Nordion.

used a similar cryogenic trap approach but with smaller diameter (0.05-mm) capillary columns and longer retention times (minutes vs. seconds) as a result of the molecular weight of the components selected for analysis. Results for a peppermint oil sample obtained using a 0.05-mm capillary column are shown in Figure 4.17. Van Es and Rijks (25) used makeup gas for the 0.05-mm columns with no adverse effect on the performance of the detector. They found a minimum detectable quantity of 0.2 pg for anthracene. This is an order of magnitude lower than packed column detection limits and about two to three times better than detection with the low dead volume cell and no makeup gas (10). Although the reduction of cell volume and detector pressure are equivalent alternatives for other concentration detectors such as the TCD, van Es and Rijks (25) found that, for the PID, the reduction of dead volume was more critical than reducing the pressure. They also found that the effect of addition of makeup gas and reduced pressure on detection limits was equivalent.

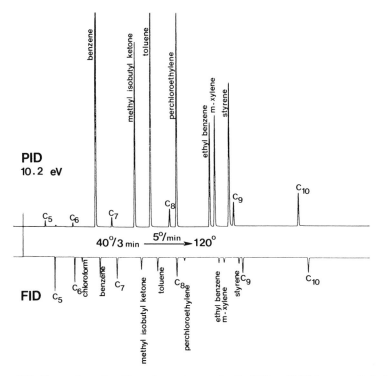

Figure 4.15. Comparison of capillary chromatograms for the FID and PID for aromatic hydrocarbons (Nordibond SE54, 25 M × 0.32 mm; temp. 40 °C/3 min, 40–120 °C at 5 °C/min). Courtesy of HNU-Nordion.

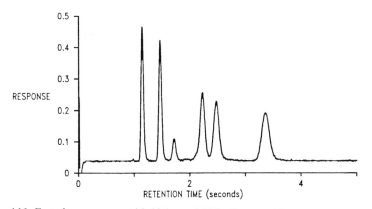

Figure 4.16. Fast chromatogram with high-speed PID. Courtesy of S. Levine, University of Michigan.

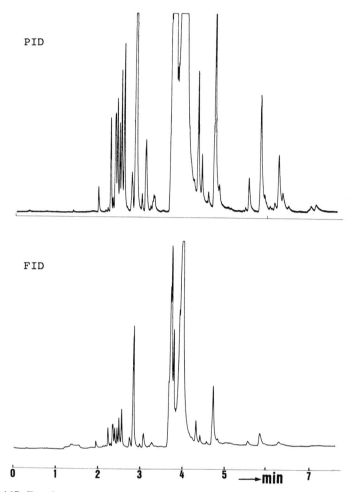

Figure 4.17. Fast chromatogram of peppermint oil using PID and FID in series (50-μM column). Courtesy of J. Rijks, University of Eindhoven.

1.5. Liquid Chromatography

Many papers have been published over the past 15 years on interfacing a PID to a liquid chromatograph (26), yet there are no commercial PIDs available for liquid chromatography (LC). This detector remains one of the more interesting detectors for HPLC and awaits development, as noted several years ago by the author (10). One interesting feature of the PID is that with

Table 4.2. Ionization Potentials of Some Common
Solvents for HPLC

Solvent	Ionization Potential (eV)
Water	12.35
Methanol	10.85
Chloroform	11.42
Carbon tetrachloride	11.47
Acetonitrile	12.22
Pentane	10.35

a 10.2-eV lamp, most of the common solvents used for reversed phase LC cannot be ionized (see Table 4.2).

Interfaces for LC have involved flowing the liquid directly into the cell or vaporizing the liquid before passing it through the cell (27). Both these approaches have drawbacks, and neither solves the problem of quenching, which limits the PID to the detection of quantities above the low-microgram (by the direct approach) or nanogram level (by the vaporization approach), in LC. On the other hand, low-picogram quantities are detected in GC. If these GC detection levels were approached more closely, this detector would become very useful for LC.

De Wit and Jorgensen (28) have described a PID that uses open tubular columns instead of conventional LC columns. The sample is vaporized before detection by the PID. Berthold et al. (29) have used laser two-photon photoionization to detect 28 polyaromatic hydrocarbons (PAHs) in an LC effluent. Typical detection limits were 50–70 pg for the PAHs compared with 6 pg detection reported by Ogawa et al. (13) for GC–PID. Note that the quenching problem described above also occurs with the laser source. A typical HPLC–PID system that vaporizes the solvent and sample before detection is shown in Figure 4.18. Selected detection limits obtained (26) by LC–PID are shown in Table 4.3. The sensitivity of this detector, and therefore the detection limits, differ considerably from those obtained by Langhorst (18) for the GC–PID described above. Why? The mechanism appears to be different. The simple mechanism described above for GC–PID [$R + h\nu \rightarrow R^+ + e^-$ (k_1)] is perturbed by the presence of high concentrations of electronegative species (solvent). In LC, where the solvent (S) is volatilized along with low levels of the solute (R), the mechanism is quite different. The reaction (k_1) occurs to a small extent because the majority of the UV radiation is absorbed by the solvent. The ionization potential of these common LC solvents is higher than the lamp energy; therefore the following indirect photoionization

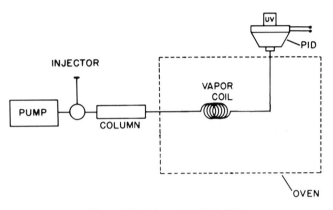

Figure 4.18. Schematic of LC–PID.

Table 4.3. Selected Detection Limits Obtained by LC–PID

Compound	Detection Limits (ng)
Bromobenzene	4
Chlorobenzene	4
Nitrobenzene	2
Phenol	3
Benzoic acid	2
Dioctyl phthalate	16
Benzene	15
Anthracene	75

mechanism will undoubtedly predominate:

$$S + h\nu \longrightarrow S^* \qquad (k_9)$$

$$S^* + R \longrightarrow R^+ + e^- + S \qquad (k_{10})$$

The quenching will be considerably greater in the LC mode than the GC mode because of the higher concentrations of electronegative species in the condensed solvents compared to the inert gases in the GC mode. This leads to the following quenching reactions:

$$S + e^- \longrightarrow S^* \qquad (k_{11})$$

$$S^* + R \longrightarrow R^+ + e^- + S \qquad (k_{12})$$

where $k_{12} \gg k_6$ owing to the increased size of S^- compared to e^-. It is clear that the mechanism and sensitivity of GC–PID and LC–PID are quite different, as seen from the above discussion and Figure 4.11. The order of response in Table 4.3 shows that strongly electronegative groups on a molecule improve the minimum detectable quantity.

It appears that the dipole, induced by the electronegative group, may prevent quenching of the ionized species before collection. The laser two-photon approach might be interesting for providing low-picogram detection of certain electronegative species such as amines or nitro compounds as a result of the improved sensitivity with HPLC–PID (as noted in Table 4.3 for anthracene). If only 2 orders of magnitude are lost by two-photon photoionization compared with 5 orders of magnitude lost for the d.c. discharge, and if similar data are observed for the electronegative species, it follows that low- or sub-picogram quantities of these species would be detected via the laser, making for a very promising technique.

We (26) have demonstrated that sub-nanogram levels of a variety of species can be detected by injecting large sample volumes (for example $50-100\,\mu L$) and that no band broadening problems occur. The problem of excess solvent quenching (reducing) the sensitivity remains the major obstacle to the widespread use of this detector in HPLC.

1.6. Supercritical Fluid Chromatography

Supercritical fluid chromatography has enjoyed a resurgence recently as a result of the work of Lee and colleagues [see Novotny et al. (30) and Wall (31)]. Some advantages of this technique, compared to GC or HPLC, include extending the molecular weight range of GC, analysis of thermally unstable compounds, and sensitive detection of compounds without spectra in the near-UV.

In one of the first papers published in this area, Gmür et al. (32) suggested that the PID was not useful for this application. However, more recent work of Sim et al. (33) and Hayhurst and Magill (34) demonstrated excellent sensitivity for the PID; these authors used a restrictor at the PID inlet, keeping the PID around ambient pressure. Gmür and co-workers (32) did not have any restrictor at the inlet and therefore ran the PID at high pressures and were unsuccessful, possibly owing to increased quenching of the response by CO_2. These authors used carbon dioxide as the supercritical fluid, which—at the wavelengths used for the PID excitation—has a minimum in its absorption spectrum.

Sim et al. (33) used typical LC columns and attained low-nanogram detection limits for a variety of polycyclic aromatic hydrocarbons (PAHs), only slightly better than the HPLC results but 2 orders of magnitude less than

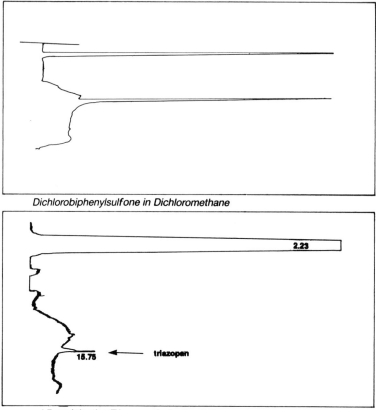

Dichlorobiphenylsulfone in Dichloromethane

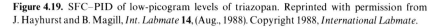

−1.7 pg Injection Triazopan in Acetronitrile

Figure 4.19. SFC–PID of low-picogram levels of triazopan. Reprinted with permission from J. Hayhurst and B. Magill, *Int. Labmate* **14**, (Aug., 1988). Copyright 1988, *International Labmate*.

results attainable by GC. The results of Hayhurst and Magill (34) show sub-nanogram levels of anthracene detected by PID using similar conditions. This suggests some quenching or loss due to the CO_2 carrier gas. Hayhurst and Magill (34) used a 6 m × 50 μm capillary column system and detected low- or sub-picogram quantities of dichlorobiphenyl sulfone and triazopan. A typical chromatogram of a low-picogram sample of triazopan is shown in Figure 4.19.

Advantages of SFC–PID include reduced analysis time by eliminating solvent tailing, use of a wider range of solvents that produce "no response" (such as methanol, methylene chloride, acetonitrile, and Freons), detection of a wide range of compounds that do not absorb in the UV, and increased sensitivity over the FID.

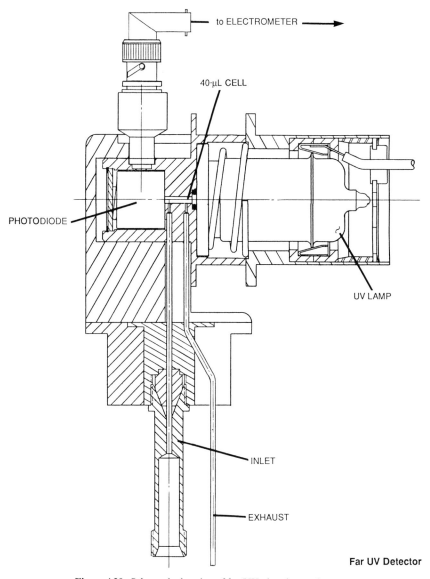

to ELECTROMETER

40-µL CELL

PHOTODIODE

UV LAMP

INLET

EXHAUST

Far UV Detector

Figure 4.20. Schematic drawing of far-UV absorbance detector.

2. FAR-UV ABSORBANCE

2.1. Introduction

The far-UV absorbance detector (FUVAD) is still fairly new to GC compared with typical GC detectors. This detector has emerged as a result of the recent development of a novel UV photodiode (35) and the availability of stable UV sources more than a decade ago (7). This detector, shown in Figure 4.20, was optimized for capillary column usage and has a 40-μL dead volume cell.

2.2. UV-Absorbing Species/Spectra

Most organic and inorganic species strongly absorb in the far-UV. Notable exceptions are the inert gases and nitrogen, which weakly absorb. Certain diatomic species (O_2, CO, etc.) that have absorption minima in the region of the lamp energy (124 nm) will have a poor response, but low-ppm levels can still be detected. A typical far-UV absorption spectrum is shown in Figure 4.21.

2.3. Applications

The far-UV detector is new to gas chromatography and frequently is compared with the thermal conductivity detector because it will respond to any compound that absorbs in the far- or vacuum-UV. The latter name is a misnomer since a carrier gas flows through the cell and a vacuum is not

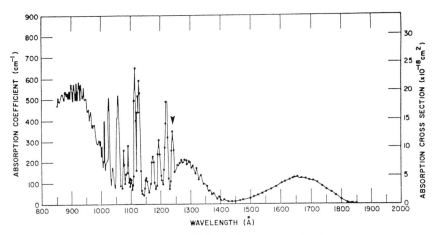

Figure 4.21. Far-UV absorption spectrum of water.

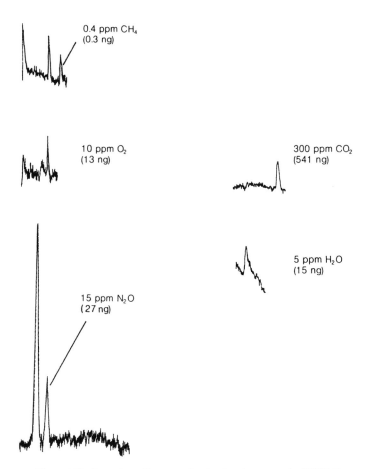

Figure 4.22. Detection of low-ppm levels of various gases by FUVAD.

needed. There are few species, notably the noble gases such as xenon, krypton, and argon, which do not absorb in this region. Thus, the detector has a response that is nearly universal, a low dead volume (40 μL), and a fast electrometer time constant (25 ms). The primary emission from this lamp is the line at 124 nm. Although there are visible lines, the photodiode is unresponsive to any long-wavelength UV or visible emission. Only the absorption at 124 nm needs to be considered. The photodiode has been described in detail elsewhere (36, 37).

Typical chromatograms for the FUVAD are shown in Figure 4.22. The detector has an excellent response to organic compounds such as methane

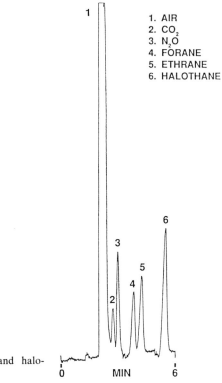

1. AIR
2. CO_2
3. N_2O
4. FORANE
5. ETHRANE
6. HALOTHANE

Figure 4.23. Detection of nitrous oxide and halothanes in air.

that absorb strongly in the far-UV. The sub-nanogram detection limit for methane is better than that achieved with the FID. For higher-molecular-weight organic compounds, the response of the FUVAD will be within a factor of 2 or 3, whereas the response of the FID increases more rapidly so that the FID detection limits will be lower than the FUVAD for high-molecular-weight hydrocarbons. This detector will still be excellent for the analysis of hydrocarbons using capillary GC where the characteristics of the FUVAD are optimized.

The data in Figure 4.22 show that at a value of two times the signal-to-noise ratio, the detection limit for oxygen is approximately 7 ppm (9.2 ng). The FUVAD is a concentration sensitive detector with a response that is inversely proportional to the flow rate of the carrier gas ($C = 1/F$, where F is the carrier gas flow rate). Low levels (ppm) of O_2 and water in gases, pharmaceuticals, and process streams are potential applications for this detector.

Low levels of atmospheric constituents such as N_2O and CO_2 can be detected. Triatomic gases such as these have band spectra in the far-UV. If a

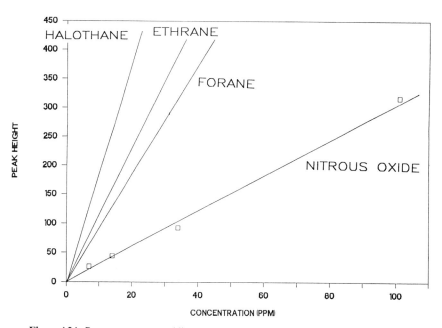

Figure 4.24. Response curves and linear range for nitrous oxide and halothanes.

minimum occurs at 124 nm, there can be several orders of magnitude difference in the observed detection limits, as seen for CO_2 and N_2O.

As expected, the chromatography of water was considerably more difficult than that of oxygen. As the first injections of 180 ppm of water on this column did not produce any sizable peaks (36), we conditioned the column for several hours with the calibration gas (180 ppm water in nitrogen). From the accumulated data, we estimate that an order of magnitude was lost because of the need to use a carrier gas flow rate of 60 mL/min. Reproducibility was good, with a coefficient of variation (CV) of 8% for the three replications. As the data in Figure 4.22 indicate, it is still possible to detect 5 ppm of water (CV = 21%), even at a carrier flow of 60 mL/min (with the 4-mL sample loop = 121 ng).

One typical application for the FUVAD involves the monitoring of anesthetic gases in operating room atmospheres. These gases have been shown to cause problems in pregnant women and, in fact, may be carcinogenic. The threshold limit values are in the low-ppm range. The only instrument currently capable of detecting these gases is the mass spectrometer, which is quite complex, expensive, and nonportable. A typical chromatogram of low-ppm levels of anesthetic gases with far-UV detection is shown in Figure 4.23. These compounds represent the majority of anesthetic gases used at present. Note that all are baseline separated and detected adequately at low-ppm levels by the FUVAD. Typical calibration curves for these species are shown in Figure 4.24. All were found to be linear over nearly three decades. This new detector appears to have many applications for the analysis of trace gases in a wide variety of sample matrices.

3. FUTURE DEVELOPMENTS

Future developments undoubtedly will involve improvements needed for capillary GC and environmental analysis. Both require improved sensitivity, faster response, and even improvements in selectivity. Such requirements could be met for the PID and FUVAD with the development of a low-cost, compact laser light source with sufficient energy to achieve two-photon absorption in the far-UV region.

REFERENCES

1. J. N. Driscoll, *CRC Crit. Rev. Anal. Chem.* **17**, 193 (1986).
2. J. E. Lovelock, *Anal. Chem.* **33**, 162 (1961)
3. D. C. Locke and C. E. Meloan, *Anal. Chem.* **37**, 389 (1965).
4. J. G. W. Price, D. C. Fenimore, P. G. Simmonds, and A. Zlatkis, *Anal. Chem.* **40**, 541 (1968).
5. J. N. Driscoll and F. F. Spaziani, *Anal. Instrum.* **13**, 111 (1974).
6. J. Sevcik and S. Krysyl, *Chromatographia* **7**, 375 (1973).
7. J. N. Driscoll and F. F. Spaziani, *Res./Dev.* **27**, 50 (1976).
8. J. N. Davenport and E. R. Adlard, *J. Chromatogr* **290**, 13 (1984).
9. J. N. Driscoll and F. F. Spaziani, U.S. Patent 4,013,913 (1976).
10. J. N. Driscoll and M. Duffy, *Chromatogr. Forum* **4**, 21 (1987).
11. D. Lubman, *Anal. Chem.* **59**, 31A (1987).
12. C. M. Klimet and J. E. Wessel, *Anal. Chem.* **52**, 1233 (1980).
13. T. Ogawa, S. Yamada, P. Zhang, S. Yoshida, and C. Sakane, *Kenkyu Hokoku—Asahi Garasu Kogyo Gijutsu Shoreikai* **46**, 65 (1985); *Chem. Abstr.* **105**(2); 17528n (1986).
14. J. N. Driscoll, C. Wood, and M. Whelan, *Am. Env. Lab.* **3**, 19 (1991).
15. J. N. Driscoll, *J. Chromatogr. Sci.* **20**, 91 (1982).
16. J. E. Arnold, *Trace Analysis of PAH, NIOSH Report*; Natl. Tech. Inf. Serv. Acces. No. PB 83 196188. NTIS, Washington DC, 1982.
17. J. N. Driscoll, *J. Chromatogr. Sci.* **20**, 91 (1982).
18. M. L. Langhorst, *J. Chromatogr. Sci.* **19**, 98 (1981)

19. J. N. Driscoll, *J. Chromatogr. Sci.* **23**, 488 (1985).

20. J. N. Driscoll, *Int. Labmate* **5**, 21 (1986).

21. J. N. Driscoll, *Am. Lab.* **18**, 95 (1986).

22. J. N. Driscoll, M. Duffy, and S. Pappas, *J. Chromatogr. Sci.* **25**, 369 (1987).

23. J. N. Driscoll, M. Duffy, and S. Pappas, *J. Chromatogr.* **441**, 63 (1988).

24. S. Levine and R. Mauridian, University of Michigan, Ann Arbor, private communication, 1988.

25. M. van Es and J. Rijks, University of Eindhoven, Eindhoven, Netherlands, private communication, 1990.

26. J. N. Driscoll, D. W. Conron, P. Ferioli, I. S. Krull, and K. H. Xie, *J. Chromatogr.* **302**, 43 (1984).

27. D. C. Locke, B. S. Dhingre, and A. D. Baker, *Anal. Chem.* **54**, 447 (1982).

28. J. S. De Wit and J. W. Jorgensen, *J. Chromatogr.* **411**, 201 (1987).

29. A. Berthod, T. Mellone, E. Voigtman, and J. D. Winefordner, *Anal. Sci.* **3**, 405 (1987).

30. M. Novotny, S. R. Springston, P. A. Peaden, J. C. Fjeldsted, and M. L. Lee, *Anal. Chem.* **53**, 407A (1981).

31. R. J. Wall, *Chromatog. Anal. (UK)* Feb., 16 (1989).

32. W. Gmür, J. O. Bosset, and E. Plattner, *Chromatographia* **23**, 199 (1987).

33. P. G. Sim, C. M. Elson, and M. A. Quilliam, *J. Chromatogr.* **445**, 239 (1988).

34. J. Hayhurst and B. Magill, *Int. Labmate* **14**, 35 (1988).

35. J. N. Driscoll, U.S. Patent 4, 614, 871 (1986).

36. J. N. Driscoll, B. Towns, and P. Ferioli, *Res. Dev.* **26**, 104 (1984).

37. J. N. Driscoll, M. Duffy, and S. Pappas, *J. Chromatogr.* **441**, 63 (1988).

CHAPTER

5

THE ELECTRON CAPTURE DETECTOR

ERIC P. GRIMSRUD

Department of Chemistry
Montana State University
Bozeman, Montana

Of the numerous schemes for gas chromatographic effluent detection that have been conceived and tested since the invention of gas chromatography (GC) in 1952, the approach based on gas-phase electron capture reactions has led to one of the most popular and powerful GC detectors in use today. The modern electron capture detector (ECD) often can provide detectable responses to picogram or even femtogram levels of specific target substances in complex samples and has become especially important for analysis problems encountered in environmental and biomedical studies. Unlike the flame ionization detector (FID), which responds with nearly uniform sensitivity to most hydrocarbons, the response of the ECD varies tremendously from one chemical class to another. Therefore, the ECD is known as a "selective" detector capable of providing extremely sensitive responses to specific sought-for substances that might be present in a sample containing a huge excess of other less interesting compounds. In many instances the ECD can be used for the direct detection of molecules that have electronegative functional groups and, therefore, attach electrons readily, without requiring prior chemical modification. Examples of these include the halogenated and the nitroaromatic hydrocarbons, two classes of compounds that have been of considerable environmental interest. By chemical derivatization procedures, the applicability of the ECD for trace analysis can be expanded further to include classes of molecules that normally do not respond strongly. An example of this type is shown in Figure 5.1, in which the detection and analysis by capillary GC–ECD of 14 aminopolycyclic aromatic hydrocarbons in a heavy distillate fraction of solvent

Detectors for Capillary Chromatography, edited by Herbert H. Hill and Dennis G. McMinn.
Chemical Analysis Series, Vol. 121.
ISBN 0-471-50645-1 © 1992 John Wiley & Sons, Inc.

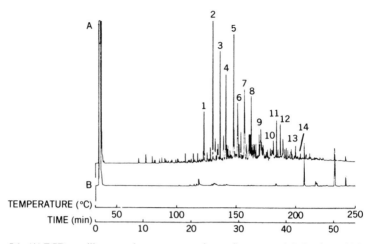

Figure 5.1. (A) ECD–capillary gas chromatogram of pentafluoropropyl derivatives of 14 amino-polycyclic aromatic hydrocarbons in a heavy distillate fraction of solvent refined coal. (B) Chromatogram observed prior to derivatization of sample. Reprinted with permission from D. W. Later, M. L. Lee, and B. W. Wilson, *Anal. Chem.* **54**, 117 (1982). Copyright 1982, American Chemical Society.

refined coal was made possible (1) by prior derivatization of the sample with pentafluoropropyl anhydride.

The development of the ECD has been a continuous process that began almost 30 years ago (2–5) following the invention of GC and still continues today. Over this period of time, hundreds of applications and instrumental modifications of the ECD have been reported and many excellent review articles have summarized these developments (6–11). In 1981 an entire monograph (12) was devoted to the ECD in which contributions from numerous experienced users were included. It is neither possible nor necessary in this chapter to provide comprehensive descriptions and discussions of all aspects of the ECD, since those can be found in the review articles indicated above and in the references listed therein. For additional reading concerning application to capillary GC, the relatively recent review articles (10, 11) are particularly recommended.

The objective of this chapter is to describe the currently most popular form of the ECD, the "^{63}Ni pulsed ECD," and to explain how this detector is expected to perform when used for capillary column gas chromatography. This objective is not a trivial one. The ECD is known to be a relatively complex transducer whose operational characteristics, including those associated with sensitivity, chromatographic resolution, and quantitative responses, are known to vary significantly with alterations in most of the experimental parameters set by the operator. Also, in recent years, new uses of the ^{63}Ni pulsed ECDs have been made possible through various schemes known collectively as the "chemically sensitized" ECD, in which chemical dopants are used to improve the response characteristics of the ECD for various classes of molecules. While these new modes of ECD operation are easily implemented, additional complexities will accompany their use. Therefore, although the ECD is recognized to be one of the most powerful tools available for the trace analysis of many compounds, the degree to which its full potential can be utilized in the laboratory depends strongly on the degree to which its basic operating principles are understood.

1. OVERVIEW OF DETECTOR OPERATION

All of the ECDs offered today by commercial suppliers use a radioactive beta emitter, usually ^{63}Ni, for the initiation of gaseous ionization. While alternate methods of ionization, not requiring a radioactive source material, have been and still are being investigated for use with the ECD, they have not yet proven as successful as the radioactive sources. With ^{63}Ni-containing foils, the problems and inconveniences often associated with radioactive materials have been minimized. These sources can be purchased from the vendors as a "sealed source" for which the vendor holds the necessary radiological license and only relatively simple radiological tests are required of the user. The ^{63}Ni sources can be safely heated up to 400 °C with no loss of activity. If such a source is accidentally overheated, loss of activity occurs by diffusion of the ^{63}Ni into the foil rather than by escape from the detector. These detectors can generally be maintained in a clean condition by periodic application of high temperature. The 93-year half-life of ^{63}Ni is conveniently long. In terms of useful lifetime, maintenance requirements, and overall cost, experience has shown that a commercial ^{63}Ni-containing ECD compares favorably with other GC detectors in common use today.

An ECD design that has enjoyed considerable popularity is shown in Figure 5.2A. Its cylindrical walls are formed by a ^{63}Ni foil of about 15 mCi total activity. A gas of moderate molecular weight (pure hydrogen or helium will not work) is passed through the detector at atmospheric pressure in order to efficiently attenuate the beta radiation, thereby creating a population of positive ions and electrons throughout the active volume of the detector. The detector gas must also be able to rapidly thermalize these secondary electrons. The aforementioned buffer gas functions are generally provided by either nitrogen or by a mixture of 5–10% methane in argon. The electron population that then exists in the cell is periodically sampled by the application of a short pulse to the detector anode, as shown in Figure 5.2A. If argon/methane buffer

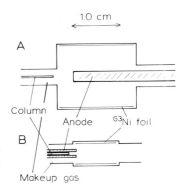

Figure 5.2. Two different designs of the pulsed ECD:
(A) internal volume = 2.5 mL, ^{63}Ni activity = 15 mCi;
(B) internal volume = 0.25 mL, ^{63}Ni activity = 3 mCi.

gas is used, the width of this pulse can be as short as 0.1 μs. If nitrogen gas is used, a width of at least 1.0 μs is typically required. The manner by which this electron current is then used for the generation of an ECD response is the subject of Section 4, below.

When the effluent of a capillary GC column is also passed through this detector, the following ionic processes (13, 14) will occur and affect the response:

$$
\begin{array}{lcr}
 & \textit{Rate} & \\
 & \textit{coefficient} & \\
\text{carrier gas} \longrightarrow \text{p}^+ + \text{e} & S & \text{(1a)} \\
\text{e} + A \longrightarrow A^- & k_{eA} & \text{(1b)} \\
\text{e} + B \longrightarrow B^- & k_{eB} & \text{(1c)} \\
\text{e} + \text{p}^+ \longrightarrow \text{neutrals} & R_e & \text{(1d)} \\
A^- + \text{p}^+ \longrightarrow \text{neutrals} & R_+ & \text{(1e)}
\end{array}
$$

These reactions represent: (1a) ionization of the detector gas by beta radiation to form positive ions and secondary electrons; (1b) either resonance or dissociative electron capture (EC) by the analyte A to form negative ions, A^-; (1c) EC by carrier gas impurities, B; (1d) ion–electron recombination; and (1e) ion–ion recombination. These reactions take place on a time scale that is fast relative to the time required for gaseous flow through the cell. Fortunately, loss of electrons by diffusion to the walls will not occur within a pulsed ECD because of a positive ion space-charge field that continuously contains the electrons within the gaseous volume and away from the walls as long as the frequency of applied pulses is sufficiently high (13). Although Reactions 1a–1e occur continuously, an ECD response to A is caused by the EC reaction (1b), which decreases the steady-state concentration of electrons as compound A passes through the cell.

2. SENSITIVITY AND RELATIVE RESPONSES

One of the requirements for successful separations by capillary GC is that the amount of sample that can be injected onto a column must be kept small to ensure that the capacity of the column is not exceeded. Since the ECD is considered to be one of the most sensitive GC detectors in existence for the detection of numerous compounds, this basic sensitivity requirement for capillary GC is easily met. In fact, with ECD detection, the maximum sample load allowed for quantitative analysis with a capillary column will often be set by the point at which the response of the detector is saturated rather than the point at which the column is overloaded.

The ECD owes its great sensitivity to the high speed with which electron capture reactions can occur. To fully appreciate this point and also the basis of the high level of chemical specificity inherent in EC reactions, let us consider the expressions given in Table 5.1 for the rate constants of a few selected classes of exothermic two-body reactions (15). The second-order rate constants are given in units that are commonly used in the gas phase literature and will be used throughout this chapter. The first reaction shown is the diffusion-limited recombination of hydronium and hydroxide ions in water and provides a reference point against which the gas phase reactions can be compared. Although this reaction is one of the fastest known in the aqueous phase, it is not fast relative to some of the gas phase reactions listed. Gas phase free radical reactions, symbolized by case 2, are typically characterized by a range of

Table 5.1. General Expressions for the Rate Constants of Various Bimolecular Exothermic Reactions

Reaction Type	Rate Constant $(cm^3 \cdot molecule^{-1} \cdot s^{-1})$
(1) $H^+(aq) + OH^-(aq) \rightarrow H_2O(aq)$	2×10^{-10}
(2) $R \cdot + M \rightarrow RM \cdot$	$1 \times 10^{-10} \exp(-E_a/RT)$
(3) $CH_5^+ + M \rightarrow CH_4 + MH^+$	2×10^{-9}
(4) $e + M \rightarrow M^-$	$4 \times 10^{-7} \exp(-E_a/RT)$
(5) $e + P^+ \rightarrow neutrals$	8×10^{-7}

activation barriers, E_a, and therefore exhibit a wide range of rate constants. Only in a few instances is this activation barrier sufficiently low so that the rate constant of a radical reaction approaches the magnitude of the preexponential factor, 1×10^{-10} (16). Case 3 represents ion–molecule reactions, on which many important mass spectrometric methods are based (17). The rate constants for these tend to a constant value of about 2×10^{-9} for exothermic reactions. Electron capture reactions are shown as case 4 of Table 5.1. The general expression for the rate constants of these includes an activation barrier. Thus, the rate coefficients for numerous compounds that undergo exothermic electron attachment but possess a nonzero activation barrier are considerably smaller than the preexponential factor, 4×10^{-7}. For these compounds, the rate of electron attachment will be increased by the use of higher detector temperatures. Some molecules, however, possess near-zero activation barriers against electron capture, and for these the rate constant for EC at all detector temperatures will be extremely fast, 1×10^{-7} and greater. Thus, fast EC reactions will occur about 2 orders of magnitude faster than ion–molecule reactions. Note that fast EC reactions are almost as fast as process 5 in Table 5.1, electron–positive ion recombination. Reactions of type 5 would be expected to occur rapidly in the gas phase owing to a coulombic attraction between the two reactants involved. It is therefore impressive, and perhaps surprising, that EC reactions, which do not enjoy this coulombic driving force, can be characterized by rate constants that are almost as great as those of electron–ion recombination reactions.

Rate coefficients for EC are well known only for a few relatively small molecules (14, 17). From these, however, the effects of various substituents on EC reactivity can be used to estimate the nature of the EC reactions of larger molecules. In Table 5.2, EC rate constants for various compounds are given. As described above, a wide range of EC rate coefficients are observed. Maximum rates approaching 4×10^{-7} are observed for several of the compounds listed, including SF_6, CCl_4, and perfluorotoluene. For many of the compounds shown in Table 5.2, the EC reaction occurs by a dissociative EC mechanism (14), shown here for the case of CCl_4:

$$e + CCl_4 \longrightarrow CCl_3 + Cl^- \tag{2}$$

Rections of this type are always irreversible, so the rate of the forward reaction alone determines the magnitude of the ECD response to the given compound. As indicated by the rate expression for EC, case 4 in Table 5.1, the rate constant for dissociative EC will increase with an increase in temperature until a limit in the range of $1–4 \times 10^{-7}$ is reached. Therefore, the EC rate coefficient given for CH_2Cl_2 or $CHCl_3$, for example, will increase significantly with an increase in detector temperature while that of CCl_4 will not.

Table 5.2. Electron Capture Rate Coefficients[a]

M	Rate Constant $(cm^3 \cdot molecule^{-1} \cdot s^{-1})$	Major Product
CH_2Cl_2	1×10^{-11}	Cl^-
$CHCl_3$	4×10^{-9}	Cl^-
CCl_4	4×10^{-7}	Cl^-
CH_3CCl_3	1×10^{-8}	Cl^-
$CH_2ClCHCl_2$	1×10^{-10}	Cl^-
CF_4	7×10^{-13}	—
CF_3Cl	4×10^{-10}	Cl^-
CF_3Br	1×10^{-8}	Br^-
CF_2Br_2	2×10^{-7}	Br^-
C_6F_6	9×10^{-8}	M^-
$C_6F_5CF_3$	2×10^{-7}	M^-
$C_6F_{11}CF_3$	2×10^{-7}	M^-
SF_6	4×10^{-7}	M^-
Azulene	3×10^{-8}	M^-
Nitrobenzene	1×10^{-9}	M^-
1,4-Naphthoquinone	7×10^{-9}	M^-

[a] Rate constants taken from summaries provided in References 14 and 17.

For some of compounds shown in Table 5.2, EC occurs by a resonance EC mechanism (14), shown here for the case of azulene:

$$e + C_{10}H_8 \rightleftharpoons C_{10}H_8^- \qquad (3)$$

Electron capture reactions of this type, in which a molecular anion is formed, can be accompanied by the reverse of Reaction 3, a process called thermal electron detachment (TED). TED will diminish or destroy the response of an ECD to the given compound if it is sufficiently fast. The rate of TED is determined largely by the electron affinity (EA) of the molecule and the temperature of the detector gas. For azulene the first-order rate constant, k_d, for TED is given by (18)

$$k_d = 1.1 \times 10^{11} \exp(-15.7/RT)\, s^{-1} \qquad (4)$$

in which the activation energy, 15.7 kcal/mol, is approximately equal to the EA of azulene. For this compound, k_d becomes significant and begins to noticeably diminish the ECD response to azulene with a detector temperature of $\sim 140\,°C$. At temperatures in excess of $200\,°C$, the ECD response to azulene is essentially destroyed. For the numerous compounds that undergo EC by the resonance process, it is helpful to know their approximate EA values, and comprehensive lists of these are available (19). If a compound undergoes EC by the resonance mechanism and its EA is greater than about $20\,kcal/mol$, TED should not noticeably diminish an ECD response with detector temperatures up to $\sim 250\,°C$. If the EA of the compound is less than $\sim 20\,kcal/mol$, care should be taken to use as low a detector temperature as the separation will allow (as a rough rule of thumb, the temperature of the ECD should be maintained at least $30\,°C$ higher than the highest column temperature used in order to maintain detector cleanliness). If the EA of the compound is less than $\sim 12\,kcal/mol$, it is unlikely that a good EC response to it can be achieved using the detector temperatures normally required for GC detection.

For some low-EA molecules, which respond primarily by the resonance EC mechanism at low temperatures, a competitive dissociative EC process may also be operative. At higher temperatures, the loss of response due to an increased rate of TED may be offset by an increase in the rate of the dissociative process. In such instances, the responses observed with low and high detector temperatures may be greater than that observed with intermediate temperatures.

3. CONTRIBUTION TO PEAK BROADENING

With use of the ECD for high-resolution capillary GC, some loss of chromatographic resolution will occur owing to the significant "mixing volume" (20–22) this type of detector adds to the total chromatographic system. The magnitude of this detrimental effect will depend on the physical design of the ECD. In Figure 5.2A a popular ECD of the pin-cup design is shown that has an internal volume of 2.5 mL. The widths of the peaks (measured at half height) observed in a chromatogram obtained by the combination of a GC with this ECD is given by (20)

$$W = (W_c^2 + W_d^2)^{1/2} \tag{5}$$

where W_c is the width of a peak produced by the column (including any contribution of the injection system), and W_d is the detector's contribution to peak broadening. For the half-height definition of width, the detector contribution is approximately $W_d = V/2F$, where $F = $ total flow rate, if the reaction vessel is "well mixed" and the detector acts as an exponential dilution vessel (20). The sample typically is carried into the detector by a flow of H_2 or He carrier gas of a few milliliters per minute combined with a much larger flow of $Ar/5\%\ CH_4$ or nitrogen makeup gas. As was explained in Section 2, the inclusion of one or the other of these common makeup gases is necessary for proper operation of the ECD. Thus, the total gas flow rate through any ECD used for capillary GC will be at least 25 mL/min. For this flow rate, the magnitude of W_d for design 5.2A becomes 3.0 s. This contribution to peak broadening would be of considerable significance in an application of high-resolution GC in which the first eluting peaks of very narrow width are of importance. For a peak of $W_c = 1.0$ s, for example, use of this ECD would provide an observed peak of $W = 3.2$ s. This is a great and probably unacceptable loss of the resolution hard won by the column. The height of this peak and therefore sensitivity would also be greatly diminished. An improvement can be achieved by increasing the makeup gas flow rate through the cell. If, say, the flow rate is increased to 50 mL/min in the foregoing example in which $W_c = 1.0$ s, the observed peak will then have $W = 1.8$ s, and if it is increased to 100 mL/min, $W = 1.25$ s. The price that might have to be paid for this solution to the resolution problem is that a corresponding decrease in sensitivity might accompany the dilution of sample in the greater makeup gas flow rates. Whether or not this sensitivity loss does indeed occur depends on other details of the EC reaction to be discussed in Section 4.

In Figure 5.2B another ECD design is shown that will reduce peak broadening by use of two factors other than increased makeup gas flow rate (21, 22). The first of these factors is its decreased internal volume, about 0.25 mL in the example shown. This will tend to decrease W_d in design 5.2B relative to that of design 5.2A by a factor of 10 for a given flow rate. The second beneficial factor inherent in design 5.2B is that the flow pattern through this ECD is "plug-like" rather than "well mixed" as in design 5.2A. With this flow pattern, the residence time of an analyte in the cell is significantly decreased and the contribution to peak broadening is then given by $W_d = V/7F$ for design 5.2B (20), rather than $W_d = V/2F$ for design 5.2A. Thus, the combination of smaller volume and plug-like flow lead to $W_d = 0.09$ s for design 5.2B using a total detector gas flow of 25 mL/min. For the example considered above of a peak for which $W_c = 1.000$ s, the observed peak using this detector will have $W = 1.004$ s. For the purpose of retaining high resolution in capillary chromatography, the ECD of design 5.2B is clearly superior to that of design 5.2A.

Another beneficial characteristic of design 5.2B is that, owing to the nature of the plug-like flow through this detector, the analyte emerging from the column will tend to pass through the central region of the active volume (21, 22), thereby minimizing contact of the analyte with the walls of the ECD. This effect will minimize surface-catalyzed degradation of the analyte and will help maintain the detector in a clean condition. Because gaseous diffusion is fast at the elevated temperatures of the typical ECD, however, analyte contact with the walls of design 5.2B is probably not completely eliminated.

With use of the smaller ECD of design 5.2B, some loss of sensitivity must be accepted because of the lower total beta-induced ionization that can be made to occur in it. With use of ^{63}Ni for ionization of the detector gas, a higher current will be measured with the larger cell owing to the greater number of ^{63}Ni atoms that can be embedded on its wall. The measured current will also tend to be greater for the larger cell because of the more effective attenuation of the beta radiation passing through this cell. The noise level associated with an ECD current measurement is determined by the random nature by which the ^{63}Ni atoms decay. The signal-to-noise ratio of the current measurement will be proportional to $N^{1/2}$, where N is the number of disintegrations per second that are determining the measured current. Since the total amount of ^{63}Ni used in design 5.2B will be about one-fifth that present in design 5.2B, the signal-to-noise ratio of the current measured from the smaller ECD will be somewhat less than half that of the larger detector.

An obvious extension of the idea behind the design in Figure 5.2B is to make an ECD even smaller, say, 50 μL in volume, and, in order to recover the loss of total current that would accompany the use of a smaller ^{63}Ni foil, a different radioisotope of greater specific activity might be deposited more densely onto the cell walls. At this point, however, no acceptable radioactive material of this type has been developed. The only other beta emitter that has seriously challenged ^{63}Ni for use with ECDs is Sc^{3}H (23). Beta-emitting foils of much higher specific activity and shorter beta penetration depths can be made using this material, and these characteristics would be well suited to a small ECD. However, owing to the lower thermal stability of Sc^{3}H, this material has not replaced ^{63}Ni as the dominant source material for ECDs. At temperatures in excess of 300 °C, the ^{3}H is lost rapidly by emanation, thereby causing a radiological safety problem as well as ruining the detector. Also, hydrogen carrier gas cannot be used with a Sc^{3}H detector because ^{3}H is then lost by chemical exchange at detector temperatures far below 300 °C.

4. NATURE OF THE QUANTITATIVE RESPONSE

4.1. The Ideal Response

Perhaps the most complicated and least understood aspect of the ECD concerns the manner in which the response changes with variations in analyte concentration. For the ideal GC detector, its response to an analyte would be given by

$$\text{response} = K \times n_0 \tag{6}$$

where K is a constant and n_0 is the concentration of the analyte in the detector gas as it enters the detector. A detector that responds in this fashion will

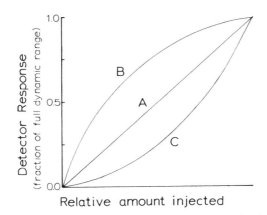

Figure 5.3. Three hypothetical response curves observed over the entire dynamic range of a detector.

provide a linear calibration curve, such as that shown as line A in Figure 5.3, if the response of the detector over its entire dynamic range is plotted against the amount of analyte injected into the GC.

The nature of the ECD response will resemble the ideal response indicated by Equation 6 and calibration curve A in Figure 5.3 only under certain operational conditions. The actual nature of the ECD's response will depend on several factors, including (a) the mode of signal processing used, (b) the electron capture rate constant of the analyte, and (c) stoichiometry of the reaction of electrons with the analyte.

4.2. Response of the Fixed-Frequency ECD (FF–ECD)

Although some commercial ECDs provide the user with both the fixed-frequency (FF) and constant-current (CC) modes of signal processing, most do not provide the FF mode. Although it is probably true that the CC mode is generally superior, the FF mode can be useful and, as will be pointed out below, should not be overlooked as a viable option under certain circumstances.

The FF mode is electronically the simplest means of obtaining a response from a pulsed ECD. It is done by the application of electron-collecting pulses at a constant frequency, f_0, while the time-averaged current, I, is measured. A simple schematic diagram of the circuitry associated with the FF–ECD is shown in Figure 5.4A. The response of the FF–ECD to a compound is taken as the decrease in current, δI, observed as the compound passes through the cell. An expression for this response to an analyte, A, can be derived in terms of Reactions 1a–1e, discussed in Section 1. The following equation describes

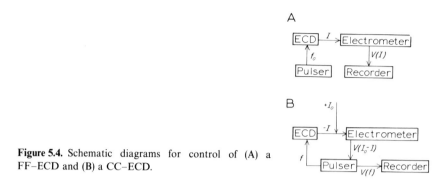

Figure 5.4. Schematic diagrams for control of (A) a FF–ECD and (B) a CC–ECD.

the instantaneous rate of change in the electron population, N_e, during the time, t, between electron-removing pulses (13):

$$dN_e/dt = S - LN_e \qquad (7)$$

where $L = k_{eA}n_A + k_{eB}n_B + R_e n_+$ is the sum of all electron-removing-pseudo-first-order rate coefficients, and n_i is the concentration of species i in the detector. In the FF–ECD, the electron population is sampled at constant intervals of time, $t = T = 1/f_0$. Thus, the time-averaged current, I, which is thereby provided by the electrometer, will be given by $I = N_e^T/T$ [a smaller positive ion contribution to the measured current (13) will be ignored here]. By integration of Equation 7 from $t = 0$ to T, the following expression for the instantaneously measured current is obtained:

$$I = \frac{S}{LT}(1 - e^{-LT}) \qquad (8)$$

An expression for the response of the FF–ECD to sample, δI, is then obtained from Equation 8 by substitution for L, with and without ($n_A = 0$) sample present, and is given by:

$$\delta I = \frac{S}{(R_e n_+ + k_{eB}n_B)T}[1 - \exp\{-(R_e n_+ + k_{eB}n_B)T\}]$$
$$- \frac{S}{(R_e n_+ + k_{eB}n_B + k_{eA}n_A)}[1 - \exp\{-(R_e n_+ + k_{eB}n_B + k_{eA}n_A)T\}] \qquad (9)$$

This awkward equation describes a complex relation between the δI response of the FF–ECD and the steady-state concentration of analyte in the detector, n_A.

This is an appropriate time to point out that n_A in Equation 9 may or may not be equal to n_A^o, the concentration of analyte introduced to the detector (24). For compounds that do not have large EC rate coefficients the consumption of A in the detector will not be significant, and for these cases n_A will be approximately equal to n_A^o. For these molecules the nature of the FF–ECD response over the full dynamic range of the detector will be described by Equation 9 and will take the shape shown as curve B in Figure 5.3. This curve indicates a near-linear relation between response and concentration only over the initial 10% of the full dynamic range of the FF–ECD.

If, on the other hand, the analyte is a molecule that attaches electrons extremely rapidly, a significant fraction of the molecules introduced to the cell may be destroyed by EC, and n_A can then be much smaller than n_A^o (24). Furthermore, the ratio n_A/n_A^o will then vary with changes in the sample concentration. For this reason the response of the FF–ECD to strongly responding compounds may not follow the relationship described by Equation 9 and curve B in Figure 5.3. While no exact mathematical expression for the FF response to strongly responding compounds has been suggested, it is qualitatively expected that the calibration curve of the FF–ECD for many of these compounds will differ from curve B in Figure 5.3 by being straighter over a greater portion of the full dynamic range. That is, for some strongly responding compounds, the FF–ECD is expected to produce calibration curves that tend toward the shape of the ideal curve A in Figure 5.3. The precise nature of the curve will depend on the magnitudes of the EC rate coefficients and the detector gas flow rate.

4.3. Response of the Constant-Current ECD (CC–ECD)

The constant-current (CC) mode of signal processing (25) is generally considered to be a superior method of controlling a pulsed ECD and is included with all modern ECDs. As shown in Figure 5.4B, a preselected level of current, I_0, is demanded by the controlling circuit of the CC–ECD, which continuously adjusts the frequency of pulsing ($f = 1/T$) to that frequency which will provide I_0 (26). Therefore, when an electron-capturing compound enters the detector, electron density will diminish and the frequency will rise. The response to sample is taken as the increase in pulse frequency, δf, which occurs as the analyte passes through the ECD. An expression for the response of the CC–ECD is readily obtained, again by consideration of Reactions 1a–1e. Equations 7 and 8, given above for the FF–ECD, also apply to the CC–ECD. Inspection of Equation 8 reveals that there is only one way that current I_0 could remain constant for all conditions of L (as analyte concentration is changed). This condition is that the product, LT, and the ratio, L/f, must remain constant (13). That is, as sample enters the detector and L increases, the frequency of pulsing is proportionately increased by the feedback circuit in Figure 5.4B. Therefore, for the CC–ECD a simple equation can be written.

$$f = KL \tag{10}$$

where K is a proportionality constant. Recalling the full definition of L provided above, with and without analyte present, we obtain the following simple expression for the CC–EDC's response to the sample:

$$\delta f = Kn_A \tag{11}$$

A comparison of Equation 11 for the CC mode of the ECD with Equation 9 given previously for the FF mode clearly reveals a significant advantage of the CC mode. The relationship between its response and the steady-state concentration of the analyte in the detector is expected to be much simpler and linear over the entire dynamic range of the detector. However, it will again be noted that n_A, and not n_A^o, is present in Equation 11. This means that a linear response of the ideal type shown in Equation 6 and curve A in Figure 5.3 will be expected only for compounds that undergo EC in a CC–ECD with moderate to low rate coefficients, so that $n_A = n_A^o$.

Since the ECD is most useful for the analysis of molecules that do attach electrons extremely rapidly, it is unfortunate that neither the FF or the CC modes described above provide linear responses to these molecules over the entire dynamic range of the instrument. The shape of the calibration curve that would be expected for many strongly responding compounds with use of the CC–ECD is shown as curve C in Figure 5.3. As the EC rate constant of a molecule is made faster, the calibration curve will increasingly depart from the ideal curve A (Figure 5.3) and will assume the character of curve C. Three different means can be attempted to eliminate or minimize this deviation from curve A.

For the CC mode, a solution has been proposed (24) for some strongly responding compounds, such as CCl_4, which are thought to undergo EC to form neutrals and ionic products that do not undergo subsequent EC reactions. The steady-state concentration of these compounds in the detector will be given by

$$n_A = n_A^o F/(\tfrac{1}{2}k_{eA}N_e + F)$$

It has been shown (24) that under this condition a response function of the following form for the CC–ECD is expected:

$$(f - f_0)(H + f)/f = Gn_A^o \tag{12}$$

where f_0 is the baseline frequency observed in the absence of analyte, f is again the instantaneous frequency, and H and G are two proportionality constants that are determined experimentally from the responses observed for at least two prepared standards. Note that this response function provides a signal linearly related to the amount of analyte introduced to the detector, rather than the steady-state concentration that is not under good experimental control. Successful applications of this extended response function for the CC–ECD responses to CCl_4 and $CFCl_3$ have been demonstrated (24). With the use of modern data systems, which frequently accompany a commercial instrument, the processing of the response of a CC–ECD in accordance with either Equation 11 or 12 or both could be easily accomplished and should lead to more linear calibration curves for many strongly responding compounds. To our knowledge, however, the vendors of GCs have not yet incorporated this software option into their standard GC data systems.

Another means by which the response of the CC–ECD to strongly responding compounds can be made more linear is by use of cell design 5.2B rather than design 5.2A. With the smaller cell and with use of "plug flow," the residence time of analyte in the cell is reduced significantly. In this cell the ratio, n_A/n_A^o, will not tend to be as greatly reduced by EC reactions as it is in a larger ECD, assuming that the same detector gas flow rates are used. Therefore, use of Equation 11 for the δf response of the CC–ECD to strongly responding compounds will be more successful (27) with this detector. For the same reasons as described earlier, the use of greater makeup gas flow rates will also tend to linearize the calibration curves of strongly responding compounds. A potential disadvantage of these approaches is that the sensitivity to moderately and weakly responding compounds will be decreased.

It was noted in the previous subsection (4.2) dealing with the FF mode of signal processing that, as the EC rate coefficient of the analyte is made larger, the nature of its expected calibration curve will change from that shown as curve B in Figure 5.3 toward the more ideal curve A. Therefore, for a given strongly responding compound, a superior and more linear calibration curve may be obtained with the FF–ECD than with the CC–ECD.

4.4. Effect of Secondary Reactions

The nature of the quantitative response of a pulsed ECD may also depend on whether or not any of the neutral products formed either in the EC reaction (1b) or in the ion–ion recombination reaction (1e) are capable of undergoing additional EC reactions. If these neutral products are EC active, and if the original analyte molecule is a strongly responding compound whose steady-state concentration is significantly diminished by the EC reaction, then these secondary reactions will also have an effect on the nature of the calibration curve observed for that substance. Because little is known of the identity of the neutrals produced by EC and recombination reactions, and because even less in known about the EC reactivity of these neutrals, it is difficult to predict what the actual calibration curves for these cases should look like. For the moment, however, we can make some reasonable guesses as to when these more complex processes might be operative and how they might affect the nature of responses.

For example, if a large polychlorinated molecule, C_xCl_y, attaches electrons by the dissociative EC mechanism to form the ion Cl^- and the neutral C_xCl_{y-1}, it seems probable that the neutral product would also readily undergo electron capture. If so, the description just given in Section 4.3 of the quantitative response of the CC–ECD to a strongly responding analyte would have to be modified. If both the original compound and its EC products attach electrons rapidly, we would expect the calibration curve to be more linear than that of curve C in Figure 5.3. Although the concentration of the original compound is depleted significantly by the EC reaction, the "effective concentration" of total EC-active species that originate from the analyte molecules is not being as significantly reduced.

Another example of important secondary reactions in an ECD is associated with the response of organic iodides (28), which typically form I^- upon EC by the dissociative mechanism. By the ion–ion recombination reaction (1e), the I^- ions can be quickly converted to molecular HI. Unlike HCl and HBr, molecular HI is known to attach thermal electrons rapidly (29), and therefore this secondary EC reaction for organic iodides will cause their ECD responses to be greater than those of organic chlorides and bromides that have the same initial EC rate constants. If the original organic iodide is a strongly responding compound, such as in the case of methyl iodide (28, 30), the nature of the calibration curve will again be affected in a manner similar to that described above for the large polychlorinated hydrocarbons. In the case of CH_3I, the quantitative response is made more linear by the secondary reaction since the sum of the two EC-active species CH_3I and HI is not so greatly reduced by EC reactions as is the parent molecule.

5.   THE CHEMICALLY SENSITIZED ELECTRON CAPTURE DETECTOR (CS–ECD)

During the last decade, it has been recognized that the ECD response to certain molecules can be greatly altered by the intentional addition of various compounds to the detector gas. These methods, which collectively might be called the chemically sensitized electron capture detector (CS–ECD), include three basically different chemical schemes in which the population of electrons within the ECD is coupled to an ion–molecule reaction of some type. The ion–molecule reaction, rather than the EC reaction, then provides the basis for an altered overall response to certain substances.

5.1. The Oxygen–Sensitized ECD (O_2S–ECD)

In using an ECD for GC detection, considerable care is normally is taken to reduce the level of oxygen in the detector gas to the lowest possible levels, about 1 ppm or less, by placement of oxygen-removing traps in the detector and carrier gas supply lines. Oxygen has a weak but positive electron affinity and can capture electrons by the following resonance EC mechanism:

$$e + O_2 \rightleftharpoons O_2^- \tag{13}$$

This reaction has the normally undesired effect of reducing the population of electrons available for reaction with the sample. It has been noticed, however, that if some sacrifice to electron population is allowed by intentionally adding a few parts per thousand (ppt) oxygen to the carrier gas, ECD responses to some compounds are significantly increased (31–34). For the case of CH_3Cl, for example, a large increase in response is observed and has been attributed to the occurrence of the following ion–molecule reaction.

$$O_2^- + CH_3Cl \longrightarrow Cl^- + \text{neutrals} \tag{14}$$

Although Reaction 14 cannot itself cause an ECD response, the concentration of the reagent ion O_2^- is reduced and it is coupled to electron population by Reaction 13, which is fast in both directions. Therefore, a loss of electrons and a normal ECD response to methyl chloride is observed that is being determined by the rate of an ion–molecule reaction rather than an EC reaction.

The O_2S–ECD can be used to improve the ECD response to certain compounds that do not normally exhibit strong EC responses. The method retains considerable chemical selectivity, however, since increased responses will be observed only for those compounds that will react rapidly with O_2^- but not with electrons. Therefore, the O_2S–ECD will not respond to saturated hydrocarbons, for example, which to not have electronegative functional groups. As a rough guide, the O_2S–ECD will provide increased responses to molecules that can be detected with the ECD but normally have poor sensitivity. In Table 5.3 the relative ECD responses and enhancements of ECD responses caused by oxygen sensitization of numerous halogenated hydrocarbons are given.

Table 5.3. Relative ECD Molar Responses[a] and Enhancements of Responses Caused by Oxygen Sensitization[b]

M	ECD Response	O_2 Enhancement
CH_3Cl	1.4	189
CH_2Cl_2	3.5	108
$CHCl_3$	420	4.8
CCl_4	10,000	1.9
CH_3CH_2Cl	1.9	228
CH_2ClCH_2Cl	4.2	161
$CH_3CHClCH_3$	1.8	195
$(CH_3)_3CCl$	1.5	95
$CH_2{=}CHCl$	0.0062	107
$CH_2{=}CCl_2$	17	1.8
trans-$CHCl{=}CHCl$	1.5	17
cis-$CHCl{=}CHCl$	1.1	20
$CHCl{=}CCl_2$	460	2.9
$CCl_2{=}CCl_2$	3,600	1.8
Ph—Cl[c]	0.026	15
Ph—CH_2Cl[c]	38	44
CF_3Cl	6.3	2.7
CHF_2Cl	1.8	190
CF_2Cl_2	160	3.4
$CFCl_3$	4,000	2.2

[a]Relative molar responses measured at 250 °C in nitrogen detector gas using a Varian 3700 GC/CC–ECD.
[b]Ratio of response observed with and without 2.0 ppt oxygen present in the nitrogen makeup gas.
[c]Ph=phenyl.

The O_2S–ECD can also be used to verify the identity of chromatographic peaks (35) even if they are unresolved, as illustrated in Figure 5.5. Two capillary GC analyses of a sample containing equimolar quantities of 1-, 2-, and 9-chloroanthracene are shown where nitrogen makeup gas was used in chromatogram 5.5 A and nitrogen containing 3 ppt oxygen was used for chromatogram 5.5 B. The arrow indicates the expected retention time of all three isomers. A doublet is observed in both chromatograms. With oxygen sensitization the first peak is increased 5.5 times and the second 21 times, so that the relative intensities of the two peaks are reversed in the two chromatograms. From parallel analyses of the pure isomers, it is known that the oxygen-caused enhancement ratios of the individual isomers are 4.0, 7.0, and 21, respectively. It is therefore clear that the first peak in the doublet shown in Figure 5.5 can be assigned to the unresolved 1- and 2-isomers and the second, to the 9-isomer.

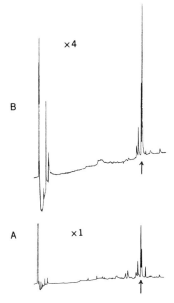

B

×4

A

×1

Figure 5.5. ECD–capillary gas chromatograms of an equimolar mixture of 1-, 2-, and 9-chloroanthracene using (A) nitrogen makeup gas and (B) nitrogen with 3‰ oxygen makeup gas, at 300 °C detector temperature. The arrow marks the expected approximate retention time of the chloroanthracenes; a doublet is observed in each case. Reprinted with permission from J. A. Campbell and E. P. Grimsrud, *J. Chromatogr.* **243**, 1 (1982). Copyright 1982, Elsevier Science Publishers.

Successful application of the O_2S–ECD generally requires that a CC–ECD be used. The useful dynamic range of the CC–ECD is much greater than that of the FF–ECD, and the initial portion of the response range is sacrificed with the addition of oxygen to the detector gas. To our knowledge, only ^{63}Ni-based ECDs have been tested for use with oxygen. No detectable damage to these has been reported.

5.2.   The Nitrous Oxide–Sensitized ECD (N_2OS–ECD)

An entirely different set of response factors is caused when a few ppm of nitrous oxide is added to the nitrogen makeup gas of an ECD (36, 37). In the absence of sample, the following reactions will occur in an ECD that contains nitrous oxide:

$$N_2O + e \longrightarrow O^- + N_2 \tag{15a}$$

$$O^- + N_2O \longrightarrow NO^- + NO \tag{15b}$$

$$NO^- + N_2 \longrightarrow NO + N_2 + e \tag{15c}$$

These reactions result is a steady-state condition in which the concentration of electrons is coupled to the concentration of O^- ions in this case, rather than to O_2^- ions as in the O_2S–ECD discussed above. Unlike O_2^-, O^- will react with almost any compound that enters the detector as shown here for the case of a hydrocarbon, C_xH_y:

$$O^- + C_xH_y \longrightarrow OH^- + C_xH_{y-1} \tag{16}$$

Because Reactions 15a–15c are very fast, the occurrence of Reaction 16 and a loss of O^- ions will cause a decrease in electron population and an ECD response. Because Reaction 16 occurs rapidly for all hydrocarbons, nitrous oxide sensitization causes an ECD to behave somewhat like an FID. That is, the N_2OS–ECD is a universal detector that will respond to almost all components of the sample. Its greatest utility at this point has been for the sensitive detection of small molecules such as hydrogen, carbon monoxide, carbon dioxide, methane, vinyl chloride, and other small hydrocarbons in gaseous mixtures and in air. These analyses can be done using GC columns made of molecular sieves or chemically bonded stationary phases, which are known to exhibit extremely low levels of column bleed. There has been less use, to date, of the N_2OS–ECD for the analysis of larger molecules using capillary columns. Perhaps this is because the more significant level of column bleed that will accompany the use of capillary columns and higher column temperatures will react with and tend to use up the finite population of O^- ions available for reaction with the sample. Nevertheless, if sufficient care is taken in the selection of low-bleed columns and in the elimination of other carrier gas impurities, the N_2OS–ECD has been shown to be compatible with capillary chromatography (11).

5.3.   The Ethyl Chloride–Sensitized ECD (EtClS–ECD)

As described in Section 2, many compounds respond in the ECD by the resonance EC mechanism shown generally as

$$e + A \rightleftharpoons A^- \tag{17}$$

Also as previously described, the ECD response to compound A may be significantly weakened by the reverse of Reaction 17, thermal electron detachment (TED), if the EA of A is relatively low and if the temperature of the detector is relatively high. For numerous compounds of this type, it has been shown that a large gain in sensitivity can be achieved by the intentional addition of an alkyl chloride, such as ethyl chloride, to the nitrogen or argon/methane makeup gas (38, 39). This gain in sensitivity is caused by the following ion–molecule reaction:

$$A^- + CH_3CH_2Cl \longrightarrow Cl^- + \text{neutrals} \tag{18}$$

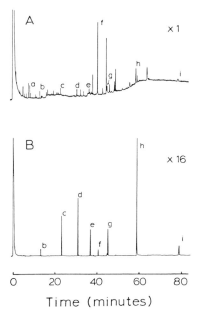

Figure 5.6. ECD–capillary gas chromatograms of a complex mixture using (A) nitrogen makeup gas and (B) nitrogen with 100 ppm ethyl chloride makeup gas, at 250 °C detector temperature. Compounds present are (a) azulene, 0.15 ng; (b) coumarin, 0.8 ng; (c) benzophenone, 4.6 ng; (d) anthracene, 11 ng; (e) 2-methylanthracene, 11 ng; (f) 1-chloroanthracene, 5 ng; (G) pyrene, 28 ng; (h) benz[a]anthracene, 29 ng; (i) benzo[e]-pyrene, 16 ng. Reprinted with permission from C. A. Valkenburg, W. B. Knighton, and E. P. Grimsrud, *J. High Resolut. Chromatogr. Chromatogr. Commun.* **9**, 320 (1987). Copyright 1986, Alfred Huethig Publishers.

in which the unstable A^- ion is converted to the stable Cl^- ion. With about 100 ppm ethyl chloride in the makeup gas, Reaction 18 is much faster than the reverse of Reaction 17 and the loss of sensitivity normally caused by TED is prevented. As shown in Figure 5.6, the EtClS–ECD is well suited to the analysis of polycyclic aromatic hydrocarbons, many of which have low EAs and undergo TED rapidly under the typical conditions of an ECD. For many of the cases shown in Figure 5.6, an increase in the signal-to-noise response of about 2 orders of magnitude has been caused by the presence of ethyl chloride. An attractive feature of the EtClS–ECD is that little damage to the standing electron population of an ECD is caused by the addition of several hundred ppm ethyl chloride to the detector makeup gas because ethyl chloride does not itself attach electrons at a significantly fast rate.

6. FUTURE DIRECTIONS

The ease of performing quantitative analyses with an ECD might be significantly increased if it could be modified so that its response were not so complicated and dependent on all of the details concerning the EC chemistry of each EC-active molecule. In order to create an ECD of this description, it would probably be necessary to replace the "reaction volume" approach

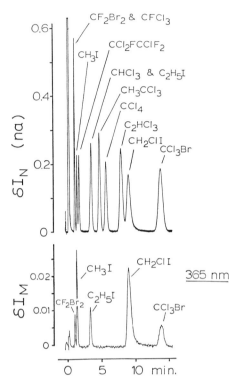

Figure 5.7. Gas chromatographic analysis of a mixture of 11 halogenated hydrocarbons with simultaneous detection by the normal (δI_N) and the photodetachment-modulated (δI_M) modes of the pulsed ECD. The light used was of wavelength 365 nm. Reprinted with permission from R. S. Mock and E. P. Grimsrud, *Anal. Chem.* **60**, 1884 (1988). Copyright 1988, American Chemical Society.

inherent in the pulsed ECD with a system in which the column effluent was exposed to a continuous "flux" of electrons and less time was allowed for events other than the initial EC reaction to occur. Approaches of this general description have been attempted. In fact, the earliest forms of the ECD used the "direct current" mode of signal processing, in which electrons produced by beta radiation were passed through the flowing detector gas by the continuous application of an electric field. In recent years, ECDs have been described in which a flux of electrons, created either by thermionic emission from a heated filament (40) or by the photoelectric effect from a low work-function surface (41), was passed across the flowing GC effluent. These latter approaches offer the added benefit of eliminating the need for radioactive materials. Although work in this area has not yet led to a product that is clearly superior, a good alternative to the ^{63}Ni pulsed ECD, offering a larger dynamic response range to all EC-active compounds, would certainly be well received.

If, on the other hand, the ^{63}Ni pulsed ECDs remain a dominant form of the ECD, additional improvements of these detectors can also be expected. The continued development of smaller pulsed ECDs will clearly be beneficial to capillary GC. Also, as more knowledge in the area of gas phase ion chemistry is acquired, parallel improvements in the pulsed ECD will follow. The three forms of the CS–ECD, discussed above, provide examples wherein new capabilities have resulted in recent years from an understanding gained of reactions that previously had constituted unknown sources of irreproducibility. In our laboratory we are currently exploring the use of photodetachment, as shown in the following reaction,

$$A^- + hv \longrightarrow A + e \qquad (19)$$

in order to further increase the response specificity of an ECD to certain substances. An application (42) of the photodetachment-modulated ECD for the specific detection of organic iodides and bromides in the presence of organic chlorides is shown in Figure 5.7. Applications of this detector to other molecules that undergo EC by the resonance mechanism also appear promising (43). Another area of major importance to the performance of a pulsed ECD, about which little is yet known (44, 45), is that concerning the fate of negative ions in Reactions 1e upon recombination with positive ions. Control and manipulation of these reactions clearly would lead to improved response characteristics (28, 45).

ACKNOWLEDGMENT

Support for research in this area by the Chemical Analysis Division of the National Science Foundation under several grants from 1977 to the present is gratefully acknowledged.

REFERENCES

1. D. W. Later, M. L. Lee, and B. W. Wilson, *Anal. Chem.* **54**, 117 (1982).
2. J. E. Lovelock and S. R. Lipsky, *J. Am. Chem. Soc.* **82**, 431 (1960).
3. J. E. Lovelock, *Anal. Chem.* **33**, 162 (1961).
4. J. E. Lovelock, *Nature (London)* **189**, 729 (1961).
5. J. E. Lovelock, and A. Zlatkis, *Anal. Chem.* **33**, 1958 (1961).
6. W. A. Aue and S. Kapila, *J. Chromatogr. Sci.* **11**, 225 (1973).
7. E. D. Pellizzari, *J. Chromatogr.* **98**, 323 (1974).
8. J. E. Lovelock, *J. Chromatogr.* **99**, 3 (1974).
9. J. E. Lovelock, and A. J. Watson, *J. Chromatogr.* **158**, 123 (1978).
10. C. F. Poole, *HRC & CC, J. High Resolut. Chromatogr. Chromatogr. Commun.* **5**, 454 (1982).

11. M. Dressler, *Selective Gas Chromatographic Detectors* Elsevier, New York, 1986.

12. A. Zlatkis and C. F. Poole, *Electron Capture, Theory and Practice in Chromatography* Elsevier, New York, 1981.

13. P. L. Gobby, E. P. Grimsrud, and S. W. Warden, *Anal. Chem.* **52**, 473 (1980).

14. W. E. Wentworth and E. C. M. Chen, in *Electron Capture, Theory and Practice in Chromatography* (A. Zlatkis and C. F. Poole, eds.), p. 27. Elsevier, New York, 1981.

15. L. J. Sears, J. A. Campbell, and E. P. Grimsrud, *Biomed. Environ. Mass Spectrom* **14**, 401 (1987).

16. C. N. McEwen and M. A. Rudat, *J. Am. Chem. Soc.* **103**, 4343 (1981).

17. A. G. Harrison, *Chemical Ionization Mass Spectrometry* CRC Press, Boca Raton, FL, 1983.

18. E. P. Grimsrud, S. Chowdhury, and P. Kebarle, *J. Chem. Phys.* **83**, 3983 (1985).

19. P. Kebarle and S. Chowdhury, *Chem. Rev.* **87**, 513 (1987).

20. D. G. Peters, J. M. Hayes, and G. M. Hieftje, *Chemical Separations and Measurements* Saunders, Philadelphia, PA 1974.

21. G. Wells, *HRC & CC, J. High Resolut. Chromatogr. Chromatogr. Commun.* **6**, 651 (1983).

22. G. Wells and R. Simon, *HRC & CC, J. High Resolut. Chromatogr. Chromatogr. Commun.* **6**, 427 (1983).

23. D. C. Fenimore, P. R. Loy, and A. Zlatkis, *Anal. Chem.* **43**, 1972 (1971).

24. W. B. Knighton and E. P. Grimsrud, *Anal. Chem.* **55**, 713 (1983).

25. R. J. Maggs, P. L. Joynes, A. J. Davies, and J. E. Lovelock, *Anal. Chem.* **43**, 1966 (1971).

26. W. B. Knighton and E. P. Grimsrud, *Anal. Chem.* **54**, 1892 (1982).

27. P. L. Patterson, *J. Chromatogr.* **134**, 25 (1977).

28. E. P. Grimsrud and W. B. Knighton, *Anal. Chem.* **54**, 565 (1982).

29. D. Smith and N. G. Adams, *J. Phys. B* **20**, 4903 (1987).

30. E. Alge, N. G. Adams, and D. Smith, *J. Phys. B.* **17**, 3827 (1984).

31. E. P. Grimsrud and D. A. Miller, *Anal. Chem.* **50**, 1141 (1978).

32. P. G. Simmonds, *J. Chromatogr.* **166**, 593 (1978).

33. D. A. Miller and E. P. Grimsrud, *Anal. Chem.* **51**, 851 (1979).

34. E. P. Grimsrud, in *Electron Capture, Theory and Practice in Chromatography* (A. Zlatkis and C. F. Poole, eds.), p.91. Elsevier, New York, 1981.

35. J. A. Campbell and E. P. Grimsrud, *J. Chromatogr.* **243**, 1 (1982).

36. M. P. Phillips, P. D. Goldan, W. C. Kuster, R. E. Sievers, and F. C. Fehsenfeld, *Anal. Chem.* **51**, 1819 (1979).

37. F. C. Fehsenfeld, P. D. Goldan, M. P. Phillips, and R. E. Sievers, in *Electron Capture, Theory and Practice in Chromatography* (A. Zlatkis and C. F. Poole, eds.), p.69. Elsevier, New York, 1981.

38. E. P. Grimsrud and C. A. Valkenburg, *J. Chromatogr.* **302**, 243 (1984).

39. C. A. Valkenburg, W. B. Knighton, and E. P. Grimsrud, *HRC & CC, J. High Resolut. Chromatogr. Chromatogr. Commun.* **9**, 320 (1987)

40. A. Neukermans, W. Kruger, and D. McManigill, *J. Chromatogr.* **235**, 1 (1982).

41. P. G. Simmonds, *J. Chromatogr.* **399**, 149 (1987).

42. R. S. Mock and E. P. Grimsrud, *Anal. Chem.* **60**, 1684 (1988).

43. R. S. Mock and E. P. Grimsrud, *J. Am. Chem. Soc.* **111**, 2861 (1989).

44. D. Smith and N. G. Adams, in *Physics of Ion-Ion and Electron-Ion Collisions* (F. Brouillard and J. W. McGowan, eds.), p.501. Plenum, New York, 1983.

45. C. A. Valkenburg, L. A. Krieger, and E. P. Grimsrud, *J. Chem. Phys.* **86**, 6782 (1987).

THE ELECTROLYTIC CONDUCTIVITY (HALL) DETECTOR

RANDALL C. HALL

Randy Hall & Associates
Cloudcroft, New Mexico

The concept of using electrolytic conductivity for the detection of gas chromatograph eluates was first reported by Piringer and Pascalau (1). These authors were interested in finding a detector that could provide a more uniform and sensitive response to hydrocarbons than that of the thermal conductivity detector, which was the most popular detector at the time. Although their results appeared promising, the electrolytic conductivity detector (ELCD) was never refined for this application, and the less complex flame ionization detector (FID) became the detector of choice for hydrocarbon analysis. The ELCD, however, has been adapted for other applications.

The ELCD detection scheme originally reported by Piringer and Pascalau consisted of several steps and on the surface appears to be quite complicated, but in practice it is quite simple and reliable. In their detection scheme, hydrocarbons were first converted to carbon dioxide by passing the GC eluate through a high-temperature reactor containing copper oxide. The carbon dioxide was extracted into a deionized water stream by passing the gas and liquid streams down a long glass capillary tube. The gas and liquid streams were then directed into a flow-through electrolytic conductivity cell where the water was separated from the gas stream and its electrolytic conductivity continuously measured, and recorded as the detector signal. This general scheme is the basis of all modern ELCDs.

Coulson (2) modified the basic concept of Piringer and Pascalau for the selective detection of halogen-, sulfur-, and nitrogen-containing compounds and made a commercial detector available in the mid-1960s. Although the Coulson detector was not compatible with capillary columns, its high selectivity

Detectors for Capillary Chromatography, edited by Herbert H. Hill and Dennis G. McMinn.
Chemical Analysis Series, Vol. 121.
ISBN 0-471-50645-1 © 1992 John Wiley & Sons, Inc.

for nitrogen- and halogen-containing compounds made it quite popular with packed-column pesticide analysts from the late 1960s to the early 1970s.

The ELCD has been extensively refined by Hall (3–5) for the selective detection of halogen-, sulfur-, and nitrogen-containing compounds, and several different versions of this detector are now commercially available for the detection of these elements. Early refinements were aimed at increasing sensitivity and selectivity, whereas later modifications addressed making the ELCD concept compatible with capillary columns.

The ELCD is used primarily to detect halogen-, sulfur-, and nitrogen-containing compounds. However, other operating modes have been reported for the selective detection of esters, nitrosamines, aliphatic halocarbons, bischloromethyl ether, and hydrogen, as well as the general detection of hydrocarbons.

Although the ELCD can be used to detect picogram quantities of most halogen-, sulfur-, and nitrogen-containing compounds, it has not been widely used as a detector for capillary column analysis except in the detection of volatile halocarbons in water, where it is specified by the U.S. Environmental Protection Agency in EPA Method 502.2. The reason for this is twofold. First, some of the commercial versions of the ELCD are not truly compatible with capillary columns, yet are sold as such. Consequently their poor capillary performance is often attributed to an ELCD incompatibility with capillary columns rather than limitations of a particular detector design. And, second, there is a general lack of user understanding as to what is required for "capillary" performance from a design and operational standpoint.

In this chapter, the operational requirements for the ELCD's three primary modes of operation and its application to the capillary column analysis of halogen-, sulfur-, and nitrogen-containing compounds are examined. In so doing, emphasis is placed on the commercially available detectors and not on experimental designs. Previously published reviews should be consulted for additional information pertaining to the historical development of the ELCD up to the early "Hall detector" and the nonstandard modes of operation (6–8).

1. BASIS OF OPERATION

The basic components of the ELCD are shown in the block diagram of Figure 6.1 and in the simplified cross-sectional view of Figure 6.2. As shown in these figures, the GC effluent is mixed with a reaction gas in the detector base and then passed through a high-temperature reactor where the heteroatoms of interest are converted to small inorganic molecules (the detection species). The reaction product stream is then directed into a flow-through electrolytic

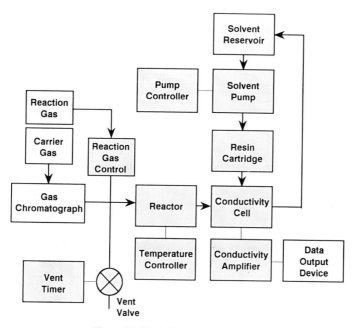

Figure 6.1. Block diagram of the ELCD.

conductivity cell where the inorganic compounds of interest are extracted into a solvent stream and the electrolytic conductivity of the solvent measured.

1.1. Detection Chemistry

The principal detection modes of the commercial ELCDs and the species detected are as follows:

- *Halogen mode*, HX (X usually $= Cl$)
- *Sulfur mode*, SO_2/SO_3 (primarily SO_2)
- *Nitrogen mode*, NH_3

The basis of detector response can be understood with the aid of Tables 6.1 and 6.2. As can be determined from these tables, response depends upon the species formed (reaction conditions), whether or not a postreaction scrubber is used, and the conductivity solvent and its pH. For example, if one wants to detect halogen compounds, a nickel reaction tube, hydrogen reaction gas, a reactor temperature of 850–1000 °C and *n*-propyl alcohol are used. Under these conditions, compounds containing chlorine will be converted to HCl and

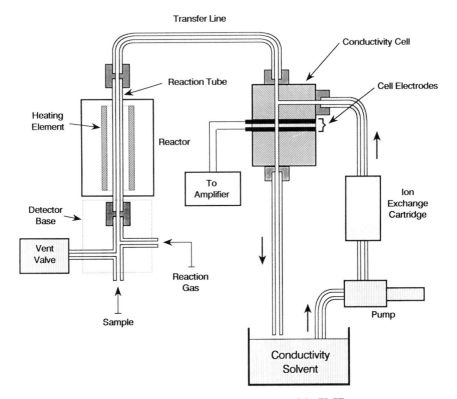

Figure 6.2. Simplified cross-sectional view of the ELCD.

other reaction products such as methane and water. The HCl will dissolve in the propyl alcohol and change its electrolytic conductivity, whereas the non-halogen products will not dissolve in the propanol and/or not change its conductivity to any significant degree. Likewise, other halogen compounds will be converted to their corresponding hydrogen halides and be detected in the same manner.

Different classes of compounds containing a given heteroatom can exhibit different dependencies on reactor temperature. For instance, aromatic chloro-carbons are more stable than vinylic chlorocarbons, which are in turn more stable than aliphatic chlorocarbons. Consequently aromatic halocarbons will require higher temperatures and/or more catalytic conditions to be converted to HX than will aliphatic halocarbons. The same is also true for sulfur compounds: mercaptans are easier to convert to SO_2 than are sulfides, which in turn are easier than disulfides.

Table 6.1. Reaction Products Produced Using a Nickel Reaction Tube under Reductive Conditions at 850–1000 °C

Compound	Main Product(s)	Comments
Halogen-containing compounds	HX	HX can be removed by N-mode scrubber and is selectively detected in X mode
Sulfur-containing compounds	H_2S	H_2S can be removed by N-mode scrubber and is poorly ionized in X mode
Nitrogen-containing compounds	NH_3	NH_3 is poorly ionized in X mode and selectively detected in N mode
Alkanes	CH_4 and lower alkanes	Products are not ionized in any mode
Oxygen-containing compounds	H_2O	H_2O gives little response in X mode and N mode

Table 6.2. Reaction Products Produced Using a Nickel or Alumina Reaction Tube under Oxidative Conditions at 850–1000 °C

Compound	Main Product(s)	Comments
Halogen-containing compounds	HX	HX can be removed by S-mode scrubber
Sulfur-containing compounds	SO_2	SO_2 is selectively detected in S mode
Nitrogen-containing compounds	N_2 and some nitrogen oxides	No response from N_2; little response from oxides
Alkanes	CO_2	CO_2 is poorly ionized in nonaqueous solvents

Differences in temperature dependency of response are more pronounced under noncatalytic conditions than under catalytic conditions. Dolan and Hall (9) used a quartz reaction tube (noncatalytic) and a reduced reaction temperature to selectively detect chlorinated hydrocarbon pesticides in the presence of polychlorobiphenyls (PCBs). Although not as great, similar enhancements in specificity can also be achieved with nickel reaction tubes.

When nickel reaction tubes are used for this purpose, caution should be exercised to make sure the catalytic activity of the tube has come to equilibrium. Nickel tubes heated to an elevated temperature and then reduced in temperature can maintain their higher temperature catalytic activity for several hours, and in some cases even overnight. Also, the catalytic activity of nickel reaction tubes will be more affected by column bleed and sample constituents at low reaction temperatures.

In the detection of sulfur compounds, the compound must be converted to SO_2. The formation of SO_2 depends on such factors as reaction temperature, composition of reaction gas, catalytic activity of the reaction tube, residence time, and the internal diameter of the reaction tube. However, in general, at temperatures around 800 °C, both SO_3 and SO_2 will be formed. The formation of SO_2 is favored up to ~ 950–1000 °C. Above this temperature SO will be formed, which does not effectively support electrolytic conductivity. The effect of reactor temperature on detector response to several sulfur compounds using an alumina reaction tube is shown in Figure 6.3. As can be seen in this figure, the alumina tube, which is fairly catalytic under oxidative conditions, moderates the effect of reaction temperature and allows a more uniform response to be obtained.

As was mentioned above, different classes of sulfur compounds can exhibit different temperature dependencies. This can be used to enhance specificity of response to mercaptans. Such schemes have not been investigated in any detail, and the same general comments apply as for the halogen schemes. Such

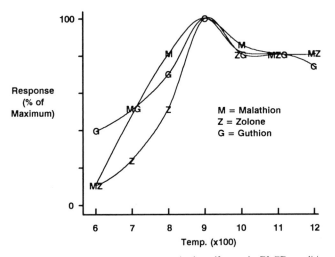

Figure 6.3. Effect of reactor temperature on response in the sulfur mode. ELCD conditions: reaction tube, 0.030-in.-i.d. alumina; reaction gas, 25 mL/min air; total detector gas flow, 50 mL/min; electrolyte, 40 μL/min 95% MeOH.

enhancement schemes, however, should not be used for routine analytical work unless thoroughly characterized for the analytes and the matrices of interest, because the matrix can change the response for certain analytes.

1.2. Solvent Systems

Nonaqueous solvents such as *n*-propyl alcohol are preferred in the halogen mode for several reasons. They increase specificity of response by reducing the ionization of interfering compounds such as H_2S and NH_3, and increase sensitivity by minimizing the detector background signal. And, although it has not been found to be a problem, they can also eliminate nonlinearities due to the autoprotolysis of water [suggested by Coulson to be a potential problem (10)]. Methyl alcohol is preferred in the sulfur mode for the same reasons.

The conductivity solvent must meet several general requirements. First, it must allow the detection species to be fully ionized. Since the detection species in the sulfur and nitrogen modes are themselves nonionic, the solvent must also allow ionic species to be formed. For example, SO_2 and NH_3 are converted to H_2SO_3 and NH_4OH, which support electrolytic conductivity. Thus for these modes, the solvent must have some water available or be capable of converting the detection species to other compounds that will support electrolytic conductivity. Methyl alcohol containing a small amount of water (1–5%) serves this need for the sulfur mode and allows the strong acid H_2SO_3 to ionize. However, methyl alcohol is not a satisfactory solvent for the nitrogen mode because it does not allow the weak base NH_4OH to fully ionize.

To achieve low detection levels, the solvent must also have a low background conductivity. Thus detection schemes for the nitrogen mode based on the reduction of electrolytic conductivity of dilute acids (11) cannot achieve the detection levels of "neutral" aqueous solvents because it is much easier to measure a small change in the electrolytic conductivity of a solution that has a low background conductance than in one that has a high background.

The pH of the solvent is another consideration in the nitrogen mode. Highly deionized solvents tend to behave as if they are slightly acidic. This usually is due to the presence of trace quantities of CO_2 that permeate into the solvent system. The net effect is that low levels of NH_3 will be neutralized. This causes low levels of nitrogen compounds to give negative peaks. Higher compound levels will exhibit peaks that have a sharp negative leading edge and a negative broader trailing edge. This can be understood by the following:

$$NH_3 + H_2O \rightleftharpoons NH_4OH \qquad \text{(solvation of } NH_3) \qquad (1)$$

$$NH_4OH \rightleftharpoons NH_4^+ + OH^- \qquad \text{(ideal ionization)} \qquad (2)$$

$$NH_4OH + HCl \rightarrow NH_4^+ + Cl^- \qquad \text{(ionization in acidic solvent)} \qquad (3)$$

Thus the result of the ionization of NH_3 in an acidic solvent is replacement of the electrolytic conductivity due to the Cl^- ion with that of the anion of the acid, and replacement of the electrolytic conductivity due to H^+ (actually H_3O^+ in water) with that of the NH_4^+ ion. Because the OH^- ion has an equivalent ionic conductance of 199, Cl^- has 76.3, NH_4^+ has 73.4, and H_3O^+ has a value of 349.8, the ionic conductance due to the HCl ions, which have a value of 426.1, will be replaced by that of the NH_4Cl ions, which have a value of 149.7. Therefore the overall ionic conductance will be decreased. Since the electrolytic conductivity is a function of the equivalent ionic conductances and their concentrations, detector response will be negative until the available acid is exhausted. This is shown pictorially in Figure 6.4.

This problem was recognized by Coulson (8) and elaborated on by Patchett (12). To prevent NH_3 neutralization, these authors used a two-bed ion exchange cartridge. The lower section was packed with a strong-base anion-resin and the upper section was packed with a neutral, mixed anion and cation exchange resin. Bleed from the anion resin sufficiently increased the pH of the water conductivity solvent to prevent NH_3 neutralization and negative peaks.

Although this technique is currently used in one of the commercial versions of the ELCD, its effectiveness is sporadic owing to inconsistencies in resin bleed and the variability of other factors such as the amount of CO_2 in the solvent stream. The reliability of this approach is not sufficient for it to be considered a solution to the NH_3 neutralization problem. For this reason another technique that can be easily reproduced and controlled on a commercial scale was developed.

The neutralization of ammonia produced in the nitrogen mode can be reliably prevented by allowing a small amount of ammonia to permeate into the solvent stream just prior to the cell (13). This is easily accomplished by passing the solvent stream through a short loop of 0.062-in. o.d. by 0.010-in.

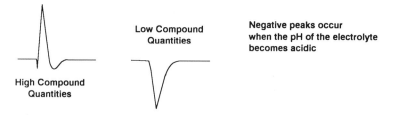

Figure 6.4. Effects of acidic electrolyte (i.e., negative peaks) on response in the nitrogen mode.

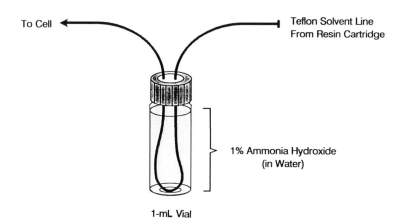

To Cell

Teflon Solvent Line
From Resin Cartridge

1% Ammonia Hydroxide
(in Water)

1-mL Vial

Figure 6.5. Permeation control system for regulating electrolyte pH in the nitrogen mode.

i.d. Teflon tubing contained in a vessel filled with dilute ammonium hydroxide. A 2–3 in. length of Teflon tubing exposed to a 1:100 dilution of concentrated ammonia hydroxide works well for this purpose. A suitable permeation system is shown in Figure 6.5. The system should be placed in a temperature-stable environment to maintain a consistent bleed of NH_3 into the solvent and to prevent baseline wandering.

Another approach that appears to work very well for preventing NH_3 neutralization is the "total" exclusion of CO_2 from the solvent stream. This is accomplished by using nonpermeable tubing in the solvent delivery system. The chromatograms shown in Figures 6.18–6.20 (see Sections 3.2 and 3.3, below) were produced in this manner.

In the sulfur mode, SO_2 is the primary sulfur species formed. The SO_2 reacts with the water present in the methyl alcohol to from sulfurous acid. Methyl alcohol is used in this mode rather than propyl alcohol because it gives a greater linear dynamic range.

Pure water is not a satisfactory solvent for the sulfur mode because of interference from carbon compounds. The interference arises from the formation of CO_2 which ionizes readily in water but not in methyl alcohol. Greater quantities of water (10–20%) in the methyl alcohol will extend the linear dynamic range of the sulfur mode (14), but specificity will suffer somewhat.

Although water is not a satisfactory solvent for the sulfur or halogen modes, it is preferred for the nitrogen modes. In this mode, nitrogen compounds are converted to NH_3, which is a weak base that requires an aqueous solvent to be fully ionized. Water containing a small amount of organic solvent (10–20%)

is frequently used in the nitrogen mode and gives greater sensitivity due to lower cell noise. One manufacturer uses a 1:1 water:propyl alcohol mixture, which is too high to allow the higher concentrations of NH_3 to fully ionize. I have found water containing 5–10% t-butyl alcohol to be an excellent solvent for the detector manufactured by O.I. Analytical.

1.3.   Reaction Gas

The general detection of nitrogen compounds requires reductive conditions and hence necessitates the use of hydrogen as the reaction gas. Since it is possible to convert some nitrogen compounds to volatile amines under neutral conditions, it should be possible to achieve additional selectivity for certain nitrogen-containing analytes by using such conditions. This, however, has not been exploited to any extent and is not currently supported by an ELCD manufacturer.

The general detection of sulfur-containing compounds requires oxidative conditions to form SO_2. There seems to be no sensitivity advantage in using oxygen gas rather than air for this purpose. In fact, oxygen is detrimental to the lifetime of nickel reaction tubes.

Detection of halogen-containing compounds requires the formation of HX, which, depending on the eluate, can be formed under either oxidative or reductive conditions. However, catalytic reductive conditions are preferred because they provide a more uniform response and greater specificity. Nevertheless, oxidative conditions may be preferred in some instances as a general screening mode for halogen- and sulfur-containing compounds. This can be achieved by simply operating the detector in the sulfur mode without the scrubber. Since some halogen-containing compounds cannot be readily detected under oxidative conditions (for example, chlorinated aromatics), detector response should be checked for each compound of interest.

The use of reaction gas doped with trace quantities of a halocarbon has been reported by Piringer and Wolff (15) to prevent peak tailing in the detection of halogen-containing compounds. I have tried various approaches based on this theme with very limited success, because of increased noise and drift due to the added halocarbon. It is much easier to prevent peak tailing by using a proper reaction tube than it is to counteract irreversible adsorption on an overly active surface.

1.4.   Reaction Tubes

Quartz reaction tubes were used in early ELCDs but fell from favor owing to their fragile nature, inconsistency of response, and poor selectivity in the halogen mode. Nickel reaction tubes, which are catalytic, are now used for all three primary modes of operation. Nickel is not an ideal material, however, and other materials may prove more useful for certain modes. For instance, gold has been reported as a good material for the selective detection of nitrosamines in the presence of other nitrogen-containing compounds (16), and an alumina reaction tube gives excellent results for the sulfur mode (see Figures 6.14–6.17 in Sections 3.1 and 3.2, below). I would expect alumina tubes to also work well for the selective detection of nitrosamines since they have low catalytic activity under reductive conditions.

Alumina tubes have a major advantage for the detection of sulfur containing compounds in that they do not eventually burn in two and therefore do not have to be replaced as often as nickel tubes. They are inexpensive and fairly resistant to breakage, and appear to be the ideal reaction tube for the sulfur mode.

Although alumina tubes are not available from the detector manufacturers at this time, they can be obtained cut to length from several manufacturers of ceramic products. I have used the 0.062-in.-o.d. by 0.030-in.-i.d. "pure" alumina tubes (AD-998), available from Coors, with excellent results. I would assume that similar tubing available from other suppliers should also work.

Nickel is currently the material of choice for the nitrogen mode and the halogen mode when a uniform response from compounds of various structures is desired. However, nickel is not suitable for the detection of iodine-containing compounds, and its response for brominated compounds is progressively poisoned by water and organic injection solvents. Deterioration in response by water is probably due to etching of the reaction tube surface, whereas organic solvents cause carbon deposition. Both effects increase the irreversible adsorption of the reaction tube surface and give rise to tailing peaks and poor sensitivity. As the reaction tube ages, the effects are first seen with brominated compounds and then with chlorinated compounds (see Figure 6.6).

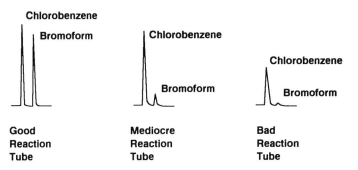

Figure 6.6. Effects of nickel reaction tube aging on response to halogen mode test compounds.

Preliminary results (17) indicate that the aging of nickel reaction tubes can be reduced by coating the inside surface of the reaction tube with a very thin layer of fused silica. The thickness of the coating will affect its catalytic reactivity, and it has been found to be difficult to produce reaction tubes of a consistent activity on a commercial scale. The layer needs to be thick enough to cover surface imperfections yet thin enough that the tube is still catalytic.

Nickel reaction tubes can be coated in the lab by simply soaking them in a 3–5% solution of methyl silicone oil, such as OV-1, in xylene. The tubes should be drained, air dried, and heated to 1000 °C for several minutes after installation. The character of such a surface is open to speculation, and some inconsistency should be expected.

Another problem with nickel reaction tubes is that at this writing there is no source of pure nickel tubing. Only nickel 200 tubing is available, which has several percent of various impurities. Also, the interior surface of such tubing is fairly rough and is variable from one lot to another. Reaction tubes having a smooth surface of pure nickel should give improved performance. An ideal reaction tube for halogen and nitrogen modes could be made by depositing a smooth layer of pure nickel on the inside of an alumina tube.

Until a better nickel reaction tube becomes available, the present tubes should be replaced periodically to maintain performance. This is easy to do in all present designs and takes only a few minutes. In addition, the nickel tubes are inexpensive. The reaction tube lifetime depends on the application. In general, the higher the reaction temperature, the shorter will be the life of the reaction tube. When the detector is operated properly, however, one should expect a tube life of approximately 1–6 months.

1.5. Postreaction Scrubbers

Postreaction scrubbers are used in the sulfur and nitrogen modes to enhance the specificity of response. The scrubbers consist of 0.062-in.-o.d. by 0.040-in.-i.d. by ~ 12-in.-long coiled copper or stainless steel tubing. The tubing contains an appropriate material that lets the monitored species pass but abstracts the unwanted components. The scrubbers are mounted at the exit port of the reactor in the sample flow path prior to the cell. Properly prepared scrubbers do not appear to increase band spreading as long as there is a total gas flow of approximately 40 mL/min.

The scrubber for the sulfur mode contains several strands of silver wire. Silver removes hydrogen halides but does not remove SO_2. The scrubber for the nitrogen mode contains several strands of quartz thread coated with KOH. There appears to be little difference between scrubbers made with copper tubing and those made with stainless steel tubing.

2. DETECTOR DESIGNS

There are two different commercial versions of the ELCD now available for which the manufacturers claim compatibility with capillary columns: one is described in U.S. Patent 3,934,193 (1976), and the other in U.S. Patent 4,555,383 (1985). The primary functional differences between them are in the designs of the cell and the electrolyte (solvent) systems. The manufacturer of the first design calls their detector a Hall detector, which is a registered trademark of Tracor, Inc. The manufacturer of the second detector refers to their ELCD simply as an ELCD. Users, however, normally refer to both detectors as the Hall detector (18). Since the two detectors have some fundamental differences that may affect performance for a given application, it is important to know which detector was actually used and to understand the differences between them.

2.1. Cell Designs

A cross-sectional view of the first cell design is shown in Figure 6.7. In this design the gas and liquid streams are mixed in a gas–liquid contactor made of an inert material such as Teflon. The gas and liquid mixture then flows into a concentric tube gas–liquid separator that also serves as a concentric electrode conductivity cell. As the liquid passes into the gas–liquid separator, it adheres

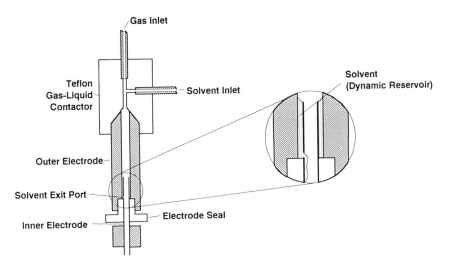

Figure 6.7. Cross-sectional view of a dynamic reservoir ELCD cell.

to the inside wall of the outer tube (electrode) and forms a dynamic reservoir of solvent between the inside wall of the outer tube and the outside wall of the inner tube. The gas simply vents through the hollow center electrode, where it is recombined with the liquid as the liquid exits from the bottom of the liquid reservoir into the center electrode through a small hole in its wall. Thus the function of the concentric tube gas–liquid separator is to isolate the liquid from the gas, thereby forming a dynamic reservoir of liquid where the electrolytic conductivity is measured.

In the second design, shown in Figure 6.8, the gas and liquid are not isolated from each other. In this case the gas and liquid are brought together in a gas–liquid contactor and the mixture is passed down an axial bore that contains the measurement electrodes. The electrodes are coplanar and form a measurement zone consisting of two closely spaced annular rings. Thus the electrolytic conductivity is measured as the gas and liquid commingle down the axial bore and through the measurement zone.

For the purpose of this discussion I will refer to the first design (Figure 6.7) as a dynamic reservoir cell (DR cell), and the second design (Figure 6.8), in which both the gas and liquid phases pass through the measurement zone, as a mixed phase cell (MP cell). The Coulson detector and the Tracor Hall detector use DR cells, whereas the O.I. Analytical ELCD uses an MP cell.

If a gas bubble passes through a DR cell, a noise spike will occur. This is due to a change in the cell constant, which takes place as the bubble passes through the measurement zone. Consequently it is essential to prevent bubbles from forming in (by solvent degassing) and/or passing through the measurement zone of all DR cells. Piringer et al. (19) reported that if the gas–liquid separator was removed in their cell (a DR cell) there was a deterioration of several orders of magnitude in the signal-to-noise ratio. Since gas and liquid

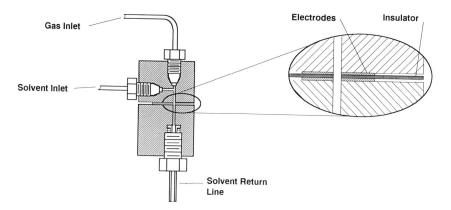

Figure 6.8. Cross-sectional view of a mixed phase ELCD cell.

are both present in the measurement zone of MP cells, bubbles are not an issue in these designs.

When the dynamic reservoir design of Figure 6.7 was developed, efficiency of gas–liquid separation, separation stability, and overall detector sensitivity were the primary concerns. At the time of this development (the early 1970s) there was a pressing need for detectors that had greater sensitivity and reliability for the analysis of halogen-, sulfur-, and nitrogen-containing pesticides, and this was the primary factor driving the work. Capillary column compatibility was not an issue, as only packed columns were then being used for pesticide analysis.

After evaluating a number of different gas–liquid separation techniques, a concentric tube gas–liquid separator was found to offer the greatest potential. However, to operate properly, the tubes had to be very close and very long compared to their separation distance. A spacing of some 0.004– 0.008 in. and a length of ~ 0.25 in. proved to offer the greatest performance.

The viscosity of the solvent moving between such closely spaced and long tubes results in a smooth gas–liquid separation, which allows much higher amplifier gains to be used than otherwise would be possible. Compared to the design described by Coulson (2), packed column detection limits for lindane (a halogen compound) were improved from ~ 1–2 ng to ~ 25–50 pg. Peak shape was also improved. Since the gas–liquid separator of this design has an extremely large surface-to-volume ratio, contamination of the stainless steel electrodes by ion exchange resin bleed and "combustion" products can result in excessive peak tailing and a loss in response. Contamination also decreases the separating effeciency of the gas–liquid separator, which increases cell noise. Thus to maintain performance the cell has to be taken apart and cleaned periodically.

The drag on the solvent passing between the inner and outer tubes of the concentric tube gas–liquid separator produces efficient gas–liquid separation, but it also results in a cell response time (dwell time) related to the viscosity of the solvent. Although the response time is fast enough for packed columns, it is too long to track accurately most capillary peaks. The response time can be decreased by increasing the distance between the electrodes or using less viscous solvents, but separation efficiency suffers. Also, the use of less viscous solvents decreases selectivity in the halogen mode. Consequently, the MP cell shown in Figure 6.8 was developed to overcome these problems. It should be noted that an MP cell design has also been reported by Piringer (15).

The internal volume of the MP cell shown in Figure 6.8 can be made quite small. For instance, the axial bore diameter of the cell shown in Figure 6.8 is 0.013 in. and the measurement zone volume is only 0.03 μL. This allows low solvent flow rates to be used (typically 20–50 μL/min), which increases the concentration of the detection species in the conductivity solvent, making it easier to detect low quantities of analyte.

A primary advantage of the mixed phase design is that the response time is not dependent upon the viscosity of the solvent (within reasonable limits). Consequently, a solvent can be used that provides optimum sensitivity and selectivity for a given mode. Another advantage of this design is that there is no separation of the gas and liquid phases. Thus the cell does not have to be cleaned periodically to maintain efficient gas–liquid separation.

2.2. Solvent Systems

The ELCD needs a low and pulse-free flow of solvent. The commercial DR cell detector available from Tracor requires a flow of ~ 0.5–$1.0 \, \text{mL/min}$, and the MP-cell detector available from O.I. Analytical requires a flow of ~ 25–$50 \, \mu\text{L/min}$. The two detectors use the same small carbon-gear pump for this purpose. Since the pump cannot operate reliably at the low flows required, the output is split after it has passed through an ion exchange resin cartridge.

Although the two detectors use similar solvent delivery systems, there are several differences. The O.I. system uses a two-cartridge ion exchange system and has a roughing resin cartridge on the inlet side of the pump and a finishing resin cartridge on the outlet side. The Tracor system uses a single cartridge. O.I. claims that periodic replacement of the resin in the roughing resin cartridge minimizes downtime since the finishing resin cartridge is on the cell side of the pump and will "buffer" any bleed from the roughing resin as it comes to equilibrium. The O.I. system also has a pneumatic pulse dampener in the finishing resin cartridge, which is said to decrease cell noise and extend the useful lifetime of the pump (5).

Another significant difference between the two systems is the way in which the solvent pH is controlled in the nitrogen mode. The Tracor system uses a resin cartridge containing two different beds of resin: the upper bed (cell side) is packed with a neutral resin, and the lower bed is packed with a basic resin. The purpose of the basic resin is to create enough bleed so that some of it makes it through the neutral resin bed and lowers the pH just enough to prevent neutralization of the NH_3 produced from the nitrogen-containing analytes. In contrast, the O.I. detector uses a patented permeation system to control solvent pH in the nitrogen mode (13).

2.3. Reaction System

Both commercial detectors use high-temperature reactors equipped with 1/4-turn quick release fittings for changing the reaction tube. The reactors in both systems have similar exterior dimensions: $\sim 1.5 \, \text{in.}$ in diameter by $\sim 3 \, \text{in.}$ long.

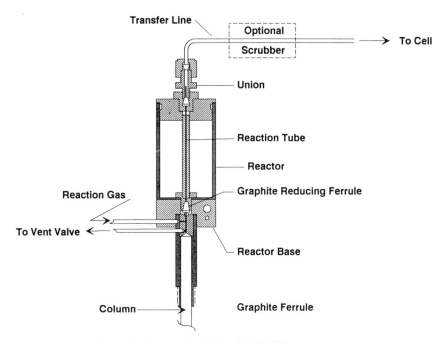

Figure 6.9. Cross-sectional view of the ELCD reactor system.

Both produce a heated zone ~ 1 in. long that is controlled within a centigrade degree. And both reactors have an upper temperature limit of 1000 °C.

The reactor base of the Tracor detector (Model 1000) is lined with glass to ensure an inert sample path. O.I. uses a reactor base made out of Nitronic 50 alloy for the same purpose. Both reactor bases have a solenoid-controlled vent for venting the injection solvent away from the reaction tube (see Figure 6.9). There are no apparent advantages of either system, and the two reaction systems can be considered to be functionally equivalent.

2.4. Conductivity Measurement Technique

Both the O.I and Tracor detectors use a square wave bipolar pulse to measure electrolytic conductivity. The bipolar pulse technique was originally described by Johnson and Enke (20) and is vastly superior to the d.c. technique used in the Coulson detector and the a.c. technique used in early Hall detectors.

In the measurement of electrolytic conductivity, there are two signal paths, a capacitive path and a resistive path. It is important to measure only the resistive component to avoid nonlinearities. The bipolar pulse technique accomplishes this by outputing only the last part of the square wave as the

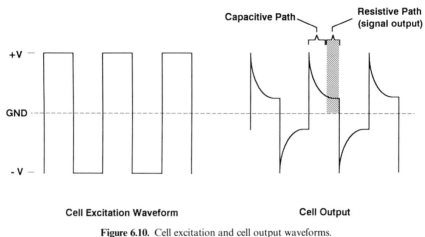

Figure 6.10. Cell excitation and cell output waveforms.

detector signal, which is free of the capacitive component. This is shown in Figure 6.10.

3. APPLICATIONS

The applications described below are for the MP cell design, and they reflect the current state of the art of the ELCD for capillary column analysis. Application of an improved DR cell ELCD for the capillary column analysis of halogen-, sulfur-, and nitrogen-containing compounds has been published previously by its manufacturer (21).

3.1. Halogen Mode

As was mentioned previously, the EPA specifies the use of the ELCD in EPA Method 502.2, which is a purge and trap procedure (dynamic headspace) employing a 0.53-mm-i.d. column. This method uses a trap that contains silica gel, which can collect approximately 10 μL of water from purging the sample. This water will be transferred to the column and into the detector unless a purge and trap concentrator that has an effective water management system is used. As was stated in the discussion on reaction tubes, water deteriorates the surface of nickel reaction tubes, which leads to peak tailing.

Since the water problem has been understood only recently and purge and trap concentrators with water management systems have only become available of late, most of the ELCD chromatograms that have been published show considerable peak tailing. This can easily lead one to conclude that the ELCD produces tailing peaks and thereby is not truly compatible with capillary columns. However, when a purge and trap concentrator that has an effective water management system is used, the ELCD exhibits no significant peak tailing.

ELCD purge and trap chromatograms of representative volatile halocarbons are compared in Figure 6.11 for purge and trap equipment with and

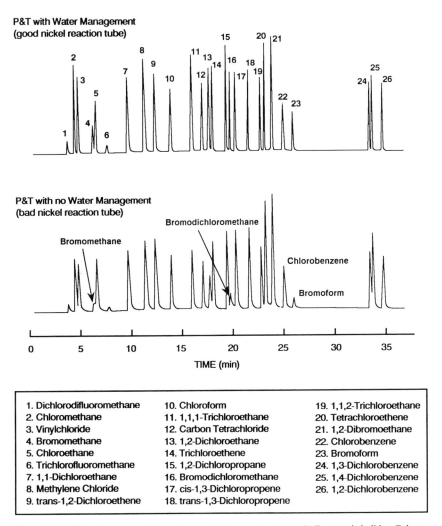

1. Dichlorodifluoromethane	10. Chloroform	19. 1,1,2-Trichloroethane
2. Chloromethane	11. 1,1,1-Trichloroethane	20. Tetrachloroethene
3. Vinylchloride	12. Carbon Tetrachloride	21. 1,2-Dibromoethane
4. Bromomethane	13. 1,2-Dichloroethane	22. Chlorobenzene
5. Chloroethane	14. Trichloroethene	23. Bromoform
6. Trichlorofluoromethane	15. 1,2-Dichloropropane	24. 1,3-Dichlorobenzene
7. 1,1-Dichloroethane	16. Bromodichloromethane	25. 1,4-Dichlorobenzene
8. Methylene Chloride	17. cis-1,3-Dichloropropene	26. 1,2-Dichlorobenzene
9. trans-1,2-Dichloroethene	18. trans-1,3-Dichloropropene	

Figure 6.11. Effect of purge and trap water on detector response to volatile organic halides. Column: Restek 502.2 FS 0.53 mm × 105 M. Program: 35 °C for 11 min, 5 °C/min to 160 °C, 4-min final hold. Carrier: helium at 8 mL/min. ELCD conditions: reactor temperature, 960 °C; electrolyte, 35 μL/min n-propyl alcohol; reaction gas, 100 mL/min H_2.

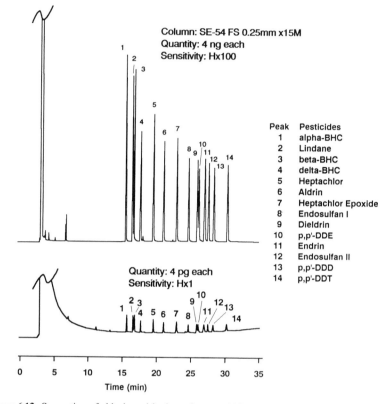

Figure 6.12. Separation of chlorinated hydrocarbon pesticides on a narrow-bore capillary column with ELCD detection. Program: 60 °C for 1 min, 30 °C/min to 150 °C/min to 235 °C. ELCD conditions: same as in Figure 6.11 except reactor temperature at 920 °C.

without a water management system. As this figure shows, when the water is prevented from being transferred to the column, excellent results are obtained. However, when no water management is used, the ELCD produces tailing peaks that get progressively worse.

Detection of chlorinated hydrocarbon pesticides and PCBs are shown in Figures 6.12–6.14. As these figures reveal, the ELCD is very sensitive to these compounds and peak shape is excellent. Low-picogram quantities of common chlorinated hydrocarbon pesticides can easily be detected using narrow-bore capillary columns (Figure 6.12). And, as shown in Figure 6.13, the ELCD with a narrow-bore column can be used to achieve fast separations of these compounds. The separation of PCBs (1254) on a relatively short 0.2-mm-i.d. column is shown in Figure 6.14. Linearity for chlorinated compounds usually

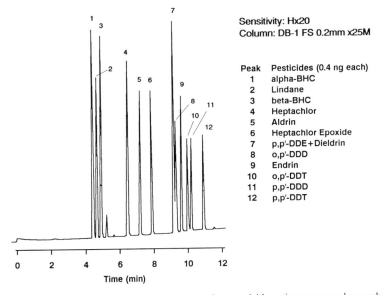

Sensitivity: Hx20
Column: DB-1 FS 0.2mm x25M

Peak	Pesticides (0.4 ng each)
1	alpha-BHC
2	Lindane
3	beta-BHC
4	Heptachlor
5	Aldrin
6	Heptachlor Epoxide
7	p,p'-DDE+Dieldrin
8	o,p'-DDD
9	Endrin
10	o,p'-DDT
11	p,p'-DDD
12	p,p'-DDT

Figure 6.13. Fast separation of chlorinated hydrocarbon pesticides using a narrow-bore column. Program: 70 °C for 0.5 min, 35 C/min to 170 °C, 10 °C/min to 240 °C. Carrier: hydrogen at 12 psig. Key BHC = benzene hexachloride; DDD = dichlorodiphenyldichloroethane; DDT = dichlorodiphenyltrichloroethane; DDE = dichlorodiphenylethylene.

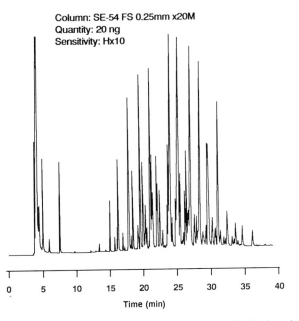

Column: SE-54 FS 0.25mm x20M
Quantity: 20 ng
Sensitivity: Hx10

Figure 6.14. Separation of PCB 1254 on narrow-bore column with ELCD detection. Program: same as in Figure 6.12. ELCD conditions: same as in Figure 6.10.

129

exceeds 5 orders of magnitude, and specificity relative to hydrocarbons is greater than 10^6 (3).

The peak shape and sensitivity of the ELCD for halogenated pesticides and PCBs are more than adequate for most capillary column analyses. Compared to the nonspecific electron capture detector, which is often used for the analysis of these compounds, the high specificity of the ELCD makes it the detector of choice for the analysis of environmental samples where the matrix is usually complex and of unknown composition.

Low ELCD sensitivity to PCBs has been a problem for some analysts. This is a particular problem when packed columns containing high-bleed silicone phases such as QF-1 or OV-210 are used. These phases bleed silicone into the reaction tube, which covers the interior of the tube with a noncatalytic layer

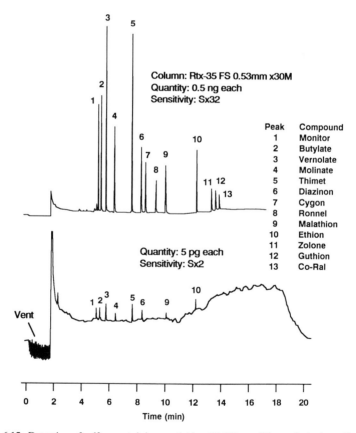

Figure 6.15. Detection of sulfur-containing pesticides. ELCD conditions: electrolyte, 40 μL/min 95% MeOH; reaction tube, 0.030-in.-i.d. alumina; reaction gas, 25 mL/min air; reactor temperature, 950 °C. Program: 65 °C for 1 min, 30 °C/min to 200 °C, 10 °C/min to 290 °C, 4-min final hold.

of glassy decomposition product. And, since a catalytic surface is required for the efficient conversion of PCBs to HCl, sensitivity is diminished. The loss in sensitivity can be partially compensated for by using a 0.040-in. i.d. nickel reaction tube rather than a 0.020-in i.d. tube. However, the best solution is to use low-bleed capillary columns and change the reaction tube when response starts to suffer.

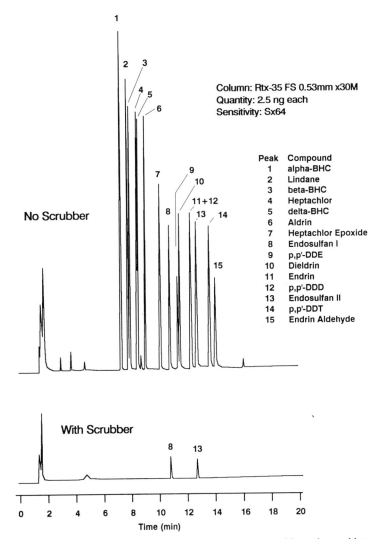

Figure 6.16. Operation of the ELCD in the sulfur mode with and without the scrubber. ELCD conditions and program: same as in Figure 6.15. Key: see Figure 6.13.

3.2. Sulfur Mode

Response of the ELCD to several sulfur-containing pesticides is shown in Figures 6.15 and 6.16. As displayed in Figure 6.15, low-piogram quantities of sulfur-containing compounds can be detected and peak shape is excellent.

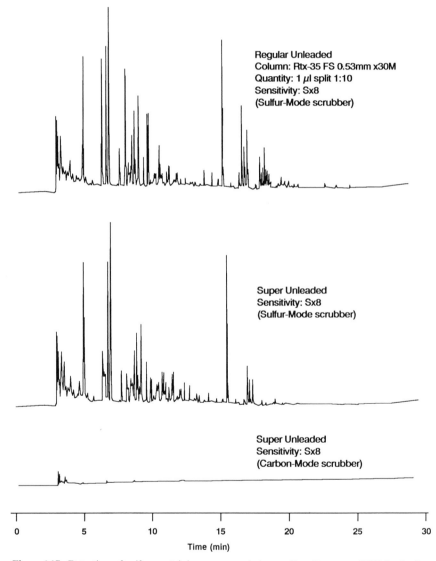

Figure 6.17. Detection of sulfur-containing compounds in gasoline. Program: 30 °C for 2 min, 10 °C/min to 290 °C. ELCD conditions: same as in Figure 6.15.

Response to the EPA 608 pesticide mixture with and without the sulfur mode scrubber is compared in Figure 6.16. When the scrubber is used, only the two pesticides containing sulfur are detected. However, when the scrubber is not used, both chlorine- and sulfur-containing compounds give a response. This could be a useful screening tool for pesticides since most of the compounds of interest contain either sulfur or halogen.

The detection of sulfur-containing compounds in two different gasolines is shown in Figure 6.17. Response to the super unleaded sample on using a carbon mode scrubber is also shown in this figure. The carbon mode scrubber contains potassium bicarbonate on quartz thread. It passes CO_2 but removes HX and SO_2/SO_3. The scrubber was conditioned for several hours with CO_2. As can be seen, there is little response with the carbon mode scrubber, which indicates that the sulfur response is due to sulfur compounds. However, selectivity in the sulfur mode for sulfur v. carbon is about 5×10^4 to 10^5, which may not be sufficient to say for certain without an independent analysis.

Response to sulfur compounds is linear from the lower limits of detection (LLD) to approximately 3 orders of magnitude above LLD. Although response tends to drop off for quantities greater than about 10 ng, response is useful

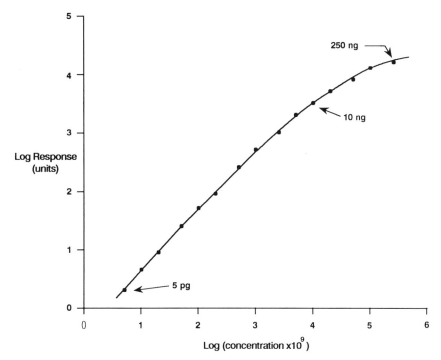

Figure 6.18. Linearity of response to diazinon in the sulfur mode. Column: Rtx-35 FS 0.53 mm × 30 M. Program and ELCD conditions: same as in Figure 6.15.

for ~ 4 orders of magnitude. Linearity of response for the sulfur-containing pesticide diazinon is shown in Figure 6.18.

3.3. Nitrogen Mode

The detection of nitrogen-containing pesticides is depicted in Figures 6.19 and 6.20. As these figures show, low-picogram quantities of nitrogen compounds can be detected. The pesticides in these figures were separated using a 0.53-mm-i.d. column, as is common in pesticide analysis in the United States. Narrow-bore columns should produce higher sensitivity. Nevertheless, quantities of only ~ 10 pg of the early-eluting carbamates, which contain ~ 10% nitrogen, can be detected, and even lower quantities of the triazines

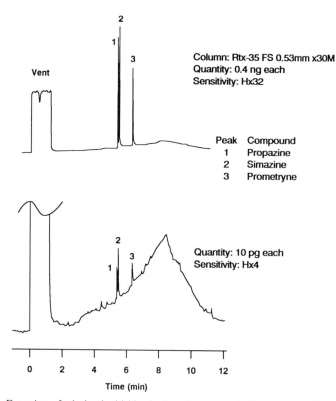

Figure 6.19. Detection of triazine herbicides in the nitrogen mode. Program: 65 °C for 1 min, 30 °C/min to 180 °C, 10 °C/min to 220 °C, 1-min final hold. GC inlet: 200 °C. GC carrier: 8 mL/min H$_2$. ELCD conditions: reactor temperature, 830 °C; scrubber, N mode; reaction gas 100 mL/min H$_2$; electrolyte: 85 μL/min 10% t-butyl alcohol; detector base, 250 °C.

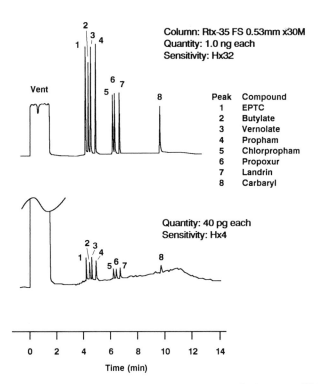

Column: Rtx-35 FS 0.53mm x30M
Quantity: 1.0 ng each
Sensitivity: Hx32

Quantity: 40 pg each
Sensitivity: Hx4

Peak	Compound
1	EPTC
2	Butylate
3	Vernolate
4	Propham
5	Chlorpropham
6	Propoxur
7	Landrin
8	Carbaryl

Figure 6.20. Detection of carbamate pesticides in the nitrogen mode. Program: 65 °C for 1 min, 30 °C/min to 170 °C, 7 °C/min to 200 °C, 2-min final hold. GC inlet: 125 °C. Other conditions: same as in Figure 6.19.

can be detected. The advantage of the ELCD for drug analyses has also been reported (22, 23).

Most common fused silica capillary columns are covered with a protective coating of polyimide. The polyimide coating is a nitrogen-containing compound that exhibits a temperature-dependent degradation and, as expected, considerable baseline is observed at high detector sensitivities (see Figure 6.20). Certain manufacturers also use a nitrogen-containing cross-linking agent in the manufacture of some of their columns, which also contributes to the baseline bleed in the nitrogen mode. Often 1 or 2 weeks of use are required before acceptable bleed levels are obtained.

Aluminum-clad columns would appear to be a solution to this problem. However, there is little comparative information, and the aluminum-clad columns are available from a limited number of suppliers. They were developed for high-temperature chromatography and consequently are available in only a few of the most temperature-stable phases.

As was mentioned earlier, the primary problem in the nitrogen mode is regulation of the pH of the solvent. This is solved by the permeation control system shown in Figure 6.5. The chromatograms reproduced in Figures 6.19 and 6.20 were obtained using this technique coupled with the "total" exclusion of CO_2.

4. CONCLUSIONS

The high selectively of the ELCD is a real advantage for the analysis of trace substances such as pesticides and drugs in complex matrices. Often cleanup procedures can be greatly abbreviated, which reduces the time and cost of analysis. When compared to other selective detectors such as the nitrogen–phosphorus detector or the electron capture detector, ELCD chromatograms typically are much cleaner.

Although the ELCD's high selectively is an advantage for most trace analyses, the ELCD is somewhat more complex and difficult to operate. There are reaction tubes, scrubbers, ion exchange resins, and solvents which must be changed from time to time in order to maintain performance. This is easy to do and is within the capabilities of any analyst who has a basic understanding of chromatographic instrumentation and can perform maintenance.

REFERENCES

1. O. Piringer and M. Pascalau, *J. Chromatogr.* **8**, 410 (1982).
2. D. M. Coulson, *J. Gas Chromatogr.* **4** (1965).
3. R. C. Hall, *J. Chromatogr. Sci.* **12**, 152 (1974).
4. R. C. Hall, U.S. Patent 3,934,193 (1976).
5. R. C. Hall, U.S. Patent 4,555,383 (1985).
6. R. C. Hall, *CRC Crit. Rev. Anal. Chem.*, **7**, 323 (1978).
7. M. L. Selucky, *Chromatographia* **5**, 359 (1974).
8. D. M. Coulson, *Adv. Chromatogr.* **3**, 197 (1966).
9. J. W. Dolan and R. C. Hall, *Anal. Chem.* **45**, 2198 (1973).
10. D. M. Coulson, *Am. Lab.* **1**, 22 (1969).
11. P. Jones and G. J. Nickless, *J. Chromatogr.* **73**, 19 (1974).
12. G. G. Patchett, *J. Chromatogr. Sci.* **8**, 155 (1970).
13. R. C. Hall and K. M. Williams, U.S. Patent 4,917,709 (1990).

14. B. J. Ehrlich, R. C. Hall, R. J. Anderson, and H. G. Cox, *J. Chromatogr. Sci.* **19**, 245 (1981).
15. O. Piringer and E. Wolff, *J. Chromatogr.* **284**, 373 (1984).
16. R. J. Anderson, U.S. Patent 4,295,856 (1981).
17. R. C. Hall, O.I. Corporation College Station, TX, unpublished results (1987).
18. F. Caron and J. R. Kramer, *Anal. Chem.* **61**, 114 (1989).
19. O. Piringer, E. Tataru, and M. Pascalau, *J. Gas Chromatogr.* **2**, *164* (1964).
20. D. E. Johnson and C. G. Enke, *Anal. Chem.* **42**, 329 (1970).
21. J. C. Butler, *Pittsburgh Conf.* Atlantic City NJ (1987). (Paper available from Tracor, Inc., Austin, TX).
22. R. C. Hall and C. A. Risk, *J. Chromatogr. Sci.* **13**, 519 (1975).
23. C. A. Risk and R. C. Hall, *J. Chromatogr. Sci.* **15**, 156 (1977).

CHAPTER

7

THE NITROGEN–PHOSPHORUS DETECTOR

PAUL L. PATTERSON

Detector Engineering and Technology
Walnut Creek, California

The nitrogen–phosphorus detector (NPD) is an ionization-type detector for gas chromatography (GC). Such devices function by ionizing the chemical compounds eluting from the GC column, and the increased ionization level produces an electrical signal of magnitude related to the quantity of the compound detected. Unlike other ionization detectors, the NPD responds mainly to organic compounds containing N or P atoms, with a level of ionization that is more than 10,000 times greater than the ionization produced by comparable amounts of hydrocarbon compounds. Hence, the NPD is also classified as an element-specific detector. Element-specific detectors are especially valuable analytical tools in applications involving complex sample mixtures.

Nitrogen is the second most commonly occurring heteroatom in organic chemistry. Consequently, there is a wide range of applications for the N specificity of the NPD. These include the following: detection of trace level drugs of abuse; detection of trace level environmental pollutants; research, development, and quality control of pharmaceuticals; research, development, and quality control of pesticides; and analysis of major and minor nitrogen constituents in petroleum samples, biological samples, and other complex organic samples. The P specificity of the NPD is used mainly for the detection of trace levels of pesticide pollutants in environmental samples, and in research, development, and quality control of organophosphorus pesticides. For both N and P compounds, the NPD is the most sensitive GC detector currently available.

The NPD evolved from an earlier type of GC detector known as an alkali flame ionization detector or AFID (1–3). The AFID used a hydrogen/air

Detectors for Capillary Chromatography, edited by Herbert H. Hill and Dennis G. McMinn.
Chemical Analysis Series, Vol. 121.
ISBN 0-471-50645-1 © 1992 John Wiley & Sons, Inc.

flame that burned near an alkali metal salt pellet so that alkali vapors were generated in the flame. Such an alkali-sensitized flame exhibited a high ionization specificity for phosphorus compounds, as well as for compounds containing halogen or nitrogen atoms under appropriately selected operating conditions. However, the AFID had the reputation of being difficult to use because of requirements for frequent adjustments of many interdependent operating parameters, unstable background signals, unstable and nonreproducible sample responses, and rapid degradation of the alkali metal salt. In 1974, Kolb and Bischoff (4, 5) described a new design for a GC detector that was specific for nitrogen and phosphorus compounds. This basic detector design (and subsequent variations) constitutes the type of detector now commonly referred to as the NPD.

The Kolb and Bischoff detector differed from prior AFID versions in three significant features: (a) the alkali-containing component of this detector was a glass bead containing nonvolatile rubidium silicate rather than a pellet of volatile alkali metal salts, as formerly used in the AFID; (b) the rubidium silicate glass bead was fused onto a platinum wire that was used to electrically heat the bead rather than being heated by a flame as in the AFID; (c) the hydrogen flow used to obtain both N and P responses was only a few milliliters per minute, which meant that there was not sufficient H_2 to support a self-sustaining flame as in the AFID.

In comparison with the AFID, the NPD of Kolb and Bischoff exhibited a much longer operating life for the alkali-impregnated component, a substantially lower background current, a more stable and reproducible response, and much better control of the key operating parameters. As a result, the NPD virtually has replaced the AFID as the detector of choice for N or P compounds.

1. TERMINOLOGY

Nitrogen–phosphorus detectors of the general type discussed in this chapter have been denominated by a variety of different names in the literature, including NP-flame ionization detector (NP–FID), alkali flame ionization detector (AFID), Rb-bead detector, thermionic ionization detector (TID), and thermionic specific detector (TSD). The names that include the word *flame* are misleading, because a distinguishing characteristic of the NPD is that a self-sustaining flame is not used. Also, the designation *Rb-bead detector* is too narrow a term, for it is now known that NP detection can be obtained without requiring Rb as the essential ingredient.

In naming GC detectors, the usual convention is to employ terminology that describes the principle of operation. In keeping with this convention, the term *thermionic ionization detector* is an appropriate description of the NPD. However, it is now known that NP detection is just one of several different modes that belong to the more general TID category of GC detectors. Hence, in the context of this chapter, NPD will refer to the NP mode of thermionic ionization detection. Furthermore, the word *thermionic* is used herein according to its strict definition, which is "the emission of electrical charge from solid surfaces that are heated" (6). In some earlier literature (2, 3) the TID terminology was more loosely used to refer to GC detectors that belong to the category identified as AFID in this work.

In accordance with the original description of Kolb and Bischoff, the active sensing element in the NPD is frequently described as a "bead." However, in recent versions of the NPD, the active element no longer has the physical appearance of a bead-like structure. Hence, the term *thermionic ionization source* is used in this chapter interchangeably with the term *bead* in referring to the active sensing element in the NPD. The thermionic ionization source is that component which defines the physical location where ions are formed in the NPD.

2.   BASIC DETECTOR COMPONENTS

Similar to a flame ionization detector (FID), the NPD requires a detector gas environment containing both H_2 and air. Consequently, NPD equipment is usually designed to mount on an existing FID-type detector base on the GC.

Figure 7.1 is a schematic illustration of the components of an NPD. The most important of these is the thermionic source, which has the shape of a bead or cylinder. This is composed of an alkali–metal compound impregnated in a glass or ceramic matrix. The body of the source is molded over an electrical heating wire connected to a precisely controlled electronic supply of heating current. By this means, the thermionic source is heated to typical operating temperatures in the range of 600–800 °C.

As shown in Figure 7.1, the thermionic source is positioned downstream of a sample inlet. Usually this inlet is just a conventional FID flame jet. As in an FID, the GC column effluent is combined with a flow of H_2 and transported into the detector through the center of the inlet structure, while air flow is provided around the (outer) periphery of this structure. An important characteristic of the NPD is that the H_2 flow to the detector is maintained in the low range of about 2–6 mL/min. As a consequence there is not enough H_2 to support a self-sustaining flame at the tip of the sample inlet structure. The magnitude of air flow supplied to the NPD is less critical than the H_2 flow and is typically in the range of 60–200 mL/min, depending on the internal volume of the detector.

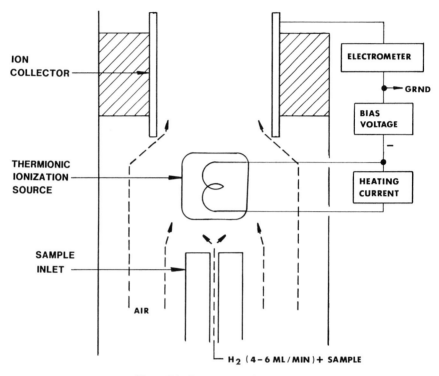

Figure 7.1. Components of an NPD.

The thermionic source is positioned such that the sample can impinge on the surface of the source en route to the detector's exit port. Located in proximity to the ionization source is an ion collector electrode that usually is cylindrical. In addition to being connected to an electronic supply of heating current, the thermionic source is also polarized at a voltage that causes ions formed at the source to move toward the ion collector. Most NPD versions are polarized to collect negatively charged ions, and this ion current is measured with an electrometer connected to the collector electrode.

Like most ionization-type detectors for GC, the NPD is a simple device from a component standpoint. The essential hardware consists of the sample inlet, the thermionic ionization source, and the ion collector, while the associated electronics are the heating current and polarization voltage supplies for the thermionic source, and the electrometer for signal measurement. In virtually all NPD versions currently in use, the thermionic source is mounted in a manner allowing it to be removed simply and replaced in the detector structure. Hence, the NPD is a detector that generally is easy to maintain in good working condition for long periods of time.

Since the NPD mounts onto an FID-type detector base, the interfacing of an NPD with a capillary column is essentially the same as a capillary–FID interface. However, some additional consideration needs to be given to the fact that the H_2 flow to the NPD is only about one-tenth that to an FID. This usually means that an inert makeup gas must be added at the detector end of the capillary column to ensure that the interior volume of the sample inlet is well purged. Also, it is common to install the capillary column so that it terminates as closely as possible to the tip of the sample inlet.

3. CONSTRUCTION OF THERMIONIC IONIZATION SOURCES

Since the original description of the NPD in 1974, most of the developments in this type of detector have focused on improving the mechanical integrity and the chemical composition of the thermionic ionization source. New developments regarding chemical composition have been directed toward improvements such as the following: extending the operating lifetime of the thermionic source; increasing the sensitivity and specificity of the detector; providing better reproducibility from one thermionic source to another; and providing better day-to-day reproducibility for any given thermionic source. The fabrication of thermionic ionization sources for the NPD requires a certain level of artistic skill, and the detailed construction techniques and chemical compositions of these sources are often treated as proprietary information by the manufacturing companies.

Figure 7.2 shows four types of structures that are currently used as thermionic ionization sources in NPDs. The structure in Figure 7.2A represents an alkali–glass bead fused onto a loop of platinum wire. This is the type of thermionic ionization source originally described by Kolb and Bischoff (4, 5). It is a small bead approximately 1 mm in diameter and usually is composed of a rubidium silicate ($Rb_2O \cdot SiO_2$) material as reported in the original description. In order to have sufficient resistance for electrical heating, the platinum wire that heats and supports the bead must have a very fine diameter. During operation both the bead and the supporting wire typically glow with a dull red orange coloration. Hence, in addition to the decomposition chemistry and ionization that occurs at the hot bead surface, there is the possibility that the hot platinum wire is contributing some catalytic effects as well.

In the detector configuration depicted in Figure 7.2A, the thermionic ionization source is positioned between the sample inlet and the ion collector. In the original NPD described by Kolb and Bischoff, the best NP response was obtained by polarizing both the bead and the tip of the sample inlet at a

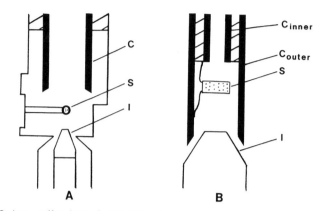

(C–ion collector, S–thermionic source, I–sample inlet)

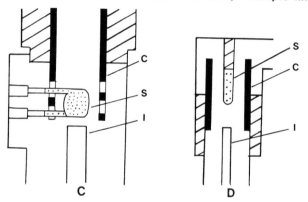

Figure 7.2. Different configurations of the thermionic ionization source, ion collector, and sample inlet used in commercially available NPDs.

negative voltage of − 180 V with respect to the ion collector. In the NPD, the efficiency of ion collection depends on the prevailing electric field between the thermionic source and the ion collector. The electric field is the ratio of polarization voltage divided by the distance separating the source and collector. Hence, with the configuration of Figure 7.2A, unit-to-unit differences in ion collection can occur unless the location of the thermionic source is maintained precisely relative to the ion collector.

Figure 7.2B represents another type of construction for the thermionic ionization source (7). In this source, a fine-diameter platinum wire is looped through a ceramic cylinder core and an alkali-glass-like material is fused over the outer surface of the ceramic. This type of thermionic source is about 2 mm in diameter by 4 mm long and is supported by attaching the ends of the platinum wire to an inner and outer collector electrode as shown. Unlike most other NPDs, this particular type of thermionic source is normally polarized to move positive rather than negative ions toward the collector. Like the thermionic source in Figure 7.2A, this thermionic source and the supporting platinum wires both glow dull red–orange under typical operating conditions. In view of Fujii and Arimoto's report (8) describing the surface ionization detector (see also Chapter 8 in this volume) that detects positive ions formed on a bare platinum wire, it is possible that the hot supporting wires in Figure 7.2B contribute some signal components in addition to the alkali–glass surface.

The usual electronic connnections to the detector configuration of Figure 7.2B also differ from the more typical electronic connections shown in Figure 7.1. In particular, the sample inlet in Figure 7.2B usually is at ground potential, whereas the thermionic source, inner and outer collectors, and electrometer are all polarized at a few hundred volts positive with respect to ground. The thermionic source heating current in this case is provided by a transformer connected between the inner and outer collectors. Although there is a large polarization voltage between the thermionic source and the sample inlet, the relevant polarization for the collection of ions formed at the surface of the thermionic source is the voltage difference between the thermionic source and the surrounding collector electrode. This voltage is just the magnitude of the voltage drop across the platinum heating wire of the thermionic source, and this is only a few volts.

The thermionic ionization source represented by Figure 7.2C is constructed of a homogeneous alkali–ceramic coating molded over a helical heater wire of nichrome (9–11). The result is a rather large thermionic source with cylindrical shape and approximate dimensions of 4-mm diameter × 5-mm height. A typical composition of the ceramic coating is 6% by weight of Rb_2SO_4 and 94% weight of a ceramic cement composed mainly of Al_2O_3. Nichrome wire has a higher resistivity than platinum, so the support wires for this source are more substantial than those of Figure 7.2A or 7.2B. Hence, in operation, only the surface of this thermionic source, and not the supporting wires, is observed to glow.

As depicted in Figure 7.2C, the cylindrical collector electrode is constructed of an open metal screen in the vicinity of the thermionic source. This allows a smooth symmetrical flow of detector gases past the surface of that source. In addition, a well-defined electrical field configuration is created between the concentric thermionic source and ion collector. As a result, there is efficient ion collection from all surfaces of this source, with minimal requirements for precise vertical or horizontal positioning. In the configuration of Figure 7.2C, the sample inlet is not polarized and the thermionic source is polarized at a modest − 4 V for optimum NP detection.

The thermionic ionization source represented in Figure 7.2D consists of a cylinder that is mounted at the downstream end of the ion collector and extends into a coaxial position in the collector (12–14). Unlike the configurations shown in Figure 7.2A, C, this mounting configuration allows an especially streamlined gas flow path in the region of ion production and collection because there are no side ports to accommodate a source mounting structure. In the construction of this thermionic source, a fine nichrome or platinum wire is routed through the interior of an alumina ceramic support such that a length of ~ 5 mm of a bare wire loop is left exposed. This bare wire is then covered with a sublayer coating consisting of a mixture of ceramic cement and a nonalkali metallic additive such as nickel. After curing, the sublayer coating is further covered by a surface layer consisting of a mixture of ceramic cement and alkali metal additives. A representative composition of the sublayer is 18% by weight of nickel in an alumina ceramic cement, and a representative composition of the surface layer is 4% by weight of Cs_2SO_4 in the alumina ceramic cement. The primary objective of the sublayer in this source construction is to minimize long-term corrosion of the hot heater wire caused by contact with alkali materials in the surface coating. The coated segment of this thermionic ionization source is ~ 1.6 mm in diameter and ~ 5 mm in length, and most of this coated segment glows a dull red–orange color during operation. For NP detection, the sample inlet in the detector represented Figure 7.2D is not polarized, and the thermionic source is polarized at -5 V with respect to the ion collector.

Thermionic ionization sources representing Figure 7.2A–D are used in NP detectors available from different manufactures. Generally, those sources formed from ceramic materials provide greater flexibility for varying the chemical composition of the source. This is because the ceramic compositions are formulated and coated from a slurry at room temperature (10, 14), whereas the glass compositions are formed in a process that proceeds through a molten glass state (15, 16). Also, the ceramic materials can tolerate much higher temperatures. At the typical source temperature of 600–800 °C, the alkali–glass formulations operate in a softened state (5) such that an accidental shutdown of a major gas supply can cause an overheating and melting of the source.

4. THEORY OF OPERATION

4.1. Hydrogen/Air Chemistry in the NPD

The NPD operates in a detector gas environment composed of a dilute mixture of H_2 in air. NP responses in this detector do not appear until the temperature of the thermionic ionization source is raised sufficiently to "ignite" the H_2/air mixture. This ignition occurs when there is enough thermal energy to dissociate H_2 molecules into reactive H atoms. The H atoms react further with O_2 molecules in a series of chain reactions that produce a highly reactive chemical environment containing H, O, OH, and H_2O in addition to the original H_2 and air. This process is analogous to that which occurs in the ignition of an H_2/air flame in a conventional FID. However, unlike an FID, there is not enough H_2 flow in the NPD to allow the active H_2/O_2 chemistry to propagate back upstream and form a self-sustaining flame attached to the end of the sample inlet structure. Instead, in the NPD, the flame-like H_2/air chemistry exists only as a gaseous boundary layer in proximity to the hot surface of the thermionic ionization source, as depicted in Figure 7.3. Furthermore, this chemically reactive boundary layer exists

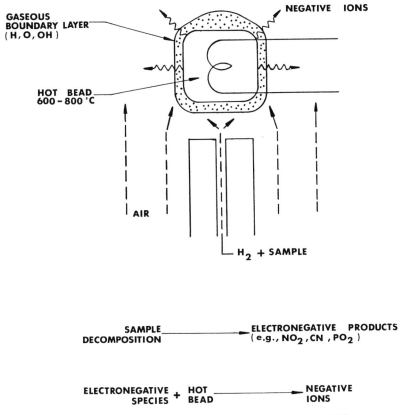

Figure 7.3. Depiction of the H_2/air chemistry required in the NPD.

only as long as heat is applied to the thermionic source. If the heating current is turned off, the boundary layer ceases to exist and the NP response disappears. Also, if the thermionic source simply is heated in an air environment in the absence of H_2, there is no initiation of the chemistry required to produce NP responses.

The NPD is known to be responsive to most organic compounds containing elemental N or P atoms. This is a consequence of sample compounds being decomposed by the hot, chemically active boundary layer, with the decomposition products being similar irrespective of the sample's original molecular structure. The detailed mechanism whereby the sample decomposition products become ionized is less clear. Two different ionization mechanisms have been proposed: one involves a gas-phase ionization process occurring in the boundary layer immediately adjacent to the hot thermionic sources; the second involves a surface ionization process occurring on the hot surface of the thermionic source. The arguments for both ionization theories are presented in the following subsections.

4.2. Gas-Phase Ionization Theory

In the original description of the NPD, Kolb and Bischoff (4, 5) viewed their detector as a modification of an AFID. Consequently, a detection mechanism was proposed that involved the gas-phase ionization of Rb atoms. Such a detection mechanism required an explanation of how Rb atoms were generated in the vapor state, as well as an explanation of how these atoms participated in the specific ionization of N or P compounds.

Kolb and co-workers (4, 17) pointed out that rubidium was present in their silicate bead as a positive ion, and they argued that the direct emission of Rb^+ into the vapor state was not possible because the bead was polarized at a negative voltage. They also argued that simple vaporization of rubidium silicate molecules was not likely at the relatively low bead temperatures of 600–800 °C. Therefore, it was concluded that the only way to get Rb into the vapor state was vaporization of Rb atoms formed on the surface of the bead by neutralization of the Rb^+ in rubidium silicate. This neutralization process was postulated to involve an electron conducted through the hot glass bead structure from the electrical heating wire. It was argued that such a vaporization of an electrically neutral atomic species could explain why the background ionization current in the NPD was several orders of magnitude lower than the background current common in previous AFIDs.

To explain the long operating lifetime of the rubidium silicate bead, Kolb et al. described an alkali recycling mechanism for the detector background current. In this recycling, vapor state Rb atoms were ionized by interactions with H atom radicals in the gaseous boundary layer of the bead and the resulting gas phase Rb$^+$ ions were drawn back into the bead surface by the negative polarization of the bead.

For the specific ionization of N and P compounds, Kolb et al. proposed a chemi-ionization reaction between gas phase Rb and electronegative radical species produced from the thermochemical decomposition of NP compounds in the reactive gaseous boundary layer. For N compounds, a likely candidate for the electronegative species was the CN radical, and the proposed ionizing reaction was

$$Rb^* + CN \longrightarrow Rb^+ + CN^- \tag{1}$$

where Rb* was a rubidium atom excited to its first energy level of 1.59 eV. An excited Rb atom was required in Equation 1, so that the electron affinity of the CN radical (3.16–3.82 eV) exceeded the ionization potential of the Rb* (2.57 eV). A similar reaction was proposed for P compounds using PO or PO$_2$ as the intermediate electronegative radical.

In view of the foregoing ionization theory, it was generally accepted for many years that Rb was an essential ingredient for the thermionic ionization source in the NPD. However, it has been shown that NP responses of comparable sensitivity and specificity are also obtained with thermionic sources composed of Cs$_2$SO$_4$ in a ceramic matrix (12, 13, 18) and LaB$_6$ in a SiO$_2$ glass matrix (19). Like rubidium silicate, neither the cesium nor lanthanum compounds are very volatile at the operating temperature of the thermionic source. Hence, if a gas-phase ionization process is to apply, there needs to be a plausible explanation of how these nonvolatile compounds get into the vapor phase.

In a recent investigation (20), laser-induced fluorescence measurements were used to determine the vapor phase density of sodium atoms in the gases immediately adjacent to a hot thermionic source made of soda glass (i.e., sodium-rich glass). These measurements detected no vapor state sodium atoms unless hydrogen was present in the gases surrounding the hot source, regardless of the temperature of the source. It was therefore concluded that sodium was not lost from the soda glass by evaporation but rather by an exchange reaction with the active hydrogen species in the gaseous boundary layer of the source. Such an exchange mechanism could explain how atoms and molecules of nonvolatile compounds could be generated in the vapor state even at temperatures too low for simple evaporation.

A basic dilemma for the gas-phase ionization theory is the explanation of why the thermionic source in an NPD has a relatively long operating life if it is continually evolving material into the vapor state. Kolb and co-workers originally addressed this issue by proposing a Rb atom recycling mechanism. However, Kolb's recycling process, as well as the chemi-ionization process in Equation 1, depend on matching energy levels of the specific reactant species. With the development of thermionic sources having compositions such as the cesium and lanthanum formulations cited previously, it becomes increasingly difficult to support arguments favoring gas-phase ionization. In order for comparable NP responses to be obtained, a gas-phase ionization theory would require that the postulated vaporization, ionization, and recycling processes occur with comparable overall efficiencies regardless of the reactant species vaporized from the thermionic source.

4.3. Surface Ionization Theory

In the theory of gas-phase ionization, a chemi-ionization reaction such as Equation 1 involves neutral reactions interacting to produce equal concentrations of positive and negative ion species. In that case, it should be possible to collect either positive or negative ions with equal efficiency with a properly constructed detector. Therefore, a graph of measured ion current vs. the magnitude of polarization voltage should be symmetrical for both positive and negative polarization voltages. As shown in Figure 7.4, this is the case for an FID, which is well known to function according to gas-phase ionization processes. Data on measured ion currents vs. polarization voltage of the NPD were reported in 1978 by Patterson (9) for a thermionic source composed of Rb–ceramic and in 1979 by Olah et al. (21) for a thermionic source composed of Rb–glass. As illustrated in the bottom half of Figure 7.4, these data produced graphs of both background and sample response ion currents that were asymmetric with respect to opposite polarities of the polarization voltage. This type of asymmetry indicated that the ion currents measured in the NPD were caused by emission of charged particles from the thermionic source itself, rather than by a gas-phase ionization process. Hence, a process of surface ionization was indicated for the NPD.

One prominent characteristic illustrated by the NPD data in Figure 7.4 is the capacity of thermionic ionization sources to emit large background currents of positive ions (22, 23). In contrast, the background current at negative polarization voltages is small. This characteristic explains why most NP detectors collect negative rather than positive ions. The few NP detectors that do collect positive ions operate at a low polarization voltage where the background current is not yet overwhelmingly high. However, small changes in the effective polarization such as might occur from detector contamination or from inaccurate positioning of the thermionic source can cause large changes and instabilities in the positive ion background current. NPDs that collect negative ions are inherently less susceptible to such background current problems.

In the theory of surface ionization in the NPD, the most important

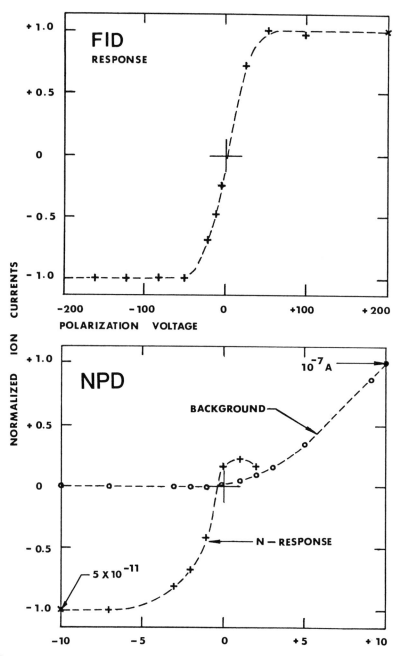

Figure 7.4. Comparison of graphs of ion current vs. polarization voltage as measured for an FID and an NPD.

characteristic of the thermionic ionization source is not its specific elemental composition but rather its bulk surface property known as the electronic work function. The "work function"of a surface is defined (24) as the amount of energy required to remove an electron from that surface. The purpose of adding alkali metal compounds to the glass or ceramic matrices of thermionic ionization sources is to lower the work function so as to more easily allow the thermionic emission of charged particles from the heated source. It has been reported (9) that the work function of the thermionic source depends on both the type and the density of the alkali metal used in the source composition. For example, if a series of sources is constructed with equal densities of either Na, K, Rb, or Cs compounds, then the resultant work functions vary in the decreasing order $Na > K > Rb > Cs$, which is the same order as the ionization potentials of the alkali metals. Similarly, for two sources composed of two different densities of the same alkali metal compound, the source with the higher density has the lower work function. It follows that there are several different ways of formulating source compositions so as to achieve a given work function. In fact, the report (19) of an NP detector that uses a thermionic source composed of $LaB_6 \cdot SiO_2$ demonstrates that alkali constituents are not necessary.

As in gas-phase ionization theory, the theory of surface ionization in the NPD assumes that NP compounds are thermochemically decomposed into electronegative products in the active H_2/air boundary layer of the thermionic source. Electronegative decomposition products such as CN, PO, and PO_2 can then be ionized by extracting negative charge from the hot thermionic surface. Since this surface is composed of a mixture of heterogeneous constituents and is immersed in a reactive gas environment containing radical chemical species, a detailed mathematical modeling of this ionization process is very complex.

One approach to a mathematical description of the surface ionization process in an NPD assumes that the thermionic surface and absorbed electronegative species are in thermal equilibrium. Then the degree of negative surface ionization is described by the classic Saha–Langmuir equation as follows (19, 25):

$$\frac{n_-}{n_0} = \frac{g_-}{g_0} \exp\left(\frac{EA - \phi}{kT}\right) \qquad (2)$$

where n_0 and n_- are the densities of neutral and negative ion species desorbed from a surface at temperature T; ϕ is the work function of the surface; EA is the electron affinity of the electronegative species; g_0 and g_- are the statistical weights of neutral and ion species; and k is the Boltzmann constant. If $n = (n_0 + n_-)$ designates the density of electronegative species formed in the thermochemical decomposition of N or P compounds, then the efficiency of surface ionization as derived from Equation 2 is as follows (8, 25):

$$\frac{n_-}{n} = \left[1 + \frac{(g_0)}{(g_-)} \exp\left(\frac{\phi - EA}{kT}\right) \right]^{-1} \tag{3}$$

Equations 2 and 3 describe mathematically the basic feature that the efficiency of surface ionization increases with decreasing work function ϕ or increasing electron affinity EA. In addition, these equations predict an increasing ionization efficiency with increasing T for the case of $\phi > EA$ or a decreasing ionization efficiency with increasing T for the case of $\phi < EA$. In the NPD, T must be at least high enough to ignite the required H_2/air chemistry. For further increases in T beyond the ignition point, most NP detectors exhibit increases in both the background and sample response currents (4, 9, 11, 19, 26). Hence the empirical data suggest that the condition $\phi > EA$ prevails in the NPD.

In an alternate mathematical description of the surface ionization process in an NPD, the general rules governing thermionic emission of electrons are cited (9) whereby the emission current density J from a surface at temperature T is expressed as follows (25):

$$J = C_1 \exp(-W/kT) \tag{4}$$

where C_1 is an empirically determined constant and W is the effective work function of the emitter. Consequently, the logarithm of J satisfies the following relationship:

$$\ln J = C_2(-W/T) \tag{5}$$

where C_2 is another constant. Patterson (9) pointed out that the surface temperature T of the thermionic source can be related to the source heating current I and the detector wall temperature T_D by using the theory of thermal conductivity detectors (27) as follows:

$$I^2 R = (C_3 \lambda + C_4 F C_p)(T - T_D) \tag{6}$$

where R is the electrical resistance of the heater wire in the thermionic source; λ is the thermal conductivity of the gas mixture around the hot thermionic source; C_p is the specific heat of the gas mixture; F is the flow rate of gases past the source; and C_3 and C_4 are constants. By rearranging terms in Equation 6, making some approximation of terms based on typical magnitudes of T and T_D, and substituting into Equation 5, Patterson arrived at the following predictions:

$$\ln J = -C_5 W/I^2 \qquad (7)$$

for variations in I with T_D held constant, and

$$\ln J = C_6 W T_D + C_7 \qquad (8)$$

for variations in T_D with I held constant, with C_5, C_6, and C_7 all constants. In a logarithmic graph of measured ion currents vs. I^{-2} for values of I greater than that required for ignition of the H_2/air chemistry and in a logarithmic graph of measured ion currents vs. T_D, Patterson deomonstrated that the background and sample response currents of a Rb–ceramic thermionic source followed the behaviors predicted by Equations 7 and 8. From the slopes of the data in these two graphs, it was determined that the effective work function W of the thermionic surface was $\sim 3.4\,eV$ in the absence of samples and $\sim 2.3\,eV$ in the presence of N or P samples. Hence, in this mathematical model, the sample response current is viewed as being caused by a reduction of the effective work functions of the thermionic source at those instances when the thermionic surface is exposed to N and P samples.

From the preceding discussion of surface ionization mechanisms, it may be concluded that the three most important parameters controlling detector response are (a) the work function of the thermionic surface, which is determined by the chemical composition of the thermionic source; (b) the temperature of the thermionic surface, determined primarily by the magnitude of electrical heating current; and (c) the composition of the gaseous environment surrounding the thermionic surface, which is dependent on the type of gases supplied to the detector as well as their chemical reactivity at the operating temperature of the thermionic source. The NPD represents a specific combination of these three operating parameters, and several other combinations are now known (13) to be possible. For example, a thermionic ionization detector has been developed (12, 18) that uses a thermionic source having a much lower work function that that normally used in an NPD. This thermionic source is composed of a high concentration of a nonvolatile Cs salt in a ceramic matrix and is operated in an inert gas environment of N_2 at relatively low temperatures of 400–500 °C. This mode of thermionic detection responds with high specificity only to sample compounds containing electronegative constituents (for example, NO_2 functional group or multiple halogen atoms) within their original molecular structure. The success of these other TID modes cannot be explained by a gas-phase ionization theory of the type proposed for the NPD. Hence, the development of multiple modes of thermionic detection through systematic changes in work function, temperature, and gas composition lends further support to the validity of the surface ionization theory for the NPD or NP mode of the TID.

5. NPD RESPONSE CHARACTERISTICS

The NPD is an example of an element-specific detector that can be optimized for both specificity and sensitivity. In such detectors, the operating conditions that provide maximum sensitivity are not always the same as those that provide maximum specificity. In evaluating the dependence of both specificity and sensitivity on various operating parameters, it is helpful to employ a test sample containing the heteroatoms in question as well as a hydrocarbon compound. An example of such a test sample for the NPD is described in Table 7.1. This sample contains nanogram levels of an N compound (azobenzene), a P compound (malathion), and an NP compound (methyl parathion), as well as a microgram level of a hydrocarbon (n-C_{17}).

Figure 7.5 shows some chromatograms of the test sample described in Table 7.1. The FID chromatogram illustrates that the hydrocarbon component is the dominant constituent of this sample and the N and P compounds are trace level constituents that are not detected at the FID sensitivity displayed. In contrast to the FID, the NPD chromatograms in Figure 7.5 exhibit prominent responses to the N and P compounds of the sample and a greatly diminished response to the hydrocarbon.

A principal difference between the NPD operating conditions and those of the FID is the magnitude of H_2 and air flows. For the FID, $H_2 = 20 \, mL/min$ and air $= 270 \, mL/min$, whereas for the NPD, $H_2 = 4 \, mL/min$ and air $= 60$ mL/min. The NPD chromatogram obtained with a Ni–ceramic thermionic ionization source demonstrates that a significant degree of NP specificity is obtained simply by generating the proper H_2/air chemistry in the boundary layer of the thermionic source, regardless of the chemical composition of the

Table. 7.1. Test Sample for the NPD

Compound	Compound Concentration (ng/μL)	Atom Concentration (ng/μL)	
Azobenzene	2.0	0.31	(P)
n-Heptadecane	4000	3400	(C)
Methyl parathion	2.0	0.11	(N)
		0.23	(P)
Malathion	4.0	0.38	(P)
Isooctane	Solvent	—	

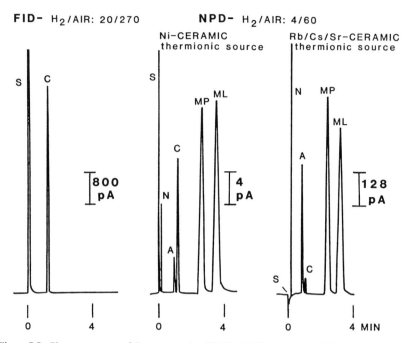

Figure 7.5. Chromatograms of the test sample of Table 7.1 illustrating the difference in response between an FID and two versions of an NPD. Sample components: S = isooctane solvent; C = *n*-heptadecane; A = azobenzene; MP = methyl parathion; ML = malathion; and N = nicotine impurity.

source. The NPD chromatogram obtained with a Rb/Cs/Sr–ceramic thermionic ionization source demonstrates how the inclusion of alkali metal constituents in the source produces substantial increases in the NP responses, as well as a significant improvement in specificity with respect to the C_{17} compound. The NPD chromatograms also reveal a nicotine impurity peak that invariably is detected in samples prepared in laboratory environments containing traces of cigarette smoke.

In most NPD versions, the P specificity is obtained readily for a wide range of operating conditions and thermionic source compositions. Therefore, the principal objective in optimizing the composition of the thermionic source usually is determining the formulation that produces the best N response. For the NPD data shown in Figure 7.5, the proportionate amounts of Rb and Cs additives in the ceramic formulation of the thermionic source were adjusted to produce the best sensitivity and specificity for the azobenzene component of the test sample. The Sr additive in the thermionic source was used to minimize the tailing of chromatographic peaks associated with the

P compounds. Chromatographic peak tailing sometimes occurs in the NPD beacuse sample decomposition products remain temporarily absorbed on the thermionic surface. Often, the tailing is reduced by operating the thermionic source at a higher temperature or with a higher H_2 flow.

In NPD chromatograms of the test sample, the quantities of interest are as follows: (a) the absolute magnitudes of response to N and P compounds, which show the absolute sensitivity of the detector; and (b) the relative magnitudes of response to N, P, and hydrocarbon compounds, which indicate the N/C, P/C, and N/P specificities of the detector. The NPD is a mass-flow-rate–sensitive detector, and its sensitivity to nitrogen atoms can be defined as follows:

$$S_N = \text{peak height/N atom flow rate} \qquad (9)$$

where the peak height of the sample compound is measured in amperes and the N atom flow rate in grams of nitrogen per second (g N/s). For a GC sample peak, the N atom flow rate is determined by dividing the N mass in the sample by the peak width at half-height. An equivalent definition of sensitivity is as follows:

$$S_N = \text{peak area/N atom mass} \qquad (10)$$

where peak area is measured in ampere-seconds or coulombs, and N atom mass in grams of nitrogen. In a similar way, the NPD sensitivities to phosphorus (S_P) and carbon (S_C) can be defined.

A unique characteristic of thermionic ionization detectors for GC is that the absolute sensitivity of the detector can be varied over a wide range, depending on the choice of operating conditions. However, the detector background and noise vary in a similar manner. Consequently, a better indicator of detector response is the detectivity, which is defined as follows:

$$D_N = (2 \times \text{noise})/S_N \qquad (11)$$

where noise is measured in amperes and the detectivity has units of g N/s. A similar definition exists for the phosphorus detectivity.

The specificity of the NPD is defined as a ratio of sensitivities. For example, the specificity of nitrogen compounds relative to hydrocarbons is defined as S_N/S_C and has the units of g C/g N. Similarly, the specificity of phosphorus compounds relative to hydrocarbons is S_P/S_C and has the units g C/g P, and the specificity of nitrogen relative to phosphorus is S_N/S_P with units of g P/g N.

Table 7.2 lists some typical values for NPD sensitivity, specificity, detectivity, and range of linear response. Generally, the NPD response begins

Table. 7.2. Typical Performance Characteristics for the NPD

Characteristic	Typical Values
Sensitivity	$S_N = 0.1$–1.0 A·s/g N $S_P = 1.0$–10.0 A·s/g P
Detectivity	$D_N = 5 \times 10^{-14}$–2×10^{-13} g N/s $D_P = 1 \times 10^{-14}$–2×10^{-13} g P/s
Specificity	$S_N/S_C = 10^3$–10^5 g C/g N $S_P/S_C = 10^4$–5×10^5 g C/g P $S_N/S_P = 0.1$–0.5 g P/g N
Linear range	10^3–10^5

deviating from a linear relationship vs. sample amounts at mass flow rates exceeding 10^{-9} to 10^{-10} g P/s or g N/s (9, 11, 27). Also, some variances of NPD response factors with molecular structure are known to exist (7, 16), although the H_2/air chemistry surrounding the thermionic source is very effective at decomposing sample compounds into common species. Those N compounds that decompose easily to yield the CN radical generally produce the highest responses, whereas amides and nitro compounds tend to produce substantially lower responses (28).

The two most important operating parameters in the NPD are the magnitude of H_2 flow to the detector and the magnitude of heating current supplied to the thermionic ionization source. Both the sensitivity and specificity of the detector can be changed substantially by changes in these key parameters. For most NPD versions, the sensitivity can be increased by as much as a factor of 10 by increasing the source heating current beyond the value required for ignition of the H_2/air chemistry. Usually, the specificity does not change much with such increases in heating current. For some NPD versions, an increase in H_2 flow in the range of 2–6 mL/min also causes a large increase in sensitivity, especially for P compounds. Such an increase in H_2 usually also changes the detector specificity in favor of P vs. N.

Since the NPD is a mass-flow-rate–sensitive detector, relatively high flows of gases can be used to purge the detector volume without causing reduced responses due to sample dilution effects. The composition and flow rates of detector gases, however, do influence the surface temperature of the thermionic ionization source. The source temperature is the result of a balance of the electrical heat input and heat losses due to conduction and convection through the gases flowing past the source. Therefore, the source temperature is dependent on the magnitude of source heating current, on the magnitude of auxiliary heating of the detector walls, on the thermal conductivity of the gas mixture flowing past the source, and on the magnitude of the total gas flow through the detector. From these considerations, the following general operating characteristics can be expected:

a. At any fixed set of gas flows and auxiliary detector wall heating, the principal means of varying source temperature is via the magnitude of source heating current.

b. If the gas flows or auxiliary heating are changed, a readjustment of the source heating current restores the source to the same surface temperature. The magnitude of detector background signal or the response to a standard sample can serve as a guide to the correct amount of readjustment.

c. Helium has a much higher thermal conductivity than nitrogen, so the use of helium as the GC carrier or detector makeup gas usually requires a higher setting of source heating current than when nitrogen is used.

d. As the total gas flow through the detector is increased, more heating current generally is required to obtain the same source temperature.

e. If auxiliary heating of the detector walls is increased, less heating current will be required to obtain the same temperature.

As a general rule, it is best to operate the auxiliary heating of the NPD detector walls as high as possible to minimize the temperature gradient between the thermionic source and the surrounding wall. This minimizes changes in source temperature that momentarily may occur owing to thermal conductivity effects as large concentrations of sample compounds pass through the detector. Also, since the thermal conductivities of most organic samples are more comparable to the thermal conductivities of nitrogen and air than of helium, source temperature changes caused by large sample concentrations are minimized if the concentration of helium is small compared to nitrogen or air in the detector volume. When connecting a capillary GC column in which a detector makeup gas is used, the above considerations suggest that nitrogen is a better choice for the makeup even if helium is the carrier gas through the column.

Two NPD versions have been described (18, 29) in which heating current to the thermionic source was controlled by a constant temperature electronic supply rather than a constant current supply. Constant temperature operation provides the convenience of an automatic electronic readjustment of source heating current whenever gas flows or detector wall temperature are changed. However, in comparison to constant current operation, constant temperature operation often produces much larger and more erratic detector signal perturbations associated with the momentary passage of large concentration sample solvent peaks through the detector.

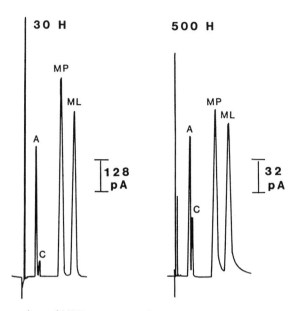

Figure 7.6. Comparison of NPD response to the test sample of Table 7.1 after operating the thermionic source continuously for the times specified. Sample components are the same as in Figure 7.5.

It is a well-known characteristic of the NPD that thermionic ionization sources lose their activity with operating time (11, 20, 26) and eventually must be replaced. Figure 7.6 shows test sample chromatograms obtained at 30 and 500 h of continuous operation of a thermionic source operated at a constant heating current magnitude just sufficient to ignite the H_2/air chemistry. In addition to about a factor of 4 decrease in absolute sensitivity to N and P compounds, the data of Figure 7.6 also illustrate about a factor of 4 degradation in the N/C and P/C specificity. All versions of thermionic sources for NPD applications exhibit losses in sensitivity and specificity with operating time, so the lifetime of the thermionic source has become one of the most important criteria of judging the quality of any given NPD version. The data illustrated in Figure 7.6 actually represent a very good performance for a thermionic source.

The thermochemical environment that the thermionic ionization source is exposed to during operation is quite harsh. Degradation of the source activity with operating time is probably due to chemical changes occurring on the surface of the source, and possibly the slow vaporization of alkali material from the surface. The exchange mechanism (20) of alkali atoms, with active hydrogen atoms, cited earlier in Section 4.2, could explain the empirical observation that thermionic sources degrade faster at higher H_2 flows. In order to prolong the life of the thermionic source, it is a good practice to turn off the H_2/air chemistry whenever the source is not being used for a few hours or more. This is accomplished by reducing the source heating current to a level insufficient to maintain the active boundary layer, or by turning off the H_2 flow.

In order to compensate for the loss of sensitivity with operating time, it is a common practice to periodically increase the source heating current (11). However, the hotter the thermionic surface, the faster it degrades. Conversely, if the thermionic source is maintained at the same heating current, its rate of decrease in sensitivity becomes less and less with increasing operating time (11, 20). Consequently, from the standpoint of preserving source life, it is better to leave the source heating constant and increase electrometer sensitivity as much as possible with increasing operating time. This was the procedure followed for the data in Figure 7.6. Standard samples should be analyzed periodically in order to monitor the decaying sensitivity.

6. APPLICATIONS

One of the most important characteristics of an element-specific detector like the NPD is that it can simplify considerably the chromatographic analyses of complex samples. An example of that situation is shown in Figure 7.7. The sample in this case was a cologne containing a complex mixture of fragrance constituents, as illustrated by the FID chromatogram. When the same chromatographic analysis was performed using an NPD, there resulted a simple chromatogram clearly revealing three prominent N compounds in this sample. As a consequence of this type of simplification of the chromatogram, an analysis can be performed in a much shorter time because there is less requirement for chromatographically resolving all the constituents of the sample. For example, the FID chromatogram in Figure 7.7 contains many unresolved peaks, whereas the NPD chromatogram reveals that even a faster analysis could have been performed before resolution of the N constituents was a problem. Short analysis times are especially valuable for laboratories handling large numbers of samples.

Figure 7.8 shows the chromatographic analysis of a sample containing 2-ng quantities of two drugs of abuse, cocaine and heroin. For the chromatographic conditions used, the FID analysis for the two drug compounds is obscured by large concentrations of other organic constituents that elute from the GC column at earlier retention imes. The NPD chromatogram, however, provides sufficient specificity to reveal prominent peaks for both

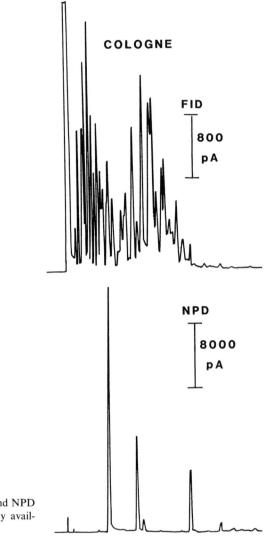

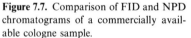

Figure 7.7. Comparison of FID and NPD chromatograms of a commercially available cologne sample.

drugs. In addition, a comparison of the electrometer sensitivities indicated on the FID and NPD chromatograms shows that the NPD produces ion current signals 100 times larger than the FID.

Figure 7.9 shows a comparison of FID and NPD chromatograms for a sample of nitro- and chlorophenols that are of concern as environmental pollutants. As in Figure 7.8, the NPD provides about 100 times more sensi-

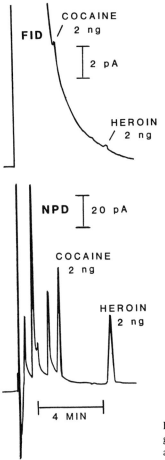

Figure 7.8. Comparison of FID and NPD chromatograms for a sample containing trace amounts of cocaine and heroin.

tivity than the FID for the N constituents of this sample. In addition, the NPD chromatogram reveals a small interference peak for pentachlorophenol. Since most NPDs measure negative ions and halogenated compounds decompose to electronegative species, it is not surprising that some polyhalogenated compounds produce interfering signals in the NPD.

Analyses of samples dissolved in halogenated solvents have been especially troublesome for the NPD (7, 17). The halogenated solvents decompose in the active H_2/air chemistry of the NPD, and the decomposition products alter the ionization characteristics of the thermionic surface for relatively long periods of time. In practice, halogenated solvents cause a large disruption

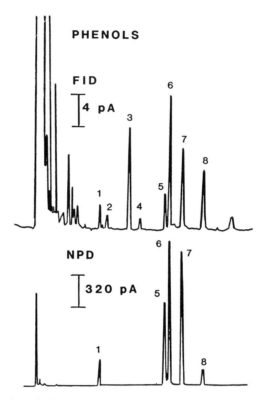

Figure 7.9. Comparison of FID and NPD chromatograms for a sample containing trace amounts of nitro- and chlorophenols. Sample components: **1** = 2-nitrophenol (6.8 ng); **2** = 2,4-dichlorophenol (6.8 ng); **3** = 4-chloro-3-methylphenol (34 ng); **4** = 2,4,6-trichlorophenol (20 ng); **5** = 2,4-dinitrophenol (20 ng); **6** = 4-nitrophenol (34 ng); **7** = 2-methyl-4,6-dinitrophenol (34 ng); and **8** = pentachlorophenol (34 ng).

of the background signal level, and the background often requires as long as an hour or more to recover fully. The specific change produced by halogenated solvents depends on the NPD version being used. Generally, NPDs with thermionic sources composed of glass matrices exhibit decreases in both background and sample response with a halogenated solvent. In contrast, thermionic sources composed of ceramic matrices generally produce increases in background and signal. In either case, the aftermath of the injection of a halogenated solvent is background and sample responses that slowly change with time back to their original levels. One way of minimizing the perturbation caused by a halogenated solvent is to turn off the H_2/air chemistry of the NPD during the time the solvent elutes through the detector. This was

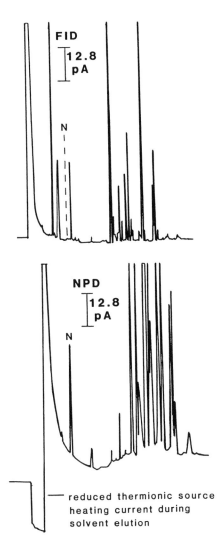

Figure 7.10. Comparison of FID and NPD chromatograms for a sample consisting of a methylene chloride extract of a commercially available shampoo; N designates the trace nitrogen constituent that was the component of interest in this sample.

done for that data shown in Figure 7.10, where the source heating current was reduced below the ignition point at the time of sample injection and then returned to an ignition level after passage of the halogenated solvent. In this case the sample was a methylene chloride extract of a shampoo product, and the objective of the analysis was to quantify a trace N constituent in the shampoo. Although the residual effects of the methylene chloride solvent still produced some background signal disruption, the magnitude of the background perturbation was substantially less than would occur if the H_2/air chemistry had not been turned off during the solvent peak elution.

Some other examples of the broad range of applications for the NPD are as follows: determination of N compounds in petroleum (30, 31); oil spill fingerprinting (32); N compounds in coal liquefication product oil (33); analyses of nitrated polycyclic aromatic hydrocarbons (34, 35); determination of nicotine in plasma (36) and air (37); determination of pentazocine and tripelennamine in blood samples of drug addicts (38); determination of aniline and substituted derivatives in wastewater (39) and N compounds in sludge extracts (49); trace detection of acrylonitrile in water, air, and other matrices (41, 42); determination of phosphoryl chloride and phosphorus trichloride in electronic grade trichlorosilane (43); determination of hydrazines in air (44); determination of trace cyanide or thiocyanate in aqueous matrices (45); determination of tetramethylsuccinonitrile in food containers and plastic packaging (46); the use of thermionic detection with liquid chromatography (47, 48); and the use of thermionic detection with supercritical fluid chromatography (49).

REFERENCES

1. A. Karmen and L. Giuffrida, *Nature (London)* **201**, 1204 (1964).

2. V. V. Brazhnikov, M. V. Gur'ev, and K. I. Sakodynsky, *Chromatogr. Rev.* **12**, 1 (1970).

3. D. J. David, *Gas Chromatographic Detectors*, Chapter 5. Wiley, New York, 1974.

4. B. Kolb and J. Bischoff, *J. Chromatogr. Sci.* **12**, 625 (1974).

5. B. Kolb and J. Bischoff, U. S. Patent 3,852,037 (1974).

6. W. C. Michels, (ed.), *The International Dictionary of Physics and Electronics*, 2nd ed. Van Nostrand, Princeton, NJ, 1961.

7. C. A. Burgett, D. H. Smith, and H. B. Bente, *J. Chromatogr.* **134**, 53 (1977).

8. T. Fujii and H. Arimoto, *Anal. Chem.* **57**, 2625 (1985).

9. P. L. Patterson, *J. Chromatogr.* **167**, 381 (1978).

10. P. L. Patterson, U. S. Patent 4,203,726 (1980).

11. P. L. Patterson and R. L. Howe, *J. Chromatogr. Sci.* **16**, 275 (1978).

12. P. L. Patterson, *Chromatographica* **16**, 107 (1982).

13. P. L. Patterson, *J. Chromatogr. Sci.* **24**, 41 (1986).

14. P. L. Patterson, U. S. Patent 4,524,047 (1985).

15. R. Greenhalgh, J. Fuller, and W. Aue, *J. Chromatogr. Sci.* **16**, 8 (1978).

16. J. A. Lubkowitz, B. P. Semonian, J. Galobardes, and L. B. Rogers, *Anal. Chem.* **50**, 672 (1978).

17. B. Kolb, M. Auer, and P. Pospisil, *J. Chromatogr. Sci.* **15**, 53 (1977).

18. P. L. Patterson, R. A. Gatten, and C. Ontiveros, *J. Chromatogr. Sci.* **20**, 97 (1982).

19. T. Fujii and H. Arimoto, *Anal. Chem.*, **57**, 490 (1985).

20. P. van der Weijer, B. H. Zwerver, and R. J. Lynch, *Anal. Chem.* **60**, 1380 (1988).

21. K. Olah, A. Szoke, and Zs. Vajta, *J. Chromatogr. Sci.* **17**, 497 (1979).

22. J. P. Blewett and E. J. Jones, *Phys. Rev.* **50**, 464 (1936).

23. D. Bombick, J. D. Pinkston, and J. Allison, *Anal. Chem.* **56**, 396 (1984).

24. R. C. Weast, ed., *CRC Handbook of Chemistry and Physics*, F-110. Chem. Rubber Publ. Co., Boca Raton, FL, 64th ed., 1983–1984.

25. E. W. McDaniel, *Collision Phenomena in Ionized Gases*, Chapter 13. Wiley, New York, 1964.

26. J. A. Lubkowitz, J. L. Glajeh, B. P. Semonian, and L. B. Rogers, *J. Chromatogr.* **133**, 37 (1977).

27. A. E. Lawson, Jr. and J. M. Miller, *J. Gas Chromatogr.* **4**, 273 (1966).

28. M. J. Hartigan, J. E. Purcell, M. Novotny, M. L. McConnell, and M. L. Lee, *J. Chromatogr.* **99**, 339 (1974).

29. B. J. Ehrlich, *Ind. Res. Dev. C*, April, p. 107 (1980).

30. D. K. Albert, *Anal. Chem.* **50**, 1822 (1978).

31. I. Dzidic, M. D. Balicki, I. A. L. Rhodes, and H. V. Hart, *J. Chromatogr. Sci.* **26**, 236 (1988).

32. G. A. Flanigan and G. M. Frame, *Ind. Res. Dev. C*, Sept., 28 (1977).

33. I. B. Rubin and C. K. Bayne, *Anal. Chem.* **51**, 541 (1979).

34. T. Ramdahl, K. Kueseth, and G. Becher, *HRC & CC, J. High Resolut. Chromatogr. Chromatogr. Commun.* **5**, 19 (1982).

35. M. C. Paputa-Peck, R. S. Marano, D. Schuetzle, T. L. Riley, C. V. Hampton, T. J. Prater, L. M. Skewes, T. E. Jensen, P. H. Ruehle, L. C. Bosch, and W. P. Duncan, *Anal. Chem.* **55**, 1946 (1983).

36. R. A. Davis, *J. Chromatogr Sci.* **24**, 134 (1986).

37. O. Grubner, M. W. First, and G. L. Huber, *Anal. Chem.* **52**, 1755 (1980).

38. M. Mackell and A. Poklis, *J. Chromatogr.* **235**, 445 (1982).

39. R. M. Riggin, T. F. Cole, and S. Billets, *Anal. Chem.* **55**, 1862 (1983).

40. V. Lopez-Avila, *HRC & CC, J. High Resolut. Chromatogr. Chromatogr. Commun.* **3**, 545 (1980).

41. T. Ramstad and L. W. Nicholson, *Anal. Chem.* **54**, 1191 (1982).

42. R. S. Marano, S. P. Levine, and T. M. Harvey, *Anal. Chem.* **50**, 1948 (1978).

43. B. K. Schulte and L. W. Shive, *Anal. Chem.* **54**, 2392 (1982).

44. J. R. Holtzclaw, S. L. Rose, J. R. Wyatt, D. P. Rounbehler, and D. H. Fine, *Anal. Chem.* **56**, 2952 (1984).

45. K. Funazo, M. Tanaka, and T. Shono, *Anal. Chem.* **53**, 1377 (1981).

46. H. Ishiwata, T. Inoue, M. Yamamoto, and K. Yoshihira, *J. Agric. Food Chem.* **36**, 1310 (1988).

47. V. L. McGriffin and M. Novotny, *Anal. Chem.* **55**, 2296 (1983).

48. F. A. Maris, R. J. van Delft, R. W. Frei, R. B. Geerdink, and U. A. T. Brinkman, *Anal. Chem.* **58**, 1634 (1986).

49. W. R. West and M. L. Lee, *HRC & CC, J. High Resolut. Chromatogr. Chromatogr. Commun.* **9**, 161 (1986).

CHAPTER

8

THE SURFACE IONIZATION DETECTOR

TOSHIHIRO FUJII

Division of Chemistry and Physics
National Institute for Environmental Studies
Tsukuba, Ibaraki 305 Japan

HIROMI ARIMOTO

Analytical Application Department
Shimadzu Corporation
Nakagyo-ku Kyoto 604 Japan

When organic molecules interact with a hot surface, thermionic emission takes place, resulting in the emission of organic ions. This was first observed in mass spectra obtained by the ionization of residual gases (1). Under the conditions used, some classes of organic molecules ionized on the surface of hot solids, with the desorption of positive ions, in the presence of weak electric fields.

Systematic studies of surface ionization (SI) of individual organics were begun in the 1960s by a group in the former Soviet Union (2). It is now established (3–8) that compounds with heteroatoms ionize on metal emitters, especially refractory metal oxides. Nitrogen-containing compounds are ionized most effectively; the extremely high ionization efficiency of amines on tungsten or rhenium oxide emitters has been demonstrated. These experimental results suggested that the SI phenomenon could be exploited in gas chromatography (GC). That is, a sensitive and selective detection of specific compounds, such as amines, eluted from a gas chromatograph may be possible by utilizing the surface ionization of the organic materials.

In this chapter, a new GC detector is described—the surface ionization detector (SID)—which employs a coiled platinum wire heated electrically in the same manner as a typical nitrogen–phosphorus detector (NPD) (9). Its development is based on the understanding (10, 11) that the detection

Detectors for Capillary Chromatography, edited by Herbert H. Hill and Dennis G. McMinn.
Chemical Analysis Series, Vol. 121.
ISBN 0-471-50645-1 © 1992 John Wiley & Sons, Inc.

mechanism involves positive surface ionization on the emitter surface, which is strongly dependent on the ionization energy (IE) of the chemical species as well as on the effective work function of the emitter surface with respect to ionization.

In the following sections we shall consider: (1) the principles of operation; (2) the instrumental description of the SID system; (3) a mass spectrometric study (12) designed to elucidate the detection mechanism, at least to the extent that the experimental results and the influence of working parameters can be understood; (4) the most important performance and response characteristics; and (5) some applications (13) in chemical analysis to demonstrate that the SID can have an important role in analytical chemistry.

1. PRINCIPLES OF OPERATION: POSITIVE
SURFACE IONIZATION

When atoms or molecules from a molecular beam or from a vapor impinge on an incandescent metal surface, they may evaporate partly as neutral atoms and partly as positive or negative ions (Figure 8.1). Kingdom and Langmuir first observed the formation of positive cesium ions on an incandescent tungsten surface (14). Subsequently, numerous studies of positive surface ionizations have been conducted, since this effect opens interesting possibilities for the analysis of chemical species with low ionization energy, for ion production, and for the detection of molecular and atomic beams.

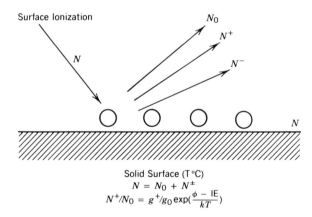

$$N = N_0 + N^{\pm}$$
$$N^+/N_0 = g^+/g_0 \exp(\frac{\phi - \text{IE}}{kT})$$

Figure 8.1. The surface with a layer of adsorbed particles of concentration, N, emit the following fluxes: N_0, of neutral particles; N^+, of positive ions; and N^- of negative ions. Under steady state conditions, the flux of particles desorbing from the surface $(N_0 + N^+)$ or $(N_0 + N^-)$ is equal to that of particles N adsorbing on the surface from the gas phase.

SI is interpreted conventionally by the use of the Saha–Langmuir equation (Equation 1), which is based upon the assumption that thermal and charge equilibria are established between the species on the surface material and the surface material itself (5). This equation describes the temperature dependence of the degree of ionization, α,

$$\alpha = N_+/N_0 = g_+/g_0 \exp\left(\frac{\phi - \text{IE}}{kT}\right) \tag{1}$$

where N_+ is the number of ions leaving the surface per unit area; N_0 is the number of neutral species emitted from the same surface in this time; ϕ is the work function of the surface at which ionization occurs at temperature T; k is the Boltzmann constant; IE is the ionization energy of the emitting chemical species; and g_+/g_0 is the ratio of the statistical weights of the ions and the neutral species.

Mass spectrometric studies on the surface ionization of organic compounds show that in most cases the compounds decompose into radicals, which have a lower ionization energy than the molecules and are ionized efficiently. For a given type of secondary species, s, formed on the surface, the resulting positive thermionic emission currents (J) and their dependence on surface temperature (T) are described by the ionization efficiency, $\beta_s(T)$, and by $Y_s(T)$, the yield of chemical reactions on the surface, such that

$$J_s(T) = N Y_s(T) \cdot \beta_s(T) \tag{2}$$

when a stationary beam of organic molecules, N, impinges from the gas phase on the surface. $\beta_s(T)$ is expressed as follows using the Saha–Langmuir equation (Equation 1) such that

$$\beta_s(T) = \frac{1}{1 + g_+/g_0 \exp[(\text{IE} - \phi)/kT]} \tag{3}$$

if equilibrium thermal ionization can be assumed. The combination of Equations 2 and 3 tells us that the emitter surface must have good pyrolytic properties [$Y_s(T)$ should be high] and the work function of the emitter must be high. Also, we see that the surface ionization process is specific, since it is strongly dependent on the IE of the species.

2.   DESCRIPTION OF THE DETECTOR

The design of the SID has the platinum emitter positioned between the quartz nozzle and the ion collector, which is in turn composed of an outer collector

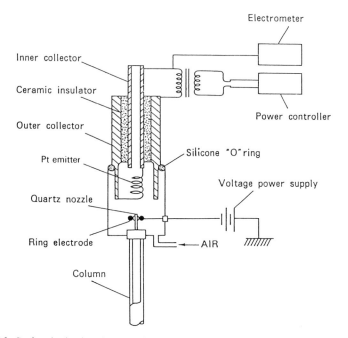

Figure 8.2. Surface ionization detector (SID) with Pt emitter. Reprinted with permission from T. Fujii and H. Arimoto, *Anal. Chem.* **57**, 2625 (1985). Copyright 1985, American Chemical Society.

and an inner collector (Figure 8.2). The ring electrode around the quartz nozzle is held at a positive potential of 200 V; the ion collector is always at a negative potential of 200 V against the ring electrode. Positive ion current directed to the collector is measured with an electrometer. Electronics is required for variable heating of the emitter filament. The emitter is a 10-turn coil of Pt (99.9% mm in diameter) capable of withstanding temperatures around 1200 °C for much more than a month in the environment of a flow of 40 mL/min of helium carrier gas mixed with dry air at 50 mL/min. Platinum was chosen as an emitter material primarily because it has a higher work function (5.65 eV) than other typical refractory metals such as tungsten and rhenium. This property easily allows positive thermionic emission under the conditions employed. Because constant carrier gas and air flow rate are important for the stability of this detector, flow controllers are recommended. In general, development of the SID has resulted in a rugged, simple, and long-lasting device.

A conventional flame ionization detector (FID) might be modified to a SID very easily if the Pt emitter is mounted in the gas flow path through the hole of the detector envelope that is used for igniting the flame.

The authors' homemade SID (10) is a structural modification of a standard NPD. It has almost the same configuration as the NPD supplied by the manufacturer except that the alkali–salt bead emitter of NPD was replaced by the Pt emitter and the applied potential was reversed.

3. MASS SPECTROMETRIC STUDIES

3.1 Atmospheric Pressure Ionization Mass Spectrometry

Although the exact identification of the charged species formed in the SID has not yet been determined, the response mechanism is assumed to involve positive surface ionization. Identification of the ions produced would assist in revealing the response mechanism as well as furthering the understanding of the detector characteristics and improving operating techniques.

Atmospheric pressure ionization mass spectrometry (APIMS) is a novel type of mass spectrometry (16) in which ionization is carried out in a reaction chamber (at atmospheric pressure) external to the low-pressure region of a mass analyzer. Ions present in the source enter the mass analyzer region through a small aperture. With this technique, informative details of various ion processes in atmospheric conditions have been possible.

Because of the demonstrated success or APIMS for the sensitive measurement of ions formed within its ECD (electron capture detector)-like source (17–19), we have taken a similar approach for the mechanistic study of the SID. The instrument used was a specialized atmospheric pressure surface ionization mass spectrometer, with an actual SID as a source (12).

We began to explore areas of application of the SID some 6 years ago. Recently we reported (11) several types of environmentally and biologically important compounds that provide a broader understanding of the potential application of the SID in chemical analysis. From these, polyaromatic hydrocarbons, antidepressant compounds, and α-pinene were chosen for the APIMS study, because the SID is clearly sensitive and thus suitable for the qualitative and quantitative analysis of these compounds. Also triethylamine (TEA), 1,3,5-cycloheptatriene, N,N-dimethylaniline, and piperidine were used. The reason these compounds were chosen is that these had been studied in detail by SIOMS (surface ionization organic mass spectrometry) in vacuum conditions (5, 7, 8). In the following subsections we will discuss: (1) the exact mass spectrometric identification of positive ions formed in the SID; (2) the effect of surface temperature, gas environments, and emitter material on the intensity and types of ions formed; and (3) the comparison of SID ions with those obtained by SIOMS.

3.2. Apparatus

The atmospheric pressure ionization mass spectrometer has been described in detail elsewhere (20). It is a modification of the apparatus specially prepared by Shimadzu Corporation as an atmospheric pressure ionization mass spectrometer for GC, which is referred to as APIMS–SID hereafter. With this instrument, the SID response can be monitored along with mass spectral measurement of the ions formed. Briefly the components are a sampling interface, a three-stage differential vacuum system, an electrostatic ion lens system and ion detection system with the channeltron electron multiplier detector operated both in the pulse mode and in the analog mode.

A detailed view of the SID atmospheric pressure ion source is shown in Figure 8.3. The ion source was an open-type modification of a standard SID for GC, with the coiled Pt emitter positioned at the midpoint between the stainless steel nozzle and entrance cone aperture of a quadrupole mass spec-

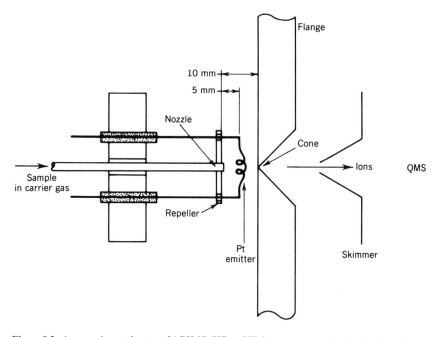

Figure 8.3. An experimental setup of APIMS–SID, a SID ion source attached to the front flange of the vacuum envelope of a quadrupole mass spectrometer. Reprinted with permission from T. Fujii, H. Jimba, and H. Arimoto, *Anal. Chem.* **62**, 107 (1990). Copyright 1990, American Chemical Society.

trometer. The repeller electrode around the nozzle was a 30-mm-diameter stainless steel disk that was 5 mm behind the emitter. All the SID ionic measurements reported here were obtained by applying + 300 V to both the repeller and the emitter.

The open-type SID ion source was positioned on the front flange of the vaccum envelope of the mass spectrometer, where a 0.5-mm-diameter orifice was drilled. This entrance cone, along with the skimmer (1 mm in diameter), provided a controlled leak of the ion source contents into the vacuum region. The grounded flange in Figure 8.3 served as the cathode to which a positively charged species would migrate.

Samples were introduced to the carrier gas stream via a temperature-controlled diffusion cell in which either a diffusion tube or a permeation tube was placed. A heated stainless steel transfer tube connected this cell with the nozzle.

SIOMS was performed with a Finnigan 3300 gas chromatograph–quadrupole mass spectrometer equipped with a home made thermionic ion source. Oxidized Re ribbon was prepared and used as a surface ionization emitter. Sample substances were admitted to the mass spectrometer from a reservoir via a variable leak valve or gas chromatograph. Full experimental details have been reported previously (5).

A Shimadzu gas chromatograph equipped with the SID, which is now commercially available, was used for the GC with SID response, GC–SID.

3.3. Identification of the Charge Carriers in the SID

Numerous experiments were conducted for each of the 10 compounds which are listed in Table 8.1 along with the formulae, molecular weight and ionization energy. Table 8.1 also includes the following experimental results: (1) the GC–SID responses, expressed in terms of sensitivity [C(coulombs)/g], under the optimum conditions for the detection limit; (2) the APIMS(SID) mass spectrum by the Pt emitter (composed of an intense peak) with the ion current (A) referenced to the sample amount of 1 g/s (determined from the height of the lines in the mass spectrum) for five volatile compounds (data were taken at the optimum emitter temperature giving the maximum total ion signal for each sample); and (3) and SIOMS mass spectrum for Re-oxide emitter with the ion current (A) referenced to a pressure of 1 Torr. Ion current expressed in amperes is the output of the ion multiplier with the gain at 5×10^3 in both APIMS–SID and SIOMS experiments.

Table 8.1 demonstrates parallels between the sensitivity of GC–SID and that obtained on APIMS(SID). The sensitivity observed by APIMS–SID is the same order of magnitude as that observed with the present SID for GC. This correlation provides support for the assumption inherent in this study—

Table 8.1. Response Comparison of APIMS–SID, SIOMS in Vacuum, and GC–SID

Compound (m/z, IE)	GC–SID (C/g)	APIMS–SID		SIOMS	
		$(A \cdot g^{-1} \cdot s^{-1})$	Mass Spectrum (m/z, %)	(A/Torr)	Mass Spectrum (m/z, %)
Triethylamine (101, 7.50) $C_2H_5)_3N$	1.6	1.2×10^{-3}	86, 100 (M–H), 63.6 (M+H), 2.6	22.8	(M–H), 100 86, 10
N,N-Dimethylaniline (121, ?) $C_8H_{11}N$	8.6×10^{-1}	1.1×10^{-3}	(M–H), 100	17.9	(M–H), 100
Piperidine (85, 8.7) $C_5H_{11}N$	3.3	3.3×10^{-3}	(M–H), 100 (M–3H), 11.9 (M+H), 8.6	45.8	(M–H), 100 (M–3H), 16 (M–5H), 14
1,3,5-Cycloheptatriene (92, 8.28) C_7H_7	2.6×10^{-2}	6.4×10^{-4}	(M–H), 100	3.3	(M–H), 100
Anthracene (178, 7.41) $C_{14}H_{10}$	2.7×10^{-1}		M, 100		M, 100
Pyrene (202, 7.48) $C_{16}H_{10}$	6.5×10^{-1}		M, 100		M, 100

Compound			(M–H), 100	(M–H), 100
Fluorene (166, 7.93) $C_{13}H_{10}$	4.3×10^{-2}			
α-Pinene (136, 8.07) $C_{10}H_{16}$	4.6×10^{-3}	1.2×10^{-5}	119, 100 105, 100 91, 51.4 121, 34.3 (M–3H), 20.0 (M–5H), 15.1 107, 15.1 95, 12 93, 12 (M–H), 11.1 (M–7H), 8.0	9×10^{-2} (M–3H), 100 (M–5H), 77.8 (M–7H), 55.6 93, 42.2 107, 38.9 105, 37.8 (M–H), 32.2 119, 26.7 91, 24.4 55, 20 43, 8.0
Imipramine(HCl) (280, ?) $C_{19}H_{24}N_2$	8.3×10^{-1}		58, 100 84, 22.5 72, 3.0	58, 100 84, 14 232, 6.3 186, 2.7 72, 2.7 70, 2.7
Lidocaine (234, ?) $C_{14}H_{22}N_2O$	2.0		86, 100 120, 23.3	86, 100 120, trace

that the mass spectrometric study explains the responses observed in an actual SID.

The positive charge carriers produced from APIMS–SID were exactly the same as the ion species obtained on conventional SIOMS using the Re oxide emitter. This agreement ensures, therefore, that the mechanisms that were considered in our initial study (10) need not be modified in order to account for these results; the mechanism for the response of SID that detects positive charge carriers involves positive surface ionization.

However, the relative intensity of the observed ion species is changed, except for those substances yielding a single peak. For example, molecular ions of $(M-nH)^+$ are dominant in the spectrum obtained by SIOMS for α-pinene (see Table 8.1). But when APIMS–SID is performed, ionic species having lower m/z values are produced more efficiently. The point is that ionization by APIMS–SID is more likely to produce extensively dissociated ion species than is that from SIOMS.

This result was expected. In the vacuum environment of the SIOMS, sample molecules interact directly with the hot emitter surface, and ions formed are the result of thermal pyrolysis. In the APIMS–SID, the hot emitter is placed in an atmospheric pressure "plasma" containing substantial concentrations of chemically active species such as H atoms and O atoms, as well as substantial amounts of water vapor. This is a thermal and combustion environment as opposed to the thermal pyrolysis conditions prevailing in the vacuum. Besides, it is certain that the variations are due partly to the emitter material, which causes a different work function and different pyrolytic properties and hence a shift in the response.

The mass spectrometric study on imipramine and lidocaine (13) is especially interesting, since the successful examination of these compounds serves as a good example of drug analysis. Hence, we believe that the SID technique may well play an important role in the analytical chemistry of many drugs in the near future. These are also interesting, from the point of view of surface ionization mass spectrometry, because they are nonvolatile compounds with rather complex structures, as shown here.

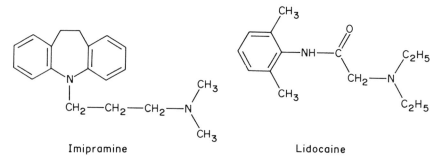

Imipramine   Lidocaine

The APIMS(SID) spectra of these compounds are characterized by DSI (dissociative surface ionization) ions, which are produced through a dissociative surface reaction (to give products with a low ionization energy) followed by surface ionization. These DSI ions are the charge carriers causing the response in the SID.

In contrast to the other substances studied, there is no indication that molecular ions are formed for either compound. But the difference in relative intensity of the mass spectrum between APIMS–SID and SIOMS was observed as in the previous cases.

The APIMS–SID spectrum of imipramine exhibits three intense peaks, m/z 58, 72, and 86, whose structures are probably $(CH_3)_2NCH_2^+$, $(CH_3)_2NCH_2CH_2^+$, and $(CH_3)_2NCH_2CH_2CH_2^+$, respectively. These are thought to be formed through a dissociative reaction after direct cleavage. The SIOMS spectrum has other peaks at m/z 232, 186, and 70. The composition of these peaks has not yet been postulated.

The mass spectrum of lidocaine comprises only two peaks at m/z 86 and 120. The most intense peak is m/z 86, presumably with the structure $(C_2H_5)_2NCH_2^+$.

3.4. Response Factors

3.4.1. The Emitter Temperature Effect

Figures 8.4 and 8.5 illustrate the response signal change caused by varying the emitter surface temperature (T) of the GC–SID and APIMS–SID. The detailed operating conditions are described in the caption. Figure 8.4 shows the temperature dependence of both the signal (i_s) and background currents (i_b) for the GC–SID. The APIMS–SID results are given in Figure 8.5. The upper graph shows the temperature dependence of the sample ion species at m/z 86 and 100 as well as ions of alkali atoms at m/z 23, 39, and 41 that result from impurities in the Pt emitter. The lower graph shows the temperature dependence of the total ion currents from both the sample substance and the impurity alkali atom (as I_s or I_b, respectively).

No difference was observed in the shape of the $i_s(T)$ curve for the GC–SID and the $I_s(T)$ curve for the APIMS–SID. Both increased with emitter temperature, and reached a maximum at a emitter temperature of $\sim 600\,^{\circ}C$, with a slight decrease at higher temperatures. Again, this agreement provides further support for the assumption that the SID mechanism involves positive surface ionization.

In the prior study (10) the background current of the SID was considered to be the appearance of Na^+ and K^+ ions from Na and K impurities in the Pt emitter. This assumption is also consistent with the present observation of

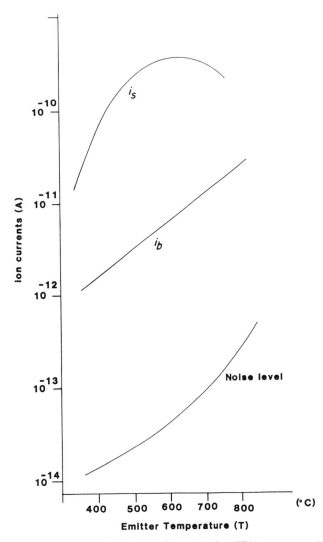

Figure 8.4. Emitter temperature dependence of triethylamine (TEA) response and background current for GC–SID. Carrier gas: He at 30 mL/min. Makeup gas: dried air at 50 mL/min. Sample: 2.6 ng of TEA in dichloromethane. Reprinted with permission from T. Fujii, H. Jimba, and H. Arimoto, *Anal. Chem.* **62**, 107 (1990). Copyright 1990, American Chemical Society.

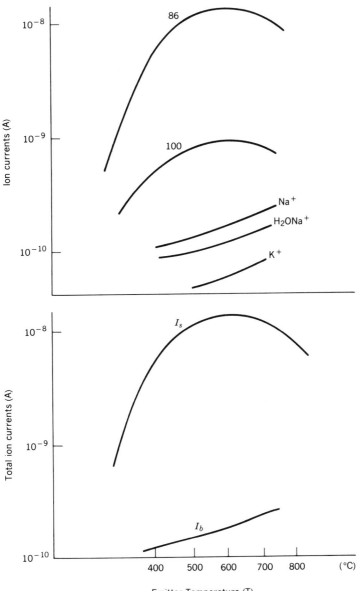

Figure 8.5. For APIMS–SID, in the lower graph, the resulting ion currents at m/e 86, 100, 23, 39, and 41 are shown as a function of the emitter temperature. TEA was introduced at the rate of 7.2×10^{-4} g/min, carried by the 50 mL/min air carrier gas. In the upper graph, curves of total ion currents either from the TEA sample or from the impurity of the Pt emitter material are drawn against the emitter temperature. Reprinted with permission from T. Fujii, H. Jimba, and H. Arimoto, *Anal. Chem.* **62**, 107 (1990). Copyright 1990, American Chemical Society.

181

the ionic species Na$^+$, K$^+$, and H$_2$OK$^+$ and H$_2$ONa$^+$, which presumably are the adduct ions associated with water vapor in the environment.

3.4.2.   Effect of a Field on the APIMS–SID Response

To determine the effects of an electric field, which is generated by applying the same voltage to the emitter and the repeller, a test sample of TEA was introduced through the diffusion tube. Response increased with applied voltage and leveled off at $\sim$ 150 V, as expected. The applied voltage affects the desorption of the ion species from the surface as well as the ion transportation to the inlet of the mass spectrometer.

3.4.3.   Emitter Material

The effect on the response caused by varying the emitter surface was studied. We examined Pt and Ir metals because both provide excellent performance as an emitter material. They were formed into 6-turn loops from 0.25-mm-diameter wire and assembled into the ion source housing to give the same length (area) and position. Since these emitter conditions were carefully controlled, it is certain that the observed large difference in $i(T)$ behavior is due to the difference in the properties of the emitter materials.

An Ir emitter gave similar results, with essentially the same i vs. T characteristics. The ion signal increased with heating current and leveled off at saturation values.

3.4.4.   Gaseous Environmental Around the Emitter

It could be concluded from the earlier study that an oxidized emitter should be used in order to improve sensitivity and stability. Such an oxidized emitter can be obtained if air (oxygen) is added to the emitter surface from another gas line. The addition of $\sim$ 50 mL/min of air to the 40 mL/min helium as a carrier gas affected the signals observed by the APIMS–SID, causing an increase in total ion intensity with a corresponding slight change in product ion distribution, which suggested that air serves to modify the surface leading to a change in the work function as well as the chemical properties of the surface.

The existence and theoretical basis of the SID mechanism is given by the mass spectrometric studies; positive surface ionization is responsible for ion formation in the SID. Thus the probability of ion formation is related to the IE of the absorbed species. In addition, on the hot surface some of the compounds decompose through consecutive series of unimolecular reactions. Some of the newly formed species at each step may be ionized. If the species has a low IE, its positive ion is formed.

The charge carriers collected at the conventional SID collector may not be those originally formed, since the ion moves through a gas in a longer electric field than the present experimental setup of the APIMS–SID and may experience charge stripping, charge transfer, ion/molecule reactions, etc. However, the SID response depends only on the total intensity of ion species initially formed.

3.5. Response and Performance Characteristics

Certain characteristics are indicative of the usefulness of the detector: (1) sensitivity; (2) selectivity; (3) noise and minimum detectable amount (MDA); and (4) linear dynamic range.

3.5.1. Sensitivity

For a given organic material, the sensitivity of the detector can be varied, depending on operating conditions such as the emitter temperature. As can be seen from Figure 8.4, sensitivity and noise both increase as emitter temperature is increased. Thus, an optimum must be determined in order to find the MDA. The sensitivity (S) of the detector can be expressed as coulombs per gram of sample. Under the optimum conditions, sensitivity was 1.58 C/g for tributylamine (TBA) and 2.98×10^{-6} C/g for dodecane.

3.5.2. Selectivity

Selectivity of the SID, which is defined as the ratio of sensitivity, was

$$S(TBA)/S(dodecane) = 5.3 \times 10^5$$

As already mentioned, sensitivity of the SID is strongly dependent on the IE of the species as well as on the yield of species generated through chemical reactions on the surface.

An experiment revealed (5) that a considerably strong signal was observed for m-xylene (M). This response is presumably due to the ionization of the (M–H) radical whose IE is 7.56 eV. Thus, it can be concluded that the conventional setup of the Pt emitter for use in the SID allows at least the detection of chemical species whose IE is less than 7.56 eV.

The relative sensitivity for different organic compounds, which is easily determined by a comparison of the signal current, varies in a wide range from sample to sample. Table 8.1 demonstrates the difference in sensitivity with the tested compounds. Therefore, for quantitative analysis, the SID requires calibration to determine correction factors.

3.5.3.   Noise Levels and Minimum Detectable Amount

At higher emitter temperatures, a decrease in the ratio of sample peak size to detector noise level occurs. This result indicates that an optimum emitter current must be found. Generally, favorable operation is achieved with an emitter temperature of 550 °C. That generates a noise level of 1×10^{-14} A. Under the optimum conditions, the MDA at a signal-to-noise ratio of 2 was measured using TBA. The result is listed in Table 8.2.

This study showed that noise levels of the SID are approximately one-tenth those of an FID under the same column conditions. While these results do not represent all the cases, this can be generalized to some extent.

3.5.4.   Linear Dynamic Range

The measurement of linear dynamic range was made with TBA as a test sample and showed that the linear portion of the working curve covers nearly 4 orders of magnitude with an uncertainty of $\pm 3.0\%$ (see Table 8.2).

3.5.5.   Baseline Stability

The stability of the baseline of the SID with a hot Pt emitter was studied. After we switched on the emitter, about 30 min was required to achieve a stable

Table 8.2. Comparison of the SID with Other Ionization Detectors

	FID	NPD	SID
Principle	H_2/O_2 flame 2000 °C plasma	Negative SI at 700 °C alkali-salt based	Positive SI at 550 °C Pt emitter
Sensitivity (C/g of X)[a]	0.015	0.2	1.58
MDA[b] (g/s of X)	3×10^{-12}	5×10^{-14}	1×10^{-14}
Selectivity (g of X/g of C)[c]		7×10^4	5.3×10^5
Linear dynamic range	2×10^6	10^5	10^4

[a] X corresponds to n-propane for FID, azobenzene for NPD, and tributylamine for SID.
[b] Minimum detectable amount.
[c] C stands for acetone in the case of NPD and dodecane in the case of SID.

baseline at a very low current level of 2×10^{-11} A. The data were obtained for an emitter after initial conditioning of 10 h at a temperature of $\sim 1000\,°C$.

3.5.6. Reproducibility

The effect of the analyte on the stable operation of the detector was investigated (11). It might be expected that temperature of the emitter surface would change with the analyte compounds because of the thermal conductivity effect. However, no significant effect was apparent from the experimental i vs. T curves, shown in Figure 8.4, which have a wide, flat portion at higher temperatures.

The second possibility is a change in the surface properties, brought about by alteration of the composition due to complex reactions between the surface material and the organic analyte. However, as already mentioned, such a poisoning of the emitter can be prevented if additional air flow is supplied to the detector.

For the Pt emitter placed in air, reproducibility tests were performed by replicate injections (20 times) of the standard sample containing 260 pg TBA. Good reproducibility was achieved; the coefficient of variation was 1.9% using peak heights (millimeters).

3.5.7. Durability

The Pt emitter is capable of withstanding more than 1 month's analytical use with no significant drift in sensitivity provided it is not heated excessively.

4. APPLICATIONS

Preliminary results suggest that the SID may be used successfully for sensitive gas chromatographic analysis of alkylamines, cyclic amines, ethanolamines, nitrosamines, aromatic hydrocarbons, polycyclic aromatic hydrocarbons (PAHs), nitro-PAHs, steroids, terpenes, and so on.

In this section, applications are illustrated using three examples to provide a broader understanding of the potential of the SID in chemical analysis.

4.1. Nitro-PAHs

Gas chromatography of PAHs was performed successfully by SID techniques and reported in detail (13). Overall the SID shows a higher response than the FID to the PAHs studied (fluorene, anthracene, fluoranthene, pyrene, chrysene, benzo[a]pyrene, benzo[e]pyrene, perylene, dibenzanthracene); perylene gives a very small peak in the FID chromatogram but a much larger peak in the

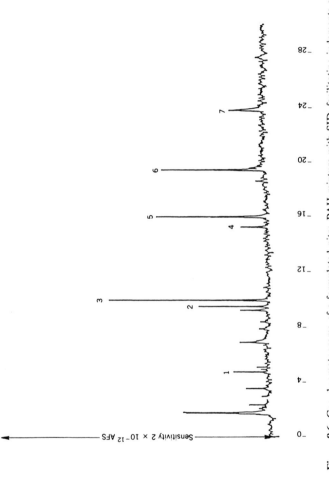

Figure 8.6. Gas chromatograms of a formulated nitro-PAH mixture with SID, facilitating judgments as to the MDA. Peaks: **1** = 1-nitronaphthalene, 820 pg; **2** = 9-nitrofluorene, 470 pg; **3** = 9-nitroanthracene, 5.7 pg; **4** = 3-nitrofluoranthene, 72 pg; **5** = 1-nitropyrene, 3.4 pg; **6** = 6-nitrochrysene, 80 pg; **7** = 1-nitrobenzo[*a*]pyrene, 4.3 pg. Analytical conditions: chemically bonded fused-silica capillary, polydimethyl-siloxane CBM1-M25-010, 25 m × 0.22 mm i.d.; injection system: A van der Berg type solventless sample injector. Carrier gas: He (4 mL/min). Column temperature: 70 °C for 2 min; then increased at 15 °C/min to 300 °C and held for 30 min. (AFS = amps full scale.)

SID chromatogram. However, the SID response is fairly unpredictable for each PAH compound because not all IE values are available. For example, it is particularly sensitive to benzo[a]pyrene and perylene, whereas the SID and FID provide almost the same response for fluoranthene.

This comparison also reveals that the SID does not give rise to peak broadening, tailing, and baseline drift, owing to its fast response characteristics. Thus the SID is compatible with capillary column techniques.

A formulated sample of nitro-PAHs, which are environmental carcinogens, was examined in order to explore another field of SID applications. Figure 8.6 corresponds to the sample, consisting of seven kinds of nitro-PAHs in tetrahydrofuran. The results illustrate a high sensitivity, with a detectability in the nanogram-to-picogram range.

A practical guideline for achieving the greatest detector stability is to operate the emitter at a slightly hotter temperature than the optimum determined for the MDA in an air environment.

4.2. Terpenes

Unlike the NPD and the ECD, which are element-selective detectors, the SID is to some extent a substance-specific detector. It is not specific elements or groups but a single compound that is responsible for detector response. In other words, a substance-specific detector would, for example, detect only a certain compound among other analogous compounds. As already mentioned, this behavior stems from the fact that the sensitivity of the SID is dependent on the IE of the species and also on the yield of the species generated through chemical reactions on the surface.

Chromatograms obtained from the analysis of a monoterpene mixture (13) clearly demonstrate these substance-specific characteristics. The SID especially detects α-pinene among the other monoterpenes of myrcene, β-pinene, d-limonene, and carene.

4.3. Drugs

Recent mass spectrometric studies of the surface ionization of individual organic compounds (7, 8) revealed that surface ionization gives greater output currents for nitrogen-containing compounds. Hence the SID appears to have great potential for use in the routine biomedical analysis of drugs, many of which possess nitrogen heteroatoms.

Recently the authors have developed a rapid, specific, and sensitive method for the determination of haloperidol in serum by GC with SID. The utilization of SID for the rapid, positive identification and determination of drugs has three advantages: (1) the possibility of interferences is greatly reduced, as this

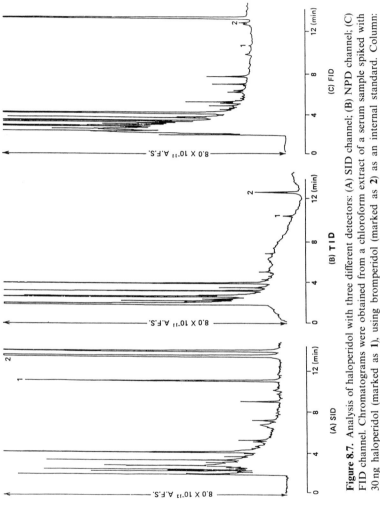

Figure 8.7. Analysis of haloperidol with three different detectors: (A) SID channel; (B) NPD channel; (C) FID channel. Chromatograms were obtained from a chloroform extract of a serum sample spiked with 30 ng haloperidol (marked as **1**), using bromperidol (marked as **2**) as an internal standard. Column: chemically bonded fused-silica capillary (polydimethylsiloxane, DB-1), 30 m × 0.31 mm i.d., 200–320 °C (6 °C/min).

188

detector shows very little response toward substances that have a high IE; (2) owing to the detector's increased sensitivity, better detection capability is achieved; and (3) cleanup procedures are usually unnecessary because of the high specificity.

Haloperidol is widely used as a neuroleptic of the butyrophenone type. Among the many methods for its determination, the reversed phase high-performance liquid chromatography method of Patlow et al. (21) may be the simplest and has been the basis of many clinical studies. However, this method lacks specificity and will produce color reactions not only with haloperidol but also with other compounds.

The sensitivity of the SID measured for haloperidol was about 0.2 C/g using a standard chloroform solution containing 100 pg of haloperidol. Considering the noise level, the MDA of haloperidol is in the picogram range, which neither conventional FID nor NPD can attain.

Figure 8.7 shows chromatograms of extracts of spiked serum containing haloperidol (100 ng/mL): (A) SID; (B) NPD; and (C) FID. The pretreatment procedure is outlined here (I). The analysis can be completed in less than 60 min.

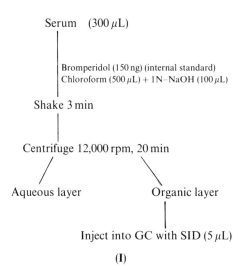

Serum   (300 μL)

Bromperidol (150 ng) (internal standard)
Chloroform (500 μL) + 1N–NaOH (100 μL)

Shake 3 min

Centrifuge 12,000 rpm, 20 min

Aqueous layer   Organic layer

Inject into GC with SID (5 μL)

(I)

The detectability of the SID is strikingly demonstrated. Another advantage is that baseline drift due to increased stationary phase bleeding in temperature programmed analysis is not significant, as the SID is not sensitive to the bulk of the liquid phase.

The overall method utilizing the SID technique and a simple pretreatment procedure may be applied successfully to the routine determination of halo-peridol in serum at concentrations as low as 3 ng/mL (at a signal-to-noise ratio of 3), provided that no other substance in the serum interferes with the haloperidol peak.

5. SUMMARY

Table 8.2 gives a comparison of performance characteristics of the SID with two other ionization detectors, the FID and NPD, in terms of principle, sensitivity, selectivity, detection limit, and linear dynamic range. The values reported for the FID and NPD are taken from the literature (22, 23). This demonstrates that the SID provides extremely high sensitivity, compared with the well-established FID and NPD.

Table 8.1 also lists the sensitivity of the SID for the compounds examined. Comparison with the FID, which shows a relatively uniform response of 0.015 C/g, provides information on which detector should be used, especially when high sensitivity is required.

The SID provides an interesting contrast to the NPD, whose mechanism involves negative surface ionization on a low work function surface (24). In the NPD environment, negative ion species CN^-, PO^-, and PO_3^- are formed predominantly when nitrogen and phosphorus compounds impact the low work function surface of the alkali–salt bead. By contrast, in the SID environment the organic molecules decompose into radicals on the hot surface. These are subsequently ionized to positive species on the high work function surface.

We believe the SID technique can play an important role in analytical chemistry and, like other GC detectors, applications to liquid chromatography and supercritical fluid chromatography may well be possible.

ACKNOWLEDGMENTS

Portions of this chapter were adapted with permission from T. Fujii, H. Jimba, and H. Arimoto, *Anal. Chem.* **62**, 107 (1990) and T. Fujii and H. Arimoto, *Anal. Chem.* **57**, 2625 (1985). Copyright 1985, 1990, American Chemical Society. Other portions were likewise adapted from T. Fujii, H. Jimba, M. Ogura, H. Arimoto, and K. Ozaki, *Analyst* **113**, 789 (1988). Copyright 1988, Royal Society of Chemistry.

REFERENCES

1. G. H. Palmer, *J. Nucl. Energy* **7**, 1 (1958).

2. E. Ya. Zandberg, U. Kh. Rasulev, and B. N. Schstrov, *Dokl. Akad. Nauk SSSR.* **172**, 885 (1967).

3. E. Ya. Zandberg and U. Kh. Rasulev, *Russ. Chem. Rev. (Engl. Transl.)* **51**, 819 (1982).

4. W. E. Davis, *Environ. Sci. Technol.* **11**, 587 (1977).

5. T. Fujii, *Int. J. Mass Spectrom. Ion Proc.* **57**, 63 (1984).

6. T. Fujii, *J. Phys. Chem.* **88**, 5228 (1984); T. Fujii, H. Suzuki, and M. Obuchi, *ibid.* **89**, 4687 (1985).

7. T. Fujii and T. Kitai, *Int. J. Mass Spectrom. Ion Proc.* **71**, 19 (1986); T. Fujii and H. Jimba, *ibid.* **79**, 221 (1987); T. Fujii, H. Ishii, and H. Jimbo, *ibid.* **93**, 73 (1989); T. Fujii, K. Kakizaki, and Y. Mitsutsuka, *ibid.* **104**, 129 (1991).

8. T. Fujii and H. Ishii, *Chem. Phys. Lett.* **163** 69 (1989); T. Fujii, K. Kakizaki, and H. Ishii, *Chem. Phys.* **147**, 213 (1990).

9. S. O. Farwell, D. R. Gage, and R. A. Kagel, *J. Chromatogr.* **19**, 358 (1981).

10. T. Fujii and H. Arimoto, *Anal. Chem.* **57**, 2625 (1985).

11. T. Fujii and H. Arimoto, *J. Chromatogr.* **335**, 375 (1986).

12. T. Fujii, H. Jimba, and H. Arimoto, *Anal. Chem.* **62**, 107 (1990).

13. T. Fujii, H. Jimba, M. Ogura, H. Arimoto, and K. Ozaki, *Analyst* **113**, 789 (1988).

14. K. H. Kingdom and I. Langmuir, *Phys. Rev.* **21**, 380 (1923).

15. E. Ya. Zandberg and N. I. Ionov, *Surface Ionization*, Israel Program for Scientific Translations, Jerusalem, 1971.

16. D. I. Carroll, I. Dzidic, E. C. Horning, and R. N. Stillwell, *Appl. Spectrosc. Rev.* **17**, 337 (1981).

17. M. McKeown and M. W. Siegel, *Am. Lab.* **7**, 89 (Nov. 1975).

18. E. C. Horning, D. I. Carroll, I. Dzidic, S. N. Lin, R. N. Stillwell, and J. P. Thenot, *J. Chromatogr.* **142**, 481 (1987).

19. E. P. Grimsrud, S. H. Kim, and P. L. Gobby, *Anal. Chem.* **51**, 23 (1979).

20. T. Fujii, M. Ogura, and H. Jimba, *Anal. Chem.* **61**, 1026 (1989).

21. P. I. Patlow, R. Miller, and M. Swigar, *J. Chromatogr.* **27**, 33 (1982).

22. B. Kolb, M. Auer, and P. Pospisil, *J. Chromatogr. Sci.* **15**, 53 (1977).

23. P. L. Patterson, *J. Chromatogr.* **167**, 381 (1978).

24. T. Fujii and H. Arimoto, *Anal. Chem.* **57**, 490 (1985).

CHAPTER

9

SULFUR-SELECTIVE DETECTORS

RICHARD S. HUTTE AND JOHN D. RAY

Sievers Research Inc.
Boulder, Colorado

1. BACKGROUND

1.1. Need for Sulfur-Selective Detection

The analysis of trace concentrations of sulfur-containing compounds is one of the most important and most difficult applications in gas chromatography. In the petroleum and petrochemical industries, the presence of sulfur compounds at parts per billion (ppb) levels can poison expensive bimetallic catalysts. At higher concentrations sulfur compounds can cause problems with corrosion and are a potential source of air pollution. Sulfur compounds are used as odorants for natural gas, liquefied petroleum gas (LPG), and other fuels. Routine determinations of the levels of these sulfur compounds are required to ensure the safety of these products.

Sulfur compounds are responsible, in part, for the distinctive taste and aroma of many foods, flavors, and beverages but can also impart undesirable tastes and odors to these products. The presence or absence of trace levels of sulfur compounds can also be used to detect alteration of essential oils and other expensive natural products.

In environmental analyses, many widely used pesticides are sulfur-containing compounds, while sulfur gases from natural and anthropogenic sources are important in acid rain and other atmospheric processes. Sulfur-selective detection is also important in soil and geological sciences.

Detectors for Capillary Chromatography, edited by Herbert H. Hill and Dennis G. McMinn.
Chemical Analysis Series, Vol. 121.
ISBN 0-471-50645-1 © 1992 John Wiley & Sons, Inc.

1.2. Difficulties in the Analysis of Sulfur Compounds

Analysis of sulfur compounds by capillary gas chromatography (GC) is complicated by their reactivity. Sorption and loss of low levels of these compounds can occur in all components of the chromatographic system, including sampling valves, injection ports, metal surfaces in detectors, and even in fused silica capillary columns. Common techniques for minimizing sorption and loss include passivation of the system using injections of high concentrations of sulfur compounds, reducing the exposure to metal surfaces, and the use of fused silica capillary columns with film thickness greater than 1 μm. The need for sensitive and selective detection of sulfur-containing compounds has led to the development of a number of detectors for gas chromatography. The principle of operation, sensitivity, and selectivity as well as selected applications for the major sulfur-selective detectors for GC are reviewed in this chapter. Other chapters in this volume should also be consulted. Each of these detectors has distinct advantages and disadvantages and, to date, no commercial sulfur-selective detector combines high sensitivity and selectivity for the measurement of sulfur compounds with low cost, easy operation, and other desirable qualities.

2. FLAME PHOTOMETRIC DETECTOR

The most widely used sulfur-selective detector for gas chromatography is the flame photometric detector (FPD). The emission of light from sulfur- and phosphorus-containing compounds in hydrogen-rich flames was first described for flame photometry (1–4), and in 1966 Brody and Chaney (5) developed the first GC detector based on this principle. In addition to detection of sulfur- and phosphorus-containing compounds, the FPD has been used for the detection of tin, germanium, selenium, tellurium, chromium, boron, and other elements (6). In this chapter, only sulfur-selective detection using the FPD will be discussed. The FPD is unique among GC detectors in that the response for sulfur compounds is nonlinear but approximately quadratic. The nonlinear response and other performance characteristics of the FPD require that care and consideration be exercised when using this detector (7–13).

2.1. Principle of Operation

Flame photometric detection is based on the formation of electronically excited diatomic sulfur (S_2^*) in a hydrogen-rich flame. The relaxation of S_2^* results in a broad band emission over the wavelength region from 320 to 460 nm, with the maximum emission at 394 nm. The chemistry of the combustion of sulfur compounds in hydrogen-rich flames has recently been reviewed (7), and the results from studies of flame chemistry have led to the development of models for the reactions of sulfur compounds in fuel-rich H_2/O_2 flames (8, 9). These models indicate that a number of sulfur-containing species, including H_2S, HS, S, S_2, SO, and SO_2, are formed in hydrogen-rich flames, with the relative concentration of these species determined by the H_2/O_2 ratio, the kinetics of the reactions, and other factors (14–20):

$$\text{sulfur compounds} \xrightarrow{H_2/O_2 \text{ flame}} S_2 + S + H_2S + HS + SO + SO_2 \qquad (1)$$

$$H_2S + H \rightleftharpoons HS + H_2 \qquad (2)$$

$$HS + OH \rightleftharpoons HS + H_2O \qquad (3)$$

$$HS + H \rightleftharpoons S + H_2 \qquad (4)$$

$$SO_2 + H \rightleftharpoons SO + OH \qquad (5)$$

$$SO + H \rightleftharpoons S + OH \qquad (6)$$

$$HS + S \rightleftharpoons S_2 + H \qquad (7)$$

Several mechanisms have been proposed for the excitation of diatomic sulfur. In one (16, 17), the excitation energy is provided by two- or three-body recombination reactions:

$$S(^3P) + S(^3P) \rightleftharpoons S_2^* \quad (B^1\Sigma g) \qquad (8)$$

$$S(^3P) + S(^3P) + M \rightleftharpoons S_2^* \quad (B^1\Sigma g) \qquad (9)$$

where M is some third body;

An alternative mechanism involves recombination of atomic hydrogen and hydroxyl radicals (4, 8):

$$H + H + S_2 \rightleftharpoons S_2^* \quad (B^1\Sigma g) + H_2 \qquad (10)$$

$$H + OH + S_2 \rightleftharpoons S_2^* \quad (B^1\Sigma g) + H_2O \qquad (11)$$

Relaxation of S_2^* results in a broad band emission in the blue and ultraviolet (UV) region of the spectrum ($\sim$ 320–450 nm), with emission maxima at 394 and 384 nm:

$$S_2^* \quad (B^1\Sigma g) \rightleftharpoons S_2^* \quad (X^1\Sigma g) + h\nu \qquad (12)$$

Emission in this wavelength region may also be observed from other sulfur species in the flame such as SO*, SH*, and SO_2^*, whereas the presence of

hydrocarbons and other non-sulfur-containing compounds in the flame results in emission from CH, C_2, CN, and other species (18–20). The excitation
reactions occur in the cooler regions above the flame, while most of the
emission from non-sulfur-containing compounds occurs in the flame itself.
Thus the FPD flame is usually shielded from view by the photomultiplier
tube. However, the emission from S_2^* is still measured against some flame
background, and therefore an optical filter is employed for sulfur-selective
detection.

2.2.   Design of Flame Photometric Detectors

There are two basic designs for the GC–FPD: single-flame and dual-flame
detectors. A generalized schematic of a single-flame FPD is shown in Figure
9.1 and consists of an enclosed H_2/O_2 (air) burner, an optical filter, and a
photomultiplier tube and associated electronics. A narrow bandpass interference optical filter with transmission at 394 nm (5–10 nm bandpass) is used
for sulfur-selective detection for most FPD designs, although FPDs with two
optical filters for simultaneous monitoring of the phosphorus and sulfur
response are also available.

2.2.1.   Single Burner

The original single-flame FPD design described by Brody and Chaney (5)
employs a diffusion flame in which the effluent from the GC column (nitrogen
carrier gas) is mixed with oxygen and delivered to the burner tip. Excess
hydrogen is not premixed with the nitrogen/oxygen stream but added to the
exterior of the burner tip. The combustion occurs inside the burner tip, thus

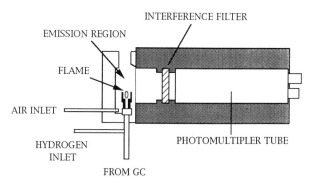

Figure 9.1. Schematic diagram of a single-flame FPD.

shielding the flame from direct view of the photomultiplier tube (PMT). With this burner design, the emission of light from hydrocarbons and other interferences occurs inside the burner tip and is not detected by the PMT, whereas the emission from sulfur and phosphorus species occurs above the shielded flame and is directly viewed by the PMT. The gas flow rates employed by Brody and Chaney (5) were 160 mL/min for the nitrogen carrier gas, 40 mL/min for the oxygen, and 200 mL/min for the hydrogen. One limitation of this design is the problem of "solvent flameout." The elution of large amounts of the solvent consumes the oxygen and extinguishes the flame, making it necessary to reignite the burner after elution of the solvent. Modified designs have been developed (21, 22) in which the hydrogen and oxygen or air inlets to the burner are interchanged, eliminating the problem of flameout.

2.2.2. *Dual Burner*

Dual-flame burner FPD designs have been proposed by several researchers (23–25) to overcome the limitations of the single-flame FPD detectors. As shown in Figure 9.2, in the dual-flame design excess hydrogen is added at the base of the detector while the GC effluent is mixed with air. The amount of hydrogen is in sufficient excess for reaction with air in the first flame, and the unreacted (excess) hydrogen is then mixed with a second air supply to produce the second flame. In operation, the top flame is ignited and ignition of the lower flame occurs as a result of a mild flashback. If the lower flame is extinguished during elution of the solvent, the flashback mechanism serves to reignite it. As will be discussed below in more detail, the use of two burners, one for the decomposition/combustion of the compounds eluting from the GC column and the other for the formation of the emitting species, overcomes

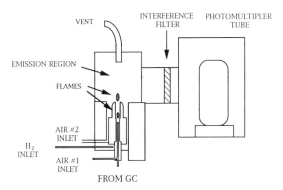

Figure 9.2. Schematic diagram of a dual-flame FPD.

some of the quenching problems observed in single-burner FPDs. The optimum flow rates for this dual burner FPD were 140 mL/min for hydrogen, 80 mL/min for air #1, 170 mL/min for air #2, and a helium carrier flow rate of 30 mL/min (23, 24).

2.2.3. Photomultiplier Tubes

Both end-on and side-view blue sensitive photomultiplier tubes are used in commercial FPDs, and the PMTs are typically operated at supply voltages ranging from 350 to 750 V. Because the PMT must be positioned near the burner assembly, there are limits to the operating temperature of the FPD, which for most applications is maintained at 150–275 °C. The FPD response to sulfur compounds decreases with increasing detector temperature (8), and PMT noise increases with increasing detector temperature; thus lower detector temperatures are preferred in sulfur-selective FPD detection. However, the need to operate the FPD at relatively low detector temperatures presents difficulties in the analysis of higher-molecular-weight sulfur compounds.

2.3.   Performance Characteristics of the FPD
for Sulfur Compounds

2.3.1. Nonlinear Response and Dynamic Range

In sulfur detection using the FPD, the species that is being measured is electronically excited diatomic sulfur (S_2^*). Thus the response of the FPD for sulfur compounds is inherently nonlinear. In practice it has been found that the intensity of the sulfur emission in the FPD is given by

$$I \propto [S]^n \qquad (13)$$

$$\log[S] \propto 1/n \log(I) \qquad (14)$$

The exponent (n) is commonly referred to as the exponential proportionality constant, the linearity factor, or simply the "n value" and is theoretically equal to 2. In practice, the value of n is dependent upon the operating conditions of the FPD and is strongly compound dependent. Values ranging from $n = 1$ to slightly greater than 2 have been reported for different compounds, different detector designs, and different operating conditions (7–13, 26–32). Several explanations have been offered to account for the observed deviation from the theoretical n value of 2, including nonoptimum flame conditions, compound-dependent decomposition, competitive flame reactions, quenching effects, and other factors (7). Calibration curves for the FPD are usually constructed by plotting the log of the response vs. the log of the sulfur concentration (Equation 14), and electronic linearizing circuitry based on this relationship is available on some commercial GC–FPD systems. Since an n value of 2 is commonly used for these linearizer circuits, errors can be introduced in the analysis (30). Corrected detector response values have also been employed. Maruyama and Kakemoto (31) suggest use of a response factor (RF)

$$RF = (\text{peak height})^{1/2} \times (\text{peak width at half height}) \qquad (15)$$

to yield a signal directly proportional to sulfur concentration, whereas Attar et al. (32) use the response factor

$$RF = \frac{\text{peak area}}{(\text{peak height})^{1/2}} \qquad (16)$$

to obtain a linear response.

A more complete discussion of variation in n values, the response of different sulfur compounds, and techniques that compensate for variable FPD response can be found in several excellent reviews of this subject (7, 8, 11, 13, 30); but for most applications and quantitative analyses using the FPD, calibration curves over the concentration range of interest must be constructed for each sulfur compound.

The reported minimum detection limits (MDL) for the analysis of sulfur compounds using the FPD range from 2 to 50 pg S/s, with the sensitivity being strongly dependent upon the hydrogen and air flow rates, the design of the FPD, and the compound used for the MDL determination (8, 13). Using log/log calibration curves, the FPD has a linear dynamic range of ~ 2.0–2.5 orders of magnitude (for example, 0.02–20.00 ng S/s). At higher concentrations, curvature is observed in the calibration curve due to quenching, self-absorption, flame effects, and other factors (7–13). The selectivity of the FPD for sulfur compounds vs. hydrocarbons has been reported to be $\sim 10^3$ g C/g S at low sulfur concentrations to 10^6 at higher concentrations using a dual-flame FPD, with selectivities of 10^4–10^5 reported for single-flame FPDs (7–13).

2.3.2. Sulfur Doping

Linearization of the FPD response for sulfur compounds and improvements in the sensitivity of the detector have been achieved by the continuous addition of a constant amount of a sulfur compound to the FPD (6, 22, 27, 33). The addition of SO_2, SF_6, or another sulfur compound results in a pseudo-first-order reaction rate for the analyte sulfur compounds. Permeation tubes, low flow rates of a pure sulfur gas, and mixtures of a sulfur compound in one of the flame gases are commonly used to deliver a constant amount of the sulfur compound to the FPD, but great care must be taken to ensure that the dopant level is constant. Detection limits as low as $2-10 \, pg \, S/s$ have been obtained using sulfur doping, with a linear response observed over $1.0-1.5$ orders of magnitude (7, 8). Although it is useful for improving the sensitivity of the FPD for sulfur detection, the doping results in a very small linear dynamic range that may not be of use for many samples containing widely varying concentrations of sulfur compounds.

2.3.3. *Quenching*

The FPD response for a sulfur compound is reduced if a non-sulfur-containing compound coelutes or is only partially resolved from the sulfur analyte. This reduction in the FPD response is attributed to nonradiative collisional quenching of S_2^* in the flame by CO_2, CH_4, and other combustion products (7–13). Although most non-sulfur-containing compounds will quench the sulfur response, hydrocarbons are particularly effective. The magnitude of this quenching can be dramatic. In many cases there is complete loss of FPD response to sulfur compounds owing to coelution of another compound.

One approach to this problem is the dual-flame FPD design (23–25), which can significantly reduce quenching. In the dual-flame FPD, the first flame is operated under strongly reducing conditions (large excess of hydrogen: $O/H = 0.1$) to produce mostly reduced combustion products (for example, H_2S or CH_4) and to lessen the competition for H atoms that occurs in single-flame FPDs. The second flame is then operated under normal FPD conditions ($O/H = 0.3$) to maximize S_2 formation.

Most researchers have found that a minimum concentration of a non-sulfur-containing compound must be present to cause quenching of the sulfur response. Typically a concentration of $> 10 \, ng \, C/s$ is reported as being required to produce significant quenching (7–13, 34–36). Recent studies have shown that coelution of a 10- to 50-fold excess of a non-sulfur-containing species is sufficient to cause significant quenching (37) and suggest that some quenching is observed whenever hydrocarbons are present in the FPD, with the magnitude increasing as the concentration of hydrocarbon in the flame increases (38).

2.4. Capillary Adaptation

Most commercial FPDs were designed for use with packed columns and therefore contain significant postcolumn dead volume, hot metal surfaces, and other design features that can lead to peak broadening or tailing. For optimum sensitivity with capillary columns, a makeup gas (usually nitrogen) is used to provide higher flow rates of an inert gas into the FPD.

Gaines et al. (28) and Barinaga and Farwell (39) have reported improved performance of commercial FPDs with capillary columns by modification of the FPD chimney. Quartz or Pyrex inserts have been positioned above the FPD burner tip to reduce the dead volume of the detector, improve flow streamlining, and minimize sorption of sulfur compounds in the detector. A column-centering device has been used to help position the fused silica capillary column end axially in the FPD column inlet (39), reducing sorption and loss of sulfur compounds in the FPD jet. These modifications improve the detection limit of the FPD, and a smaller range of n values (1.8–2.0) has been reported for the analysis of low-molecular-weight sulfur compounds using an improved fused silica capillary GC–FPD system (28, 39).

Despite the less than ideal performance characteristics of the FPD (nonlinear compoud-dependent response, quenching, etc.), the FPD has proven to be a valuable sulfur-selective detector for GC.

3. HALL ELECTROLYTIC CONDUCTIVITY DETECTOR

3.1. Principle of Operation

The Hall electrolytic conductivity detector (ELCD) is a selective detector for GC that can be operated in either a halogen-, nitrogen-, or sulfur-selective mode (40). A generalized schematic of the ELCD is shown in Figure 9.3 and consists of the following: (1) a reaction gas; (2) a high-temperature furnace and reaction tube; (3) a chemical scrubber; (4) a gas–liquid contactor; (5) a gas–liquid separator; (6) a conductivity cell; (7) a solvent reservoir; (8) a circulating pump; and (9) an ion exchange resin bed. In operation, the effluent of the GC column is mixed with a reaction gas and enters the high-temperature furnace for oxidation (sulfur) or reduction (halogen or nitrogen) of the sample depending on the species being detected. The reaction products are then dissolved in the deionized conductivity solvent in the gas–liquid contactor and undissolved gases are removed via the gas–liquid separator. In the solvent, the reaction products ionize, resulting in an increase in the conductivity of the solvent that is measured in the conductivity cell using a bipolar pulsed technique. A solenoid vent valve is positioned between the GC and the reactor to permit venting of the solvent to minimize interference and prevent fouling of the reaction tube. (Chapter 6 in this volume contains additional details on the ELCD.)

For sulfur-selective detection, the reaction gas is air or oxygen and a nickel reaction tube operated at 750–1000 °C is used (41, 42). Sulfur compounds are

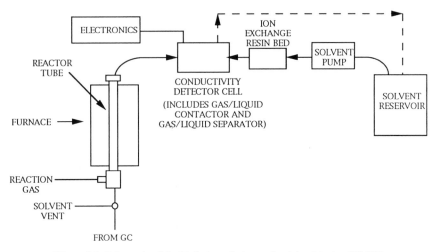

Figure 9.3. Schematic of the Hall electrolytic conductivity detector (ELCD).

oxidized to produce SO_2 and/or SO_3, while hydrocarbons are converted to CO or CO_2. The conductivity solvent is usually pure methanol or a methanol/water mixture (7% H_2O), which permits ionization of SO_2/SO_3 but minimizes the ionization of CO_2, thus reducing hydrocarbon interferences (42). The presence of halogenated compounds in the samples results in the production of strong acids (HX), and therefore a halogen scrubber consisting of a silver wire in a copper tube is usually positioned between the furnace and the gas–liquid contactor.

3.2.   Sensitivity, Selectivity, and Linearity

A study on the optimization of ELCD for surfur-selective detection (41) found the optimum flow rate for the methanol solvent was 1.9 mL/min. However, operation at these flow rates was not practical owing to the formation of bubbles in the gas–liquid separator, and lower flow rates (0.6–1.4 mL/min) were used for routine operation. The optimum air flow rate was 20 mL/min, and the furnace was operated at 800 °C. Using these conditions, the minimum detection limit of the ELCD for diethyldisulfide [signal-to-noise ratio $(S/N) = 2.5$] was 0.85 pg S/s, which was 20 times better than the MDL obtained using an FPD operated under similar chromatographic conditions. Detection limits were determined for several low-molecular-weight sulfur compounds and ranged from 0.21 pg S/s for CH_3SO_4 to 1.3 pg S/s for COS. The response of the ELCD was linear over 2–4 orders of magnitude.

The selectivity of the ELCD for saturated hydrocarbons was $> 10^9$ S/g C, whereas the selectivity for acetone was 2×10^4 S/g C. The selectivity for sulfur compounds vs. chlorinated hydrocarbons was only 6.1 g S/g Cl without the Ag scrubber, and it increased to 5×10^3 g S/g Cl using the scrubber. Sulfur response and peak shape (packed column) was not affected by use of the scrubber.

Another study directly compared the sulfur detection using an ELCD and an FPD (41). The operation conditions of the ELCD were as follows: reactor temperature, 1000 °C; air flow rate, 55 mL/min; and solvent flow rate, 4.6 mL/min. Pure methanol or 7% water/methanol were used as the solvents. Minimum detection limits for the ELCD ($S/N = 2$) were 1 pg S/s for COS, 1.1 pg S/s for thimet, and 1.2 pg S/s for parathion, with pure methanol as the solvent. The MDL for the ELCD using 7% water/methanol was 5.2 pg S/s for H_2S. The MDLs for COS and H_2S using the FPD in this study were 4.0 and 5.6 pg S/s respectively.

The linear range of the ELCD with pure methanol was ~ 2 orders of magnitude, whereas the use of 7% water/methanol extended the linear range to ~ 4 orders of magnitude. In contrast, the linear range of the FPD, using square root electronic linearization, was 2.5 orders of magnitude. The selectivity of the ELCD and FPD were reported to be $> 10^5$ g S/g C, with the ELCD offering the potential for some control of the selectivity by selection of furnace temperature. Interference in the ELCD from hydrocarbons can be reduced by operating at lower furnace temperatures; however, some loss in oxidation efficiency for sulfur compounds may be observed if the temperature is too low.

Analysis of sulfur compounds in hydrocarbon matrices (42, 43) revealed that little or no quenching of the sulfur response is observed using an ELCD whereas the FPD response was decreased. The conclusion of Ehrlich et al. (42) is that the ELCD offers some advantages over the FPD for sulfur-selective detection, including a linear response, better sensitivity, and the absence of quenching.

While the ELCD is used most frequently for halogen-selective detection, it provides high sensitivity and selectivity when operated in the sulfur-selective mode and offers advantages over the FPD for the detection of sulfur compounds, particularly in hydrocarbon matrices. The major disadvantages of the ELCD as a sulfur-selective detector for GC include compound-dependent response factors, operational complexity (solvent pumps, gas–liquid interface, etc.), potential fouling of the nickel catalyst tube, and relatively high maintenance requirements.

Commercial versions of the ELCD have been improved to permit operation with capillary columns without substantial loss of efficiency or peak tailing even with the relatively large postcolumn dead volume of the ELCD (43).

4. CHEMILUMINESCENCE DETECTORS

Several different chromatography detection methods have been developed in recent years based on chemical reactions that produce light (44), a process referred to as chemiluminescence (CL). Reaction is generally fast but of low efficiency in the various molecules producing light. However, because the light is emitted against a dark background, CL detectors are among the most sensitive. Low light levels produced in the visible or near-visible range can be measured using photomultiplier vacuum tubes that amplify the photocurrent by 10^6 or more. The nanoampere signals are further amplified to give usable millivolt signals. Because of the inherently low noise of many CL reaction cells, the large amplification factors still produce usable signals.

There are several complications in CL detection methods. The light detection and amplification equipment can be bulky compared to the commonly used FID or TCD detectors. There are not, as yet, solid state photodetection devices with the required sensitivity. The reaction cell is also relatively large, which can cause chromatographic peak distortion or broadening. Reagents other than those in the chromatographic effluent are also required. Often these reagents are corrosive and require additional pressured cylinders of gases. In practice, however, each of these complications can be handled reasonably well.

4.1. Chemiluminescence Detector Based on Ozone Reactions

4.1.1. Principle of Operation

Ozone reacts with a variety of organic and inorganic compounds to produce light (45, 46). This chemiluminescence is the basis for a number of atmospheric ozone monitors (47–49). When these detectors are operated in reverse, that is, with excess ozone and hydrocarbon samples, they can detect various unsaturated organic compounds (50). At higher temperatures both saturated and aromatic hydrocarbons exhibit chemiluminescence. Chromatograms comparable to those obtained with a commercial FID can be obtained at temperatures greater than 250 °C (50). Detection limits are reported in the nanogram range for most compounds and, as long as the ozone is in large excess, the reaction is a pseudo-first-order and response is linear. Emissions are believed to be from glyoxals, excited-state formaldehyde, CH radicals, and hydroxyl radicals.

The chemiluminescent reaction of NO and O_3 to produce light in the red region of the spectrum is the preferred method for measuring nitrogen oxide air pollutants. A GC detector (51, 52) used a postcolumn gold catalyst reduction of nitrogen dioxide and the NO/O_3 CL reaction to detect reducing compounds, including sulfur compounds. Thus H_2S, CS_2, SO_2, sulfides, and thiols could readily be detected. Compounds that could reduce NO_2 were detected. However, in general, the chromatograms were less complicated than those produced using an FID.

An analogous CL reaction is that of SO to produce light in the blue region (280–420 nm):

$$SO + O_3 \longrightarrow SO_2^* + O_2 \tag{17}$$

$$SO_2^* \longrightarrow SO_2 + hv \tag{18}$$

Halstead and Thrush (53) studied this reaction and recorded a broad peak at 340 nm. Akimoto et al. (54) observed that O_3 reacted with several sulfur species including SO, CH_3SH, CH_3SCH_3, and H_2S to produce the identical SO_2^* emission spectrum. It was concluded that SO was an intermediate in all of these reactions. A hydrogen flame produces a number of sulfur species when a sulfur-containing compound is burned. In fact, SO occurs in the hydrogen flame in higher concentrations than S or S_2, species that are the basis of the FPD. Benner and Stedman (55) reported a method for sampling a hydrogen flame with a quartz probe and subsequently detecting the SO in a reaction cell. Optimum sensitivity was obtained at short flame residence times (2.5 ms) and with hydrogen 80% greater than what was needed for combustion. A linear response was obtained based on the sulfur content of the species introduced into the flame. Relative response of different sulfur compounds was the same within the precision of the calibrations. Thus, a highly selective method for sulfur detection was available with linear response based only on the number of sulfur atoms present. The method was also effective when an FID flame was sampled.

4.1.2. Sulfur Chemiluminescence Detector Design

A commercially available sulfur chemiluminescence detector (SCD) uses a variation on the Benner and Stedman method. A high-purity ceramic probe is used to sample the hydrogen flame from a normal GC–FID. The SO produced in the flame is transferred in a Teflon line to a reaction cell where O_3 is added (Figure 9.4). The detector signal is amplified and converted to a voltage signal by an electrometer. The output signal can be connected to a recorder, integrator, or GC data system.

The sampling probe for the SO causes some loss in sensitivity of the FID, but allows both the FID signal and the sulfur signal to be viewed simultaneously. Under normal operation the ceramic probe sits about 4 mm above the hydrogen jet (Figure 9.5) and draws 80–90% of the total gas flow into the SCD. Optimum response is obtained by adjusting the total gas flow and the

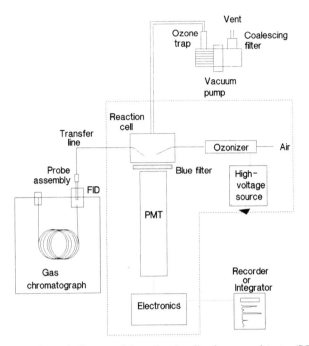

Figure 9.4. Schematic diagram of the sulfur chemiluminescence detector (SCD).

fuel/air ratio. Tailing is minimized by the low pressure in the transfer line and the short residence time. Under some circumstances, heating the ceramic probe can improve the sensitivity. Heating or shortening the transfer line was found to have minimal effect.

Recent improvements in the SCD may lead to enhanced sulfur detection. A dedicated burner can give greater sensitivity than using the flame of an existing FID. Quartz, metal, and ceramic burners have been shown to work well. Detection limits have been lowered by more than a factor of 10 over the standard instrument by using all ceramic burners with enclosed combustion at reduced pressure. Detection limits in the 10–100 fg sulfur range have also been obtained using specially configured SCDs for direct atmospheric monitoring (56) and as a GC detector (57).

There is some trade-off between maximum sensitivity and maximum selectivity depending on the H_2/air ratio. Reducing the air from the typical ratio of 200 mL/min H_2 and 400 mL/min air will increase the selectivity. More critical for sensitivity is the probe position. In addition to the height adjustment, the probe must be centered within the flame or a significant loss in signal can

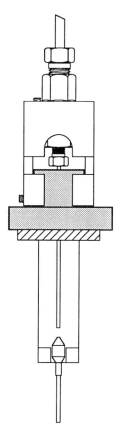

Figure 9.5. Schematic diagram of the SCD sampling probe interface with an FID. The transfer line is attached at the top. The ceramic probe is centered just above the hydrogen jet.

occur. Apparently the SO concentration is dependent on location, the localized temperature, and the reducing character of the flame. The best sensitivity is obtained when the tip of the ceramic probe glows bright red. This provides a visual indication of correct positioning.

The size of the flame is also important. Greater combustion gas flows increase the flame height and change the flame shape, requiring that the probe be repositioned. If the flows are low, ambient air is drawn into the SCD along with the sample. This leads to dilution of the sample and possible noise from ambient sulfur compounds.

For most applications the flame source for the SCD is a conventional FID operated under hydrogen-rich conditions, and it is possible to obtain simultaneous FID and SCD chromatograms from a single injection. The flow rate of hydrogen needed for the SCD results in an FID sensitivity loss of about 1 or 2 orders of magnitude. The SCD peaks are slightly delayed compared to the FID response and the peaks are slightly broadened owing to the electronics and dead volume in the transfer line and reaction cell. The simultaneous chromatograms greatly assist in peak assignment.

4.1.3. Sensitivity, Linearity, and Selectivity

The SCD is highly specific for sulfur compounds and detects both oxidized and reduced sulfur species. Hydrogen sulfide, carbonyl sulfide, sulfur dioxide, and carbon disulfide can be detected equally as well as the organosulfur compounds. The selectivity (as sulfur to hydrocarbons) has been determined to be greater than 10^7 (57). Thus, the detector is not subject to quenching or interference from coeluting compounds at the usual GC sampling volumes.

Calibration curves obtained by repeated injection of standards are linear over 5 orders of magnitude. The sulfur detection limit, based on the sensitivity and 3 standard deviations above the rms (root mean square) noise, is < 1 pg S/s. SCD systems used for direct sampling of air have obtained detection limits 10–100 times less than that (56), and greater sensitivity for chromatography may be available in the future.

Table 9.1. Comparison of SCD Relative Response for Different Sulfur-Containing Compounds

Compound	Relative Response	Standard Deviation (%)
Ethanethiol	1.0	15.9
Dimethylsulfide	1.0	4.6
Iso-propanethiol	1.1	8.0
t-Butanethiol	1.1	9.0
n-Propanethiol	1.0	6.5
Methylethylsulfide	0.8	13.4
s-Butanethiol	1.0	9.4
Isobutanethiol	1.1	3.5
Diethylsulfide	1.0	4.2
n-Butanethiol	1.0	7.1
Dimethyldisulfide	1.0	6.9
Diethyldisulfide	1.0	7.0

Table 9.2. Sensitivity of the SCD Normalized to SO$_2$ for
Several Sulfur Compounds

Compound	Response
SO$_2$	1.0
SF$_6$	0.98
CH$_3$SCH$_3$	1.2
C$_2$H$_5$SH	0.72
H$_2$S	1.1

Source: From Benner and Stedman (55).

Unlike other sulfur detectors, the response of the SCD is nearly equimolar for different sulfur compounds. Tables 9.1 and 9.2 list response factors for a variety of sulfur compounds. To within the errors of the experiments, the sensitivity is equimolar. This can be an advantage when the sulfur content of a sample is desired, but not all of the peaks have been identified. A single representative calibration curve is often suitable for a wide range of sulfur compounds.

4.2. Other Chemiluminescent Detectors for GC

4.2.1. Fluorine SCD

A selective detector for reduced sulfur compounds based on the fluorine-induced chemiluminescence reaction has been used in chromatography (58). The effluent from the GC is mixed postcolumn with 5% F$_2$ in helium to form vibrationally excited hydrogen fluoride. Emission is monitored between 660 and 740 nm using optical filters and a red-sensitive PMT.

This detection method does not respond to all sulfur compounds. In general, sulfur-containing organics such as sulfides, thiols, and disulfides are detected. Inorganic sulfur compounds do not chemiluminesce with fluorine, so SO$_2$, COS, H$_2$S, and CS$_2$ give no response. This can be a benefit in some applications, but it can lead to errors if total sulfur is the objective. The selectivity of this detector vs. saturated hydrocarbons is $> 10^6$, but certain alkenes and aromatic hydrocarbons chemiluminesce with fluorine and are detected with the same sensitivity as reduced organosulfur compounds.

Relative sensitivity varies by sulfur compound; some selected sulfur compounds are listed in Table 9.3. Detection limits are in the 20–200 pg range. The response is linear over 3 orders of magnitude.

Table 9.3. Fluorine Chemiluminescence Method: Detection Limits and Relative Response Factors for Sulfur Compounds

Compound	Detection Limit (pg)	Response Factor
Allyl sulfide	24	8.3
Ethanethiol	24	4.6
Ethyl sulfide	46	6.0
Butanethiol	35	4.6
n-Butyl disulfide	73	4.3
Hexanethiol	84	2.5
n-Butyl sulfide	180	1.4
Isoamyl disulfide	200	1.8
Octanethiol	257	1.0

Source: From Nelson et al. (58).

4.2.2. Sulfur Detection by Direct Ozone Reaction

Sulfur-containing organic compounds also react with ozone but produce light in the blue and UV spectral region. Kelly et al. (59) used this CL reaction (with an optical filter to selectively view wavelengths < 400 nm) as a GC detector for reduced sulfur compounds. The detection limit was in the low-ppb range (tens of nanograms) with a selectivity over alkenes of 10^3. Each sulfur compound had a different sensitivity, and oxidized sulfur compounds were not observed. In general, the response decreased in the order $CH_3SH > CH_3SCH_3 > H_2S >$ thiophenes and the calibration was slightly curved. It is likely that SO is being produced during the ozone oxidation of the organic components and that the chemiluminescence is the source of the signal (54).

5. OTHER SULFUR-SELECTIVE DETECTORS

5.1. Atomic Emission Detector

Principle of Operation

The atomic emission detector (AED) is an element-selective detector for GC based on the formation of electronically excited atoms in a atmospheric pressure microwave-excited helium plasma and monitoring of the atomic emission using a spectrometer equipped with a silicon photodiode array (60–66). It is covered in more detail in Chapter 10 of this volume. The AED provides selective detection for C, H, O, N, Cl, Br, S, P, Si, and many other metals and nonmetals. For sulfur-selective detection, emission in the vacuum UV (180.7 nm) is monitored and, with the photodiode array, the response from C, S, and N can be monitored simultaneously. Reagent gases are commonly used with the AED to minimize the formation of soot in the microwave plasma discharge tube (in addition to venting the solvent). A mixture of oxygen and hydrogen was reported to provide optimum peak shape and selectivity for sulfur detection. Response of the AED for sulfur is linear, with a minimum detection for sulfur ($S/N = 2$) of 1 to 1.7 pg S/s, a selectivity of 1.5×10^5 mol S/mol C and a dynamic range of 2×10^4 (60, 64). In principle the response of the AED should be independent of the compounds, and the prospects for elemental formula determination using the AED have been discussed (66).

While the AED is an extremely powerful tool for chromatographers, the detector is expensive and may require highly skilled personnel for operation. In addition, the AED may not be suited for all sulfur measurements (for example, analysis of thiophene in benzene) owing to problems with soot formation in the discharge tube. Thus, the AED may not be the ideal instrument for all sulfur-selective analysis, but it is a valuable GC detector for research applications.

5.2. Electron Capture Detector

Principle of Operation

The electron capture detector (ECD) is highly sensitive for a number of electronegative elements, including sulfur. The basic principles and operation of the ECD are covered elsewhere in this volume (Chapter 5). Its use in measuring sulfur compounds deserves mention here, however. Reduced sulfur compounds can be detected at the ppbv (parts per billion by volume) level using an ECD. The optimum detector temperature and sensitivity varies by compound. Compared to an FPD, the ECD was as sensitive for disulfides, less sensitive for mercaptans, and much less sensitive for the others. A variation on the ECD has been reported that can detect reduced sulfur species in GC effluents at the femtomole level (67). This extraordinary sensitivity is obtained by fluorination of the reduced sulfur compounds with a heated AgF_2 catalyst. The resulting SF_6 is fairly inert and has an extremely low minimum detectability on an ECD. Differences in sensitivity are due mostly to the relative conversion efficiency of the catalytic fluorinator.

6. APPLICATIONS

Applications for sulfur-selective detection cover a wide array of industrial, environmental, and other samples. The selection of which sulfur-selective GC detector to use will depend upon several factors including sample matrix, desired sensitivity and selectivity, cost, and the desire to detect other elements. A comparison of the sensitivity, selectivity, and dynamic range of the six major sulfur detectors discussed above is given in Table 9.4. The values listed are general ranges and, as already noted, the sensitivity of some detectors depends on the compound being analyzed. Two detectors, the ECD and the fluorine–sulfur chemiluminescence detector (F_2–SCD) have responses to other compounds that are as large as the response to sulfur compounds. Finally, the dynamic range of the detectors is reported using square root linearization for the FPD. All the detectors listed in Table 9.4 have found practical applications in GC analyses. Some of these are discussed below.

In the petroleum and petrochemical industries important applications include the determination of trace sulfur content in process feedstocks and petroleum products such as the measurement of carbonyl sulfide in propylene (8); measurement of odorant concentrations in natural gas and LPG (68); and determination of individual sulfur compounds, total sulfur content, and boiling range distributions in crude oils, distillation cuts, and other refinery products (69–72). Applications in the chemical and pharmaceutical industries include determination of sulfur compounds in plastics (73) and detection of sulfur-containing amino acids (74) and other natural products.

Another example of sulfur-selective detection in the petroleum industry is the determination of sulfur compounds in gasoline. Current and pending state and federal regulations limit the amount of sulfur-containing compounds that can be present in automotive fuels. Figures 9.6 and 9.7 show the analysis of a commercial gasoline using the SCD. Since the SCD uses a flame ionization detector (FID) for the production of sulfur monoxide, both sulfur and hydro-

Table 9.4. Comparison of Sulfur-Selective Detectors for GC

	Sensitivity (g S/s)	Selectivity (g S/g C)	Linear Dynamic Range
FPD	5×10^{-12}–5×10^{-11}	6×10^4–1×10^5	1×10^2–5×10^2
SCD	$< 5 \times 10^{-12}$	$> 10^6$	10^4 to 10^5
F_2–SCD	1–5×10^{-12}	Variable	10^4
ELCD	2×10^{-12}–5×10^{-11}	10^4 to $> 10^6$	10^4
ECD	7×10^{-11}–5×10^{-12}	Variable	10^4
AED	1×10^{-12}–2×10^{-12}	1×10^5	2×10^4

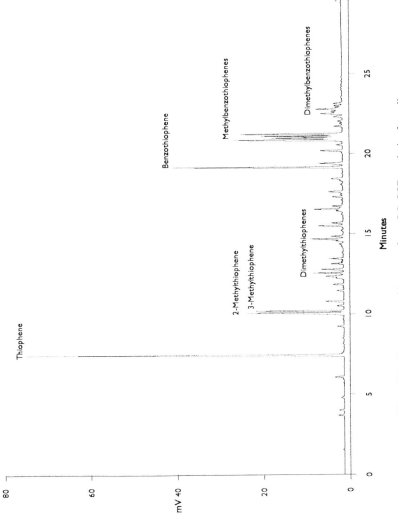

Figure 9.6. Chromatographic output from GC–SCD analysis of gasoline.

213

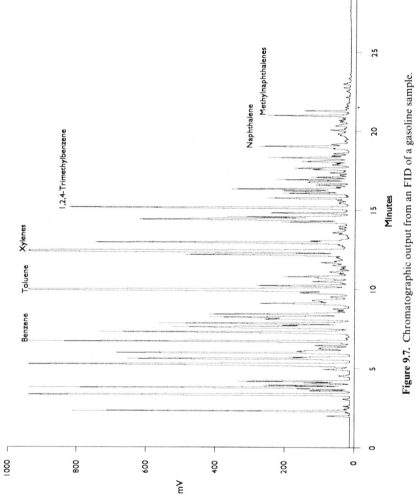

Figure 9.7. Chromatographic output from an FID of a gasoline sample.

carbon chromatograms are obtained from a single injection. The sample (1 μL of gasoline split 100:1) was analyzed using a 30-m $\times$ 0.32-mm-i.d. SPB-1 fused silica capillary column with a 4-μm film thickness. The column temperature program was 1 min at 35 °C to 300 °C at 10 °C/min, with the injection port operated at 250 °C and the FID detector base operated at 300 °C. The FID response shows the typical hydrocarbon components of gasoline, with aromatic, aliphatic, and alkene hydrocarbons. The SCD response shows the presence of thiophene, C_1- and C_2-alkyl thiophenes, benzothiophene, C_1- and C_2-alkyl benzothiophenes, and other thiols, sulfides, and disulfides. The equimolar response of the SCD also permits determination of the total sulfur content of this fuel by summing the concentrations of the individual components, $\sim$ 200 mg S/L for this particular sample.

Applications of sulfur-selective detection in the foods, flavors and beverage industries span the range from sulfur compounds in beer, wine, coffee, citrus products, and carbonated beverages to volatile sulfur compounds in dairy products, essential oils, and artificial flavors.

Environmental applications for sulfur-selective detection include the determination of pesticides and pesticide residues and decomposition products in foods, water, and soils (75, 76) and the measurement of atmospheric sulfur gases using the FPD (77). Two of the common chemical warfare agents, mustard gas and the nerve agent VX, are sulfur-containing compounds and sensitive detection of these compounds is performed using the GC detectors described above (78, 79).

Biogenically produced dimethyl sulfide (DMS) has been proposed as a regulator of global climate by the production of fine sulfuric acid particles that change the albedo of clouds (80). The principal source of DMS is thought to be phytoplankton in the oceans. Direct measurement of DMS at a few parts per trillion by volume (pptv) is not possible; thus some method of preconcentration from up to 100 L of air is used before GC. Both FPD and ECD detectors are currently being used with either packed columns (81) or capillary columns (82). The preconcentration step is critical to this analysis. Cryogenic trapping in a cooled loop or trapping on gold wool at ambient temperature has been widely used (83, 84). The detection limit of a GC–FPD with the hydrogen doped with SF_6 is a few picograms of sulfur in the concentrated 1-L air sample. With a capillary column the equivalent DMS measured is in the low-pptv range.

These methods all suffer from problems due to the preconcentration step. Ozone reacts with sulfur species to oxidize them, and water interferes with the analysis. None of the techniques is close to being a real-time measurement, and the multiple handling steps require diligence on the part of the operator. The SCD offers advantages in detection limit, linearity, and universal response. We may well see wider use of the SCD in this field in the future.

REFERENCES

1. H. Draeger and B. Draeger, Ger. Patent 1,133,918 (1962).
2. W. L. Crider, *Anal. Chem.* **37**, 1770 (1965).
3. R. M. Dagnall, K. C. Thompson, and T. S. West, *Analyst* **92**, 506 (1967).
4. A. Syty and J. A. Dean, *Appl. Opt.* **7**, 1331 (1968).
5. S. S. Brody and J. E. Chaney, *J. Gas Chromatogr.* **4**, 42 (1966).
6. W. A. Aue and C. G. Flinn, *J. Chromatogr.* **158**, 161 (1978).
7. S. O. Farwell and C. J. Barinaga, *J. Chromatogr. Sci.* **24**, 483 (1988).
8. M. Dressler, *Selective Gas Chromatography Detectors*, p. 133. Elsevier Science Publishing, Amsterdam, 1986.
9. S. O. Farwell and R. A. Rasmussen, *J. Chromatogr. Sci.* **14**, 224 (1978).
10. L. S. Ettre, *J. Chromatogr. Sci.* **16**, 396 (1974).
11. S. O. Farwell, D. R. Gage, and R. A. Kagel, *J. Chromatogr. Sci.* **19**, 358 (1981).
12. H. V. Drushel, *J. Chromatogr. Sci.* **21**, 375 (1983).
13. J. Sevcik, *Detectors in Gas Chromatography*, p. 145. Elsevier, Amsterdam, 1976.
14. C. H. Muller, III, K. Schofield, M. Steinberg, and H. P. Broida, *Symp. Int. Combust.* [*Proc.*] **17**, 867 (1979).
15. T. M. Sugden, E. M. Bulewicz, and A. Demerdache, *Chemical Reactions in the Lower and Upper Atmosphere*, p. 99. Wiley, New York, 1862.
16. T. Sugiyama, Y. Suzuki, and T. Takeuchi, *J. Chromatogr.* **77**, 309 (1973).
17. T. Sugiyama, Y. Suzuki, and T. Takeuchi, *J. Chromatogr.* **80**, 61 (1973).
18. C. H. Muller, III, K. Schofield, and M. Steinberg, *ACS Symp. Ser.* **134**, 103 (1980).
19. C. F. Cullis and M. F. R. Mulcahy, *Combust. Flame* **18**, 225 (1972).
20. P. T. Gilbert, in *Nonmetals in Analytical Flame Spectroscopy* (R. Mavrodineau, ed.), p. 281. Springer-Verlag, New York, 1970.
21. C. A. Burgett and L. E. Green, *J. Chromatogr. Sci.* **12**, 356 (1974).
22. W. L. Crider and R. W. Slater, Jr., *Anal. Chem.* **41**, 531 (1969).
23. P. L. Patterson, R. L. Howe, and A. Abu-Shumays, *Anal. Chem.* **50**, 339 (1978).
24. P. L. Patterson, *Anal. Chem.* **50**, 345 (1978).
25. W. E. Rupprecht and T. R. Phillips, *Anal. Chim. Acta.* **47**, 439 (1969).
26. H. A. Moye, *Anal. Chem.* **41**, 1717 (1969).
27. T. J. Cardwell and P. J. Marriott, *J. Chromatogr. Sci.* **20**, 83 (1982).
28. K. K. Gaines, W. H. Chatham, and S. O. Farwell, *J. High Resolut Chromatogr.* **13**, 489 (1990).
29. B. Wenzel and R. L. Aiken, *J. Chromatogr. Sci.* **17**, 503 (1976).
30. C. H. Burnett, D. F. Adams, and S. O. Farwell, *J. Chromatogr. Sci.* **15**, 230 (1977).
31. M. Maruyama and M. Kakemoto, *J. Chromatogr. Sci.* **16**, 1 (1978).
32. A. Attar, R. Forgey, J. Horn, and W. H. Corcoran, *J. Chromatogr. Sci.* **15**, 222 (1977).

33. J. M. Zehner and R. A. Simonaitis, *J. Chromatogr. Sci.* **14**, 348 (1976).

34. E. Mangani, F. Bruner, and N. Penna, *Anal. Chem.* **55**, 2193 (1983).

35. H. W. Grice, M. L. Yates, and D. J. David, *J. Chromatogr. Sci.* **8**, 90 (1970).

36. D. A. Ferguson and L. A. Luke, *Chromatographia* **12**, 197 (1979).

37. J. Efer, T. Maurer, and W. Engewald, *Chromatographia* **29**, 115 (1990).

38. G. H. Liu and P. R. Fu, *Chromatographia* **27**, 159 (1989).

39. C. J. Barinaga and S. O. Farwell, *HRC & CC, J. High Resolut. Chromatogr. Chromatogr. Commun.* **9**, 388 (1986).

40. R. C. Hall, *J. Chromatogr. Sci.* **12**, 152 (1974).

41. S. Gluck, *J. Chromatogr. Sci.* **20**, 103 (1982).

42. B. J. Ehrlich, R. C. Hall, R. Anderson, and H. G. Cox, *J. Chromatogr. Sci.* **19**, 245 (1981).

43. R. G. Schiller and R. B. Bronsky, *J. Chromatogr. Sci.* **15**, 541 (1977).

44. R. S. Hutte, R. E. Sievers, and J. W. Birks, *J. Chromatogr. Sci.* **24**, 499 (1986).

45. D. H. Stedman and M. E. Fraser, in *Chemi- and Bioluminescence* J. G. Burr, ed.), p. 439. New York, 1985.

46. J. W. Birks, ed., *Chemiluminescence and Photochemical Detection in Chromatography*. VCH Publishers, New York, 1989.

47. C. W. Nederbragt, A van der Horst, and T. van Duijn, *Nature (London)* **206**, 87 (1965).

48. B. J. Finlayson, J. N. Pitts, Jr., and R. Atkinson, *J. Am. Chem. Soc.* **26**, 5356 (1974).

49. J. D. Ray, D. H. Stedman, and G. J. Wendel, *Anal. Chem.* **58**, 598 (1985).

50. W. Bruening and F. J. M. Concha, *J. Chromatogr.* **112**, 253 (1975).

51. S. A. Nyarady, R. M. Barkley, and R. E. Sievers, *Anal. Chem.* **57**, 2074 (1985).

52. R. E. Sievers, S. A. Nyarady, R. L. Shearer, J. J. DeAngelis, R. M. Barkley, and R. S. Hutte, *J. Chromatogr.* **349**, 395 (1985).

53. C. J. Halstead and B. A. Thrush, *Proc. R. Soc. London* **295**, 380 (1966).

54. H. Akimoto, B. J. Finlayson, and J. N. Pitts, Jr., *Chem. Phys. Lett.* **12**, 199 (1971).

55. R. L. Benner and B. H. Stedman, *Anal. Chem.* **61**, 1268 (1989).

56. R. L. Benner and D. H. Stedman, *Environ. Sci. Technol.* **24**, 1592 (1990).

57. R. L. Shearer, D. L. O'Neal, R. Rios, and M. D. Baker, *J. Chromatogr. Sci.* **28**, 24 (1990).

58. J. K. Nelson, R. H. Getty, and J. W. Birks, *Anal. Chem.* **55**, 1767 (1983).

59. T. J. Kelly, J. S. Gaffney, M. F. Phillips, and R. L. Tanner, *Anal. Chem.* **55**, 138 (1983).

60. B. D. Quimby and J. J. Sullivan, *Anal. Chem.* **62**, 1027 (1991).

61. J. J. Sullivan and B. D. Quimby. *Anal. Chem.* **62**, 1034 (1990).

62. L. Ebdon, S. Hill, and R. W. Ward, *Analyst* **111**, 1113 (1986).

63. E. A. Estes, P. C. Uden, and R. M. Barnes, *Anal. Chem.* **53**, 1829 (1981).

64. P. L. Wylie and B. D. Quimby, *J. High Resolut. Chromatogr.* **12**, 813 (1989).

65. P. C. Uden, *Chromatogr. Forum* **1** (4), 17 (1986).

66. J. J. Sullivan and B. D. Quimby, *J. High Resolut. Chromatogr.* **12**, 283 (1989).

67. J. E. Johnson and J. E. Lovelock, *Anal. Chem.* **60**, 812 (1988).

68. K. J. Rygle, G. P. Feulmer, and R. F. Scheidemann, *J. Chromatogr. Sci.* **22**, 514 (1984).

69. C. Bradley and D. J. Schiller, *Anal. Chem.* **58**, 3017 (1986).

70. I. Dzidic, M. D. Balicki, I. A. L. Rhodes, and H. V. Hart, *J. Chromatogr. Sci.* **26**, 236 (1988).

71. J. L. Buteyn and J. J. Kosman, *J. Chromatogr. Sci.* **28**, 19 (1990).

72. T. R. McManus, *Anal. Chem.* **63**, 48R (1991).

73. C. G. Smith, R. A. Nyquist, P. B. Smith, A. J. Pasztor, and S. J. Martin, *Anal. Chem.* **63**, 11R (1991).

74. S. I. Bonvell and R. H. Monheimer, *J. Chromatogr. Sci.* **18**, 18 (1980).

75. W. A. Aue, *J. Chromatogr.* **13**, 329 (1971).

76. P. L. Wylie and R. Oguchi, *J. Chromatogr.* **517**, 131 (1990).

77. S. O. Farwell and R. A. Rasmussen, *J. Chromatogr. Sci.* **14**, 224 (1976).

78. M. W. Ellzy, F. E. Ferguson, and L. G. Janes, *U.S. Army CRDEC Rep.* **CRDEC-TR-88168**, Order No. AD-A200908 (1988).

79. R. E. Sievers and R. S. Hutte, in *Proceedings of Scientific Conference on Chemical Defense Research*, Aberdeen Proving Grounds, Aberdeen, MD, 1990.

80. R. J. Charlson, J. E. Lovelock, M. O. Andreae, and S. G. Warren, *Nature (London)* **326**, 655 (1987).

81. E. S. Saltzman and D. J. Cooper, *J. Atmos. Chem.* **7**, 191 (1988).

82. P. D. Goldan, W. C. Kuster, D. L. Albritton, and F. C. Fehsenfeld, *J. Atmos. Chem.* **439**, 467 (1987).

83. M. O. Andreae, H. Berresheim, T. W. Andreae, M. A. Kritz, T. S. Bates, and J. T. Merrill, *J. Atmos. Chem.* **6**, 149 (1988).

84. H. Berresheim, M. O. Andreae, G. P. Ayers, R. W. Gillett, J. T. Merrill, V. J. Harris, and W. L. Chameides, *J. Atmos. Chem.* **10**, 341 (1990).

CHAPTER

10

ATOMIC PLASMA ELEMENT-SPECIFIC DETECTION

PETER C. UDEN

Department of Chemistry
University of Massachusetts at Amherst
Amherst, Massachusetts

High-resolution analytical chromatography, particularly capillary column instrumentation and methodology, places great demands upon the devices that are utilized as eluate detectors. This is especially so when sophisticated characterization detectors are employed rather than simpler monitoring devices. Present-day instrumentation relies heavily on the conceptual integration and physical interfacing of what may seem, at first sight, to be quite dissimilar and independent techniques.

Chromatographic procedures have as their basis the transformation of a complex multicomponent matrix into a time-resolved separated analyte stream, observable in the analog differential signal mode. The chromatographic sample is thus distinctive in that analytes are changing in nature and with time. In high-resolution chromatographies, these changes occur rapidly and thus more rigorous demands are placed on the analytical detection system than for other types of sample input. It is also likely that sample component levels will be reduced in capillary chromatography, placing an additional sensitivity burden upon the detection device. There can be two quite different perspectives on an instrument that interfaces the chromatograph with a complex "sample characterization device" like the atomic emission spectrometer. From the chromatographer's viewpoint the spectrometer is a sophisticated chromatographic detector, but from the spectroscopist's point of view the chromatograph is a component-resolving sample introduction device. Both observations are valid; emphasizing that this mode of chemical analysis must involve optimization of both the separation and the detection process, as well as the interface between them.

Detectors for Capillary Chromatography, edited by Herbert H. Hill and Dennis G. McMinn.
Chemical Analysis Series, Vol. 121.
ISBN 0-471-50645-1 © 1992 John Wiley & Sons, Inc.

1. THE SELECTIVE CHROMATOGRAPHIC DETECTOR

A detection device for qualitative and quantitative determination of the components resolved by the column should respond immediately and predictably to the presence of solute in the mobile phase. One group of detectors, the bulk property detectors, respond to changes produced by eluates in a characteristic property of the mobile phase itself. A second group, the solute property detectors, measure some physicochemical property of the eluates themselves directly.

An important class of solute property detectors comprises those giving selective or, under optimal circumstances, specific information on the eluates. Spectral property detectors such as the mass spectrometer, the infrared (IR) spectrophotometer, and the atomic emission spectrometer fall in this class. Such detectors may be element selective, structure or functionality selective, or property selective (1).

1.1. The Element-Selective Detector

The major objectives of element-selective chromatographic detection are to acquire qualitative and quantitative determination of eluates, frequently in *interfering* background matrixes, by virtue of their elemental constitution. Further, just as in classical elemental microanalysis, simultaneous multi-element detection can enable empirical formulas of eluates to be determined. Element-selective gas chromatographic (GC) detectors in common use include the alkali flame ionization detector (AFID), often known as the nitrogen–phosphorus detector (NPD), selective for these elements; the flame photometric detector (FPD), selective for sulfur and phosphorus, and the Hall electrolytic conductivity detector (ELCD), for halogen, nitrogen, and sulfur. The widespread adoption of these detectors shows the value of element-selective detection, but they are too limited for general empirical formula determinations. In high-performance liquid chromatography (HPLC) and supercritical fluid chromatography (SFC), no element-specific detectors are in general use.

Multielement chromatographic detection is a worthwhile technical objective to complement molecular and structural specific detection by interfaced mass spectroscopy and Fourier transform IR spectroscopy. Atomic emission spectroscopy is a natural choice for such interfaced detection in view of its capacity to monitor all elements. The rebirth, during the past decade, of analytical atomic emission spectroscopy, particularly using plasma excitation sources, has refocused the efforts of chromatographers to employ its capacity in on-line detection.

2. ATOMIC EMISSION SPECTROSCOPIC DETECTION (AESD) IN CHROMATOGRAPHY

Element-selective chromatographic detection by interfaced atomic spectroscopy has involved primarily atomic absorption (AAS) and atomic plasma emission (APES), with flame emission (FES) and atomic fluorescence (AFS) being used to a smaller extent (2, 3). APES has the greatest capabilities for simultaneous multielement measurement, coupled with wide dynamic measurement ranges, good sensitivities, and selectivities over background elements. The major advantages of interfaced chromatography–atomic plasma emission spectroscopy (C–APES) are as follows: (a) monitoring eluted species directly for their elemental composition with high elemental sensitivity; (b) measurement of the selected element with high selectivity over coeluting elements; (c) toleration of nonideal chromatography and incomplete chromatographic resolution from complex matrices; (d) detection of molecular functionality through derivatization with element-tagged reagents; and (e) simultaneous multielement detection to make empirical and molecular formula determination.

2.1. Classes of Atomic Plasma Emission Chromatographic Detectors

Emission spectral excitation sources utilize various energy sources to transform solid, liquid, or gaseous samples into energetic plasmas of electrons, atoms, ions, and radicals. When the excited states of atoms and ions deactivate they generate light quanta to give an elemental emission spectrum. The principal plasma emission sources that have been used for GC detection have been the microwave-induced and sustained helium or argon plasma (MIP), operated at atmospheric or reduced pressure, and the direct-current argon plasma (DCP). The inductively coupled argon plasma (ICP) has been little used for GC, but it and the DCP have been used effectively as HPLC detectors. Alternating-current and capacitively coupled radio-frequency (RF) plasmas have been developed, as have detectors utilizing discharges in helium, nitrogen, or other gases. Their major characteristics are given in the following subsections.

2.1.1. The Microwave-Induced Electrical Discharge Plasma Detector

The principal atom–ion reservoir plasma excitation systems used for GC detection have been microwave-induced electrical discharge plasmas (MIPs).

An argon or helium plasma is initiated and sustained in a microwave cavity that serves to focus power from a microwave source, typically at 2.45 GHz, into a discharge cell made of quartz or some other refractory material. Such microwave plasmas may be operated at atmospheric or reduced pressures in various cavities (4, 5). Power levels for analytical microwave plasmas are usually much lower, $\sim$ 50–100 W, than for the DCP or the ICP, but power densities are similar owing to the smaller size of the MIP. Plasma temperatures are not as high as in some other plasmas, but high electron temperatures are available, notably in the helium plasma, producing intense spectral emission for many elements, including nonmetals, which respond poorly in the argon ICP or DCP. MIP efficiency depends on the discharge cavities and waveguides, metal tubes that transfer power from a microwave generator to the plasma support gas. An interruption in the waveguide causes total reflection of energy traveling along it, setting up standing waves and forming a resonant cavity. Risby and Talmi compared microwave cavities in their general review of GC–MIP (6).

The most widely used cavity for reduced pressure helium or argon plasmas has been the 3/4 wave cavity described by Fehsenfeld et al. (7), but the cavities that have been most widely adopted for GC-MIP are based upon the TM_{010} cylindrical resonance cavity developed by Beenakker (8) (Figure 10.1), which can sustain argon or helium atmospheric pressure discharges at relatively low power levels.

In the Beenakker cavity, light emitted is viewed axially, and not transversely through cavity walls whose properties change with time. The discharge tube may be made from opaque materials such as alumina or boron nitride,

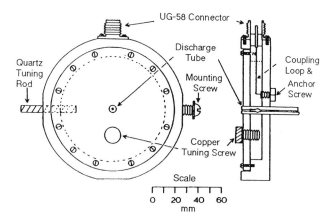

Figure 10.1. The Beenakker microwave plasma cavity. Reproduced by permission from S. A. Estes, P. C. Uden, and R. M. Barnes, *J. Chromatogr.* **239**, 181 (1982). Copyright 1982, Elsevier Science Publishers, Physical Sciences & Engineering Division.

which have been used effectively in GC–MIP. The instrumental advantages of atmospheric pressure operation greatly simplify GC detection, and the cavity configuration allows capillary GC columns to be terminated a few millimeters from the plasma, minimizing any degradation of chromatographic peak efficiency in the interface.

Another class of atmospheric pressure microwave plasma cavity that has been used successfully in GC–MIP application is the Surfatron, which can sustain a discharge over a wide pressure range. It operates by surface microwave propagation along a plasma column that may be viewed axially or transversely since it extends outside of the plasma structure (9).

2.1.2. The Direct-Current Argon Plasma (DCP) Discharge

The DCP is an electrical discharge sustained by a continuous d.c. arc and stabilized by flowing inert gas (10). For chromatographic detection, a cathode jet is placed above two symmetrically placed anode jets in an inverted Y configuration (11). Flowing argon causes vortices around the anodes, and a thermal pinch gives an arc column having a power level between 500 and 700 W at an operating potential of 40–50 V. Solutions are introduced from a nebulizer–spray chamber, or vapor-phase samples are directly channeled, into the junction of the two columns; there analyte spectral emission is observed from the exhaust plume of the discharge, which constitutes the excitation source. The DCP has been interfaced with both HPLC and GC. Since, unlike the MIP, the DCP is not constrained in a discharge tube, additional makeup argon must be introduced annularly around the chromatographic column effluent stream to direct it into the excitation region without loss of chromatographic efficiency.

2.1.3. The Inductively Coupled Argon Plasma Discharge

The ICP is the most widely used general analytical emission spectrochemical source (12), the discharge resulting from interaction of an RF field (usually 27 or 41 MHz) on argon flowing through a quartz tube within a copper coil. The RF field creates a varying magnetic field in the gas, which in turn generates a circulating eddy current in the heated argon. A stable, spectrally intense plasma discharge is produced with temperatures as great as 9000 K. Liquid samples are introduced through a nebulizer–spray chamber from which aerosol is carried by the argon into the discharge where solvent is evaporated and the analyte atomized. The ICP is a natural complement for liquid chromatography, and HPLC–ICP procedures have been quite widely adopted. It has gained less prominence as a GC detector but may be useful for those elements, mainly metals, whose sensitivity is high in the argon plasma discharge. A major problem is in maintaining capillary column efficiency during introduction of eluate into the large volume plasma region.

2.1.4. Other Atomic Plasma Chromatographic Detectors

A 60-Hz alternating-current helium plasma (ACP) has been employed as a GC detector. It acts as a stable, self-maintaining emission source that requires no external initiation and does not extinguish under high solvent loads (13).

Atmospheric pressure capacitively coupled plasma (CCP) emission discharges have been developed as sources for atomic spectroscopy (14, 15), and reduction in volume of this plasma has allowed its use for capillary GC without appreciable band broadening. It can sustain a stable plasma over a wide range of input power (10–500 W) and frequencies (200 kHz to 30 MHz) and carrier gas flow rates down to 20 mL/min (16). A series of studies on low-frequency, high-voltage electrodeless discharges sustained in argon, nitrogen, and helium showed analytically useful emission from the atmospheric pressure afterglow region for GC detection. The helium system has been the most extensively developed because the metastable energy carriers have the capability of the highest collisional energy transfer and thus the best ability to excite other elements to emission (17). Since chromatographic effluents are introduced directly into the afterglow discharge region, extinguishment and contamination of the primary discharge are eliminated. A stabilized capacitive plasma (SCP) operating with a RF at 27.12 MHz and with up to 200 W of RF energy has been shown to exhibit long-term stability and minor matrix interferences (18).

2.2. Analytical Information from Chromatography–Atomic Emission Detection (C–AED)

A number of different capabilities of plasma–AED make it a valuable tool for elemental speciation in a wide variety of samples. Complex analytes such as environmental, petrochemical or biological materials have many components that complicate chromatography. Unresolved interferences that may be at much greater levels than the targeted analyte may make it impossible to quantify or even to identify the eluate. Element-selective atomic plasma emission detection can reduce or eliminate such interferences.

2.2.1. Interelement Selectivity

The ability to detect the target element without interference from signals of other elements present in the plasma is most important. Selectivity depends on emission properties of the element and of possible interferences, and on the resolution of the spectroscopic measurement system. Some ultraviolet (UV)–visible spectral regions are less prone than others to interference due to emission from the plasma background or from line or band spectra of carbon, nitrogen, oxygen, or molecular combinations. A useful measure of interelement selectivity, at the measured emission wavelength, defines it as the peak area response per mole of analyte element divided by the peak area response of the background element per mole of that element. Selectivity against carbon is most frequently determined, but other elemental background matrices necessitate separate selectivity criteria. Since selectivities vary greatly among elements, between plasmas and with instrumental conditions, calibration is necessary.

2.2.2.   Elemental Sensitivity and Limits of Detection

The sensitivity for an element in the AED depends on its spectral intensity at the measured wavelength. Each element has many possible wavelengths for determination, and the best must be chosen for sensitivity and selectivity since lines exhibit different sensitivities in different plasmas. Sensitivity, defined by the slope of the response curve, is less often used in C–AED than detection limits, expressed as absolute values of element mass (in a resolved peak) or in units of mass flow rate. The latter measurement allows direct comparison with other mass-flow-rate–sensitive detectors. Detection limits for different elements differ by 2 or 3 orders of magnitude, and this affects interelement selectivity if spectral overlap is present.

2.2.3.   Dynamic Measurement Range

Linear dynamic ranges of response in capillary GC–AED typically extend from the upper linear analyte-carrying capacity of the columns, around 100 ng, to the detection limit of the target element (1–100 pg). Chemical, gas dopant, and plasma–wall interaction effects modify the limits.

Table 10.1 summarizes elemental detection limits, selectivities, and linear dynamic ranges for atmospheric pressure microwave-induced helium plasma capillary GC detectors.

2.2.4.   Simultaneous Multielement Detection

Multielement monitoring is accomplished in various ways. Rapid sequential switching between elemental wavelengths of a monochromator is useful if timing is compatible with peak elution rates (19), but this is more difficult for rapidly eluting capillary peaks. The most widely used multielement detection has been by a direct reading polychromator that can display up

Table 10.1. Selected Detection Limits and Selectivities for Atmospheric Pressure Helium Microwave Plasma GC Detection

Element	Wavelength (nm)	Detection Limit [pg/s (pg)]		Selectivity vs. C	Dynamic Range
Carbon (b)	247.9	2.7	(12)	1	1,000
Hydrogen (b)	656.3	7.5	(22)	160	500
Deuterium (b)	656.1	7.4	(20)	194	500
Boron (b)	249.8	3.6	(27)	9,300	500
Chlorine (a)	479.5	40		3,000	10,000
Bromine (a)	478.6	60		2,000	1,000
Fluorine (a)	685.6	50		20,000	2,000
Sulfur (a)	180.7	2		8,000	10,000
Phosphorus (a)	177.5	1		5,000	1,000
Silicon (b)	251.6	9.3	(18)	1,600	500
Oxygen (a)	777.2	120		10,000	5,000
Nitrogen (a)	174.2	50		2,000	20,000
Germanium (b)	265.1	1.3	(3.9)	7,600	1,000
Tin (b)	284.0	1.6	(6.1)	36,000	1,000
Arsenic (b)	228.8	6.5	(155)	47,000	500
Iron (b)	259.9	0.3	(0.9)	180,000	1,000
Lead (b)	283.3	0.17	(0.71)	25,000	1,000
Mercury (b)	253.7	0.6	(60)	77,000	1,000

Sources: Compiled from (a) Firor (25) and (b) Estes et al. (27).

to 12 monitoring wavelengths simultaneously (5), used with either reduced pressure (20) or atmospheric pressure MIP detection (21, 22). Diode array detection has shown considerable versatility and sensitivity (23), and a commercial GC instrument has been developed (24, 25) incorporating an atmospheric pressure helium MIP with a water-cooled discharge tube to maximize signal to background signals, and a movable photodiode array detector measuring from 170 to 780 nm. Simultaneous detection of up to four elements and display of element-specific chromatograms is possible. The array range is approximately 25 nm, which determines what combinations of elements can be measured in a single chromatogram. Some analytical figures of merit of this instrument are included in Table 10.1.

Among the valuable features of multielement GC–AED detection is the ability to carry out quantitative ratioing of elements to determine the empirical formulas of eluates. Such determinations do not achieve the accuracy of conventional microanalysis, and it must be supposed that there is no dependence of response upon molecular structure, a factor that has been questioned. However, measurements may be obtained on eluant peaks at sample levels up

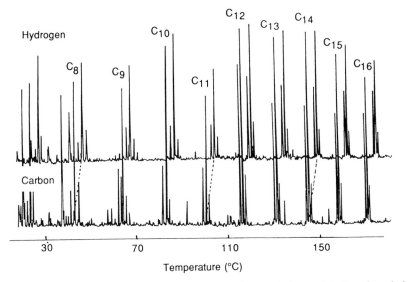

Figure 10.2. Section of GC pyrograms of simultaneous carbon and hydrogen detection of pyrolysis products of HDPE. Pyrolysis conditions: 20 s at 800 °C for ∼ 500-mg sample. Column: 25-m DB5 fused silica capillary. Temperature: 7 min isothermal at 30 °C, programmed at 4 °C/min to 240 °C. Reproduced by permission from H. J. Perpall, P. C. Uden, and R. L. Deming, *Spectrochim. Acta* **42B**, 243 (1987). Copyright 1987, Pergamon Press.

to 6 orders of magnitude below classical methods. An approach to obtain higher precision and accuracy uses a multireferencing method employing pyrolysis products from standard polymers to provide a reproducible reference range; results with relative errors less than 1% are usually obtained (22). Figure 10.2 is such a chromatogram from the pyrolysis of high-density polyethylene (HDPE), showing the well-documented pattern corresponding to a mechanism in which homolytic chain scission is followed by simultaneous saturation and unsaturation at the site of the scission, giving an α,ω-diolefin, α-olefin, and *n*-paraffin. This triplet pattern formation is noted at carbon numbers up to about 36 in this case, and some representative data are shown in Table 10.2.

3. PLASMA SAMPLE INTRODUCTION FROM CAPILLARY GAS CHROMATOGRAPHS

Since eluant from GC columns is at atmospheric pressure, simpler interfacing configurations are possible with atmospheric pressure plasmas than for

Table 10.2. Overall Mean Molecular Formulas for C_8–C_{20} Alkenes Calculated from Replicate Pyrograms of HDPE

Molecular Formula	Mean Calculated Molecular Formula	S/X^a (%)	Relative Error (%)
C_8H_{16}	$C_8H_{13.7}$	5.2	2.1
C_9H_{18}	$C_9H_{18.8}$	2.4	4.4
$C_{10}H_{20}$	$C_{10}H_{20.9}$	2.0	4.5
$C_{11}H_{22}$	$C_{11}H_{22.9}$	2.7	4.1
$C_{12}H_{24}$	$C_{12}H_{24.8}$	2.2	3.3
$C_{13}H_{26}$	$C_{13}H_{26.4}$	1.4	1.5
$C_{14}H_{28}$	$C_{14}H_{28.8}$	2.2	2.9
$C_{15}H_{30}$	$C_{15}H_{30.4}$	4.3	1.3
$C_{16}H_{32}$	$C_{16}H_{31.9}$	2.0	0.3
$C_{17}H_{34}$	$C_{17}H_{33.7}$	3.0	0.9
$C_{18}H_{36}$	$C_{18}H_{38.2}$	2.9	0.5
$C_{19}H_{38}$	$C_{19}H_{37.8}$	1.5	0.5
$C_{20}H_{40}$	$C_{20}H_{39.0}$	0.8	2.5

Source: From Perpall et al. (22).

$^a S/X$ = Relative standard deviation of the mean.

reduced pressure plasmas. Interfacing of reduced pressure MIPs involves evacuating a silica sample chamber contained within the MIP cavity to a pressure of ~ 1 Torr (26). With packed columns, little degradation in peak efficiency is evident, but the volume of the cavity leads to some degradation of capillary peak efficiency. The atmospheric pressure cavities such as the TM_{010} are simple to interface with capillary GC columns since the latter can be terminated within a few millimeters of the plasma, giving minimal transfer volume (27). Heating is needed to prevent analyte condensation along the interface. Helium makeup gas or other reactant gases can be introduced within the transfer line to optimize plasma performance and minimize peak broadening. A typical GC–MIP interface is illustrated in Figure 10.3 (28).

Improvement in the performance of the GC–MIP has been obtained with a threaded tangential flow torch (TFT) (29, 30) to give a self-centering plasma that can give enhanced emission and better stability. The plasma loses relatively little energy to the walls; thus atom formation and excitation appear to be enhanced by comparison with the straight capillary torch. A disadvantage, however, is the high volume (liters per minute) of helium flow gas required. Detection limits of 30 pg/s for carbon, 150 pg/s for iodine, and

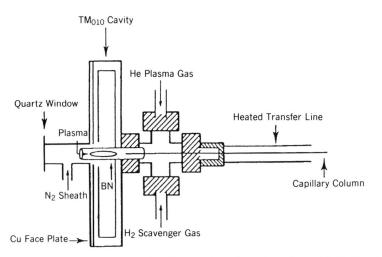

Figure 10.3. GC–MIP interface. Reproduced by permission from C. Bradley and J. W. Carnahan, *Anal. Chem.* **60**, 858 (1988). Copyright 1988, American Chemical Society.

300 pg/s for phosphorus were obtained with linear calibration over some 3–4 decades. Alternative torch designs have also been investigated.

The MIPs have found much greater use in GC than in HPLC interfacing, although the application of the direct injection nebulizer (DIN) for microbore column effluent flow rates may expand the latter's potential (31).

4. ANALYTICAL CAPILLARY GC APPLICATIONS

Carbon selective detection may be considered as a universal mode of detection for organic compounds. The most frequently used emission wavelength is from the carbon ion line at 247.9 nm, but other spectral emission features corresponding to CN, CH, etc. may be used (32). This AED mode response is analogous to flame ionization but is more completely independent of the carbon atom environment and exhibits as great or greater sensitivity (5). Figure 10.4 shows carbon selective and lead selective responses by atmospheric pressure MIP for a leaded gasoline, obtained under identical chromatographic conditions (33).

4.1. GC–AED Detection of Nonmetallic Elements

The helium MIPs have been used most often for nonmetals detection, since for many of these elements argon metastable energy carriers show insufficient

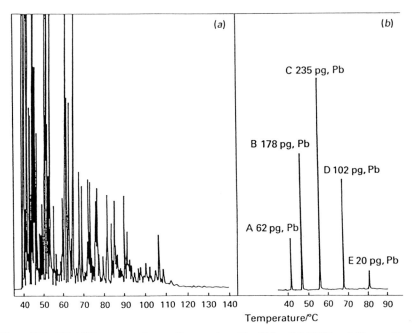

Figure 10.4. GC–MIP responses for gasoline samples with a 12.5-m SP-2100 fused silica capillary column, an inlet split of 1:100, and a temperature program from 40 to 100 °C at 5 °C/min: (a) Carbon-specific response for 0.10 mL at 247.86 nm; (b) lead-specific response at 283.3 nm. Peak designations: **A**, tetramethyllead; **B**, trimethylethyllead; **C**, dimethyldiethyllead; **D**, methyltriethyllead, **E**, tetraethyllead. Reproduced by permission from P. C. Uden, *Anal. Proc.* **18**, 189 (1981). Copyright 1981, Royal Society of Chemistry.

collision energy transfer for adequate excitation. The argon MIP, DCP, and ICP have shown some utility, however, for elements such as fluorine, phosphorous, and silicon.

4.1.1.   GC–MIP Detection

Reduced Pressure Plasmas. The analytical performance of a multichannel instrument utilizing a polychromator configuration with Rowland Circle optics was reported by Brenner (20). With a helium plasma at 0.5–3.0 Torr and oxygen or nitrogen scavenger gas, detection limits ranged from 0.02 ng/s for bromine to 4 ng/s for oxygen. Reproducibility of 95–98% and linearity of 3–4 decades were seen, and selectivities against carbon obtained with background correction were in the range 500–1000 for a number of elements. Hagen et al. (34) used elemental derivatizing reagents such as chlorofluoroacetic anhydride to tag the analyte molecules, thereby permitting detection of acylated amines through their fluorine and chlorine content. Developments by Zeng et al., with a similar reduced pressure system, focused upon oxygen-specific detection (35). Highly purified plasma gases, and rigorous exclusion of air, improved the limit of detection to 0.3 ng/s with a linear dynamic range of 3 decades. Oxygen-specific analysis is becoming more important with the increasing use of oxygenates in fuel oils, and oxygen fingerprints of such materials have potential in environmental analyses. The absence until recently of effective oxygen-specific detectors for GC has emphasized the value of such analyses.

Complementary information on elemental composition of capillary GC eluates has been derived from MIP emission and mass spectral data (36). GC effluent was split inside the oven with part passing to a quadrupole mass spectral detector and part to a spectrometer having a low-pressure Evenson cavity and a 0.75-m monochromator. Output was monitored for C, N, H, F, S, Cl, and O from molecules of differing structure in the molecular weight range from 200 to 400. Emission responses were calibrated from a small set of molecules. If an accurate mass was available, the molecular formula could be calculated using a subset of element ratios, but for low-resolution mass spectrometry (MS) all possible measured elemental ratios were required for the determination. H/C and other element ratios were almost always found to be structurally independent. With high-resolution MS and multielement AED data, the number of possible formulas fitting the data for larger molecules could be drastically reduced. Figure 10.5 shows six individual GC–MIP responses obtained in this study for the compounds noted.

Sklarew et al. (37) observed that although the low-pressure Evenson source provides the greatest excitation energy for nonmetals, its sensitivity is inherently less than that of the Beenakker cavity because of its transverse viewing geometry. The atmospheric pressure Beenakker cavity provides a more efficient way to couple microwave energy into the plasma, and its axial geometry provides inherently more sensitivity; however, its high pressure causes tube erosion, at least if the discharge tube is not cooled, thus creating reliability problems. Sklarew and co-workers devised a modified Beenakker cavity operating at reduced pressure, 20 Torr, which was effective for sulfur-specific detection, and empirical formula determination in complex matrices such as an oil shale retort off-gas.

Olsen et al. (38) compared reduced pressure and atmospheric pressure MIP systems for Hg, Se, and As detection in shale oil matrices and found the latter system to be better both for detection limits and for selectivity. Figure 10.6 shows comparison chromatograms for low pressure and atmospheric pressure helium GC–MIPs of dialkylmercury compounds in a shale oil matrix.

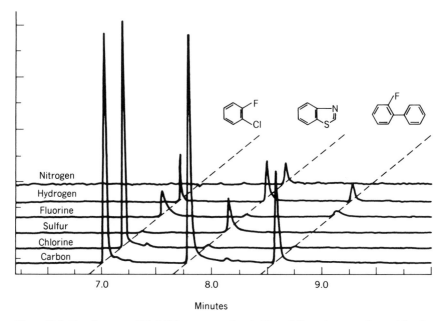

Figure 10.5. Simultaneous GC–MIP responses for 1-chloro-2-fluorobenzene, benzothiazole, and 2-fluorobiphenyl with the low-pressure Evenson cavity. Reproduced by permission from D. B. Hooker and J. DeZwann, *Anal. Chem.* **61**, 2207 (1989). Copyright 1989, American Chemical Society.

Atmospheric Pressure Plasmas. Improved transfer efficiency of microwave power to the plasma discharge by means of such cavity structures as the Beenakker TM_{010} allows plasmas to be maintained at atmospheric pressure at power levels similar to those possible for reduced pressure cavities. Another advantage is the fact that light emitted from the plasma can be viewed axially, rather than transversely through a quartz discharge tube wall subject to variations in spectral throughput because of contamination and corrosion from extended use.

Flexible fused silica open tubular (FSOT) columns can be interfaced to within a few millimeters of the plasma, and such direct interfaces have been widely used in capillary GC–MIP, although there are some advantages in incorporating a gas switching device to introduce additive gases or to purge segments of the chromatographic eluates. A system of the latter kind, incorporating a deactivated valveless fluidic logic device facilitating solvent venting and addition of dopant gas, was used successfully to determine chemically active and thermally sensitive trialkyl lead chlorides at the sub-nanogram level (39).

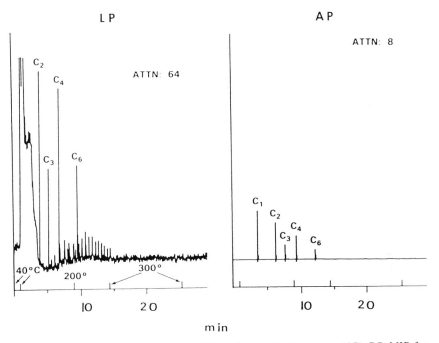

Figure 10.6. Comparison of low-pressure (LP) and atmospheric pressure (AP) GC–MIP for dialkylmercury compounds in a shale oil matrix. Mercury-specific detection at 253.6 nm: C_1, C_2,... refer to dimethylmercury, diethylmercury, etc. Reproduced by permission from K. B. Olsen, D. S. Sklarew, and J. C. Evans, *Spectrochim. Acta* **40B**, 357 (1985) Copyright 1985, Pergamon Press.

A number of developments show the versatility of GC–MIP research and applications; widespread adoption of the technique for more routine use is now feasible with the introduction of the generally available instrumental system mentioned earlier (24, 25). Initial reports have focused upon instrument design features, notably quantitative and qualitative aspects of C, H, N, and O detection including sensitivity, precision, peak shapes induced by the plasma, and response variation between compounds. Precision ranged from 2% (C, H, N) to 7% (O). Valid element ratios for O and N are reported and empirical formulas calculated with an accuracy to one atom in most cases. Figure 10.7 shows a cutaway view of the microwave cavity used in this system.

Muller and Cammann used a helium TM_{010} MIP in combination with oscillating narrow band interference filters, their system having the capacity for multielement detection by using several filters and a beam-splitting light-guide optic (40). Their data is exemplified in Figure 10.8, which shows

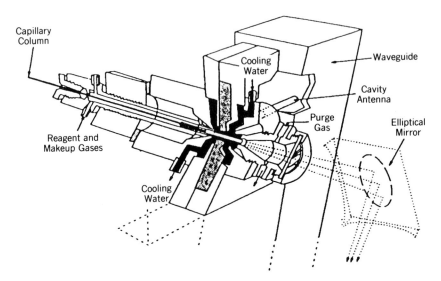

Figure 10.7. Cutaway view of cooled microwave cavity. Reproduced by permission from J. J. Sullivan and B. D. Quimby, *J. High. Resolut. Chromatogr.* **12**(5), 282 (1989). Copyright 1989, Dr. Alfred Huethig Publishers.

simultaneous bromine, chlorine, and carbon detection of compounds at around the microgram level.

A rapid-scanning spectrometer has also been applied for multielement detection, molar elemental response for carbon, chlorine, and bromine being found to be independent of molecular structure despite the low power (50–60 W) used (19).

For halogen detection in haloorganics, the advantage of the MIP over the other principal GC detectors used, the electron capture detector (ECD) and the Hall electrolytic conductivity detector (ELCD), are noteworthy. Although the MIP does not possess the extreme sensitivity of the ECD for polyhalogenated compounds, it has the advantage of virtually uniform response to element content for each halogen, independent of analyte molecular structure. A major advantage over the ELCD is the element specificity for individual halogens, which is not possible in the halogen-specific mode of that detector. An instructive halogen-specific application of GC–MIP operated with glass capillary columns was in the examination of chlorinated products of natural humic materials in drinking water sources (41). Such humic substances—i.e., amorphous, hydrophilic, acidic polydisperse substances with molecular weights of several hundreds to tens of thousands derived from soil, vegetation, and algae—are found in most natural freshwater sources. Their structures consist

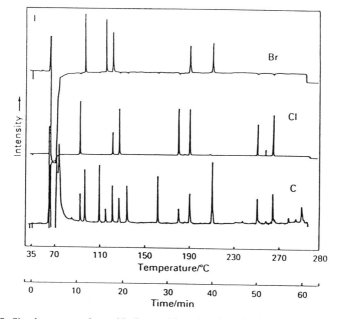

Figure 10.8. Simultaneous carbon, chlorine, and bromine detection by GC–MIP using oscillating interference filters. Reproduced by permission from H. Muller and K. Cammann, *J. Anal. At. Spectrom.* **3**, 907 (1988). Copyright 1988, Royal Society of Chemistry.

of aromatic polyhydroxy, polymethoxy, and polycarboxylic acid structures, with smaller amounts of carbohydrates, nitrogen bases, and nucleotide residues. The range of potential products formed upon chlorination during water treatment procedures is large, and Figure 10.9 shows an example of analytical GC–MIP analysis. The most distinctive feature of the upper chromatogram of a nonmethylated chlorinated humic acid extract is the broad chlorine-containing band designated **2**. Upon methylation with diazomethane, peak **2** in the lower chromatogram is produced and shown to be methyltrichloracetate, derived from trichloracetic acid. The amount of chlorine detected in the upper chromatographic band emphasizes the relative abundance of this chlorination product.

As noted earlier, among the capabilities of multichannel AED is its capacity for quantitative determination of element ratios to obtain empirical formulas of chromatographic eluates. An example of such a determination is shown for the chromatographic separation of methylated fulvic acid chlorination products in Figure 10.10 (42). Simultaneous carbon, hydrogen, and chlorine detection is displayed, methyltrichloroacetate (D in Figure 10.10) being assigned

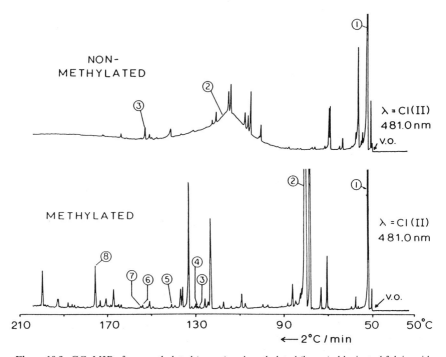

Figure 10.9. GC–MIP of nonmethylated (upper) and methylated (lower) chlorinated fulvic acid extracts. Peak identities: upper, chloroform (**1**), trichloracetic acid (**2**), and 1-chlorophenol (**3**); lower, chloroform (**1**), methyltrichloracetate (**2**), 2,4,6-trichlorophenol methyl ether (**3**), methyl-2-chlorobenzoate (**4**), methyl-3,5-dichlorobenzoate (**5**), 1-chlorophenol (**6**), 2-chlorophenol (**7**), and pentachlorophenol methyl ether (**8**). Reproduced by permission from B. D. Quimby, M. F. Delaney, P. C. Uden, and R. M. Barnes, *Anal. Chem.* **52**, 259 (1980). Copyright 1980, American Chemical Society.

by retention time standardization as a reference to compute empirical formulas of other major components. Unknowns were identified as chloroform (A in Figure 10.10), trichloracetaldehyde (B), and methyldichloracetate (C) from the stoichiometries listed in Table 10.3. Such determinations do not attain the accuracy of those obtained from conventional microanalysis, but measurements are made directly on GC eluant peaks at sample levels up to 6 orders of magnitude smaller!

Oxygen-selective detection with an oxygen-to-carbon selectivity of 10^2 and a linearity of 3 decades has been reported by Bradley and Carnahan (43) with a TM_{010} cavity in a polychromator system. Background oxygen spectral emission from plasma gas impurities, leaks or back-diffusion into the plasma was minimized to give oxygen sensitivities between 2 and 500 ppm in complex

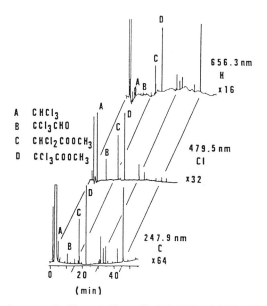

Figure 10.10. Simultaneous C-, H-, and Cl-specific GC–MIP of fulvic acid chlorination and methylation products. Reproduced by permission from P. C. Uden, K. J. Slatkavitz, R. M. Barnes, and R. L. Deming, *Anal. Chim. Acta* **180**, 401 (1986). Copyright 1986, Elsevier Science Publishers, Physical Sciences & Engineering Division.

petroleum distillates. The selective detection of phenols in a light coal liquid distillate is seen in Figure 10.11.

MIPs produced by Surfatron structures have been used as element-selective detectors for capillary GC (9, 23). Several parameters such as line selection, pressure, input power, helium flow rate, tube configuration, and the role of added amounts of nitrogen have been studied to optimize determinations of

Table 10.3. Empirical and Molecular Formula Determinations of Chlorination Products of Fulvic Acid by GC–MIP

Compound	Formula	RSD[a] (%)	
		H	Cl
$C_3H_3Cl_3O_2$ (methyltrichloroacetate)	Reference		
$C_3H_4Cl_2O_2$ (methyldichloroacetate)	$C_3H_{3.9}Cl_{2.0}O_x$	1.8	3.6
$C_2H_1Cl_3O_1$ (chloral)	$C_2H_{0.9}Cl_{2.9}O_y$	8.7	1.1
$C_1H_1Cl_3$ (chloroform)	$C_1H_{1.1}Cl_{2.9}$	11.1	0.4

Source: From Uden et al. (42).

[a] RSD = relative standard deviation.

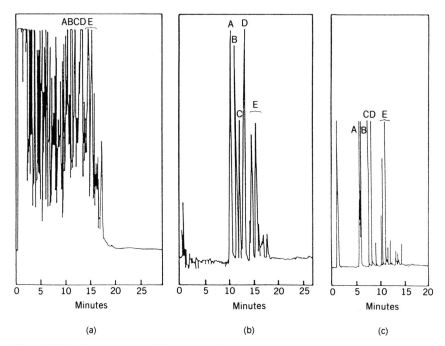

Figure 10.11. Chromatograms of light coal distillate: (a) carbon MIP; (b) oxygen MIP; (c) FID trace of phenolics concentrate of the same distillate. Peaks. phenol (**A**), o-chlorophenol (**B**), o-cresol (**C**), m- and p-cresols (**D**), C_2-phenols (**E**). Temperature program: GC–MIP, 30 °C (1-min hold) to 250 °C at 10 °C min; CG–FID, 50–250 °C at 10 °C min. Reproduced by permission from C. Bradley and J. W. Carnahan, *Anal. Chem.* **60**, 858 (1988). Copyright 1988, American Chemical Society.

phosphorus, sulfur, chlorine, and bromine in pesticides (44). With nonspecific detection, the detection limit of carbon is in the range of 2–10 pg/s. With specific detection, detection limits of elements present in compounds are between 3 and 60 pg/s. In contrast with other elements, signals obtained for sulfur seemed to depend on the chemical formulas of the pesticides. This study also compared results obtained over a pressure range from 20 to 760 Torr, concluding that the best results were obtained at 50 Torr.

A Fourier transform (FT) spectrometer has been coupled with a Surfatron MIP for nonmetals GC detection, taking advantage of multielement determination through its multiwavelength capability (45). Figure 10.12 shows spectrochromatograms for the separation and detection of two isomers, isoflurane ($CF_3CHClOCHF_2$) and enfurane (CHF_2OCF_2CHFCl), covering the range 680–1080 nm, to permit simultaneous detection of fluorine, oxygen, chlorine, and carbon. The near-IR region was selected because of high halogen sensitivity,

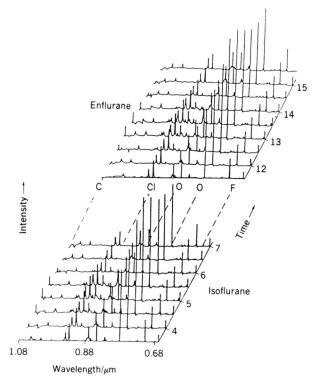

Figure 10.12. GC–Surfatron MIP-FTIR spectrochromatograms of isoflurane and enflurane. Reproduced by permission from C. Lauzon, K. C. Tran, and J. Hubert, *J. Anal. At. Spectrom.* **3**, 901 (1988). Copyright 1988, Royal Society of Chemistry.

simplicity of background, and absence of interferences. Acquisition time per interferogram was 0.58 s for spectral resolution of 16 cm^{-1}. Detection limits were from 1 to 3 orders of magnitude poorer than with single element detection with a dispersive system, but methods to gain improved performance are noted.

4.1.2. Radio-Frequency and Afterglow Plasma GC Detectors

Rice et al. developed a high-voltage, electrodeless discharge-afterglow sustained in helium at atmospheric pressure (17). Incident power was 45 W at a frequency of 26–27 kHz. Elements determined were F, Cl, Br, I, C, P, S, Si, Hg, and As, with limits of detection from 0.5 to 50 pg, linear ranges from 1 to 7×10^3, and selectivities from 20 (As) to 7000 (Hg). Skelton et al. have described an RF plasma detector (RPD) for sulfur-selective capillary GC of fossil fuels (46). Sulfur selectivity at 921 nm, greater than 10^3, was obtained in petroleum distillates and coal extracts. Oxygen was doped in at $< 0.05\%$. Power levels were 50–80 W at about 335 kHz, and the detection limit was 0.5 pg/s with 4 decades linear range. Substituent phenyl groups on the thiophene rings had

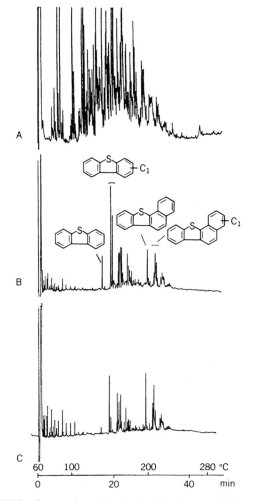

Figure 10.13. GC–RPD of aromatic and isolated sulfur heterocyclic fractions of an Upper Freeport coal extract: (A) H detection at 656 nm of aromatic fraction; (B) S detection at 921 nm of the aromatic fraction; (C) S detection of the isolated sulfur fraction. Reproduced by permission from R. J. Skelton, H.-C. K. Chang, P. B. Farnsworth, E. K. Markides, and M. L. Lee, *Anal. Chem.* **61**, 2292 (1989). Copyright 1989, American Chemical Society.

minimal effect on sulfur signal intensity, and hydrocarbon coelution affected sulfur response only at very high levels. Figure 10.13 depicts RPD detection for hydrogen and sulfur in aromatic and isolated heterocyclic fractions of a coal extract.

4.2. GC–AED Detection of Metals

GC applications for metallic compounds are less common than for nonmetals, but many volatile organometallic and metal chelate compounds can be quantitatively gas chromatographed (47). GC–AED detection methods are valuable in confirming elution and acquiring analytical data, since metal elemental detection is usually more sensitive and selective than for nonmetals.

4.2.1. GC–MIP Detection

As shown in Table 10.1 (see Section 2.2.3, above), GC–MIP data have been obtained for many transition and main group metals. Lead and mercury in particular have been the subject of a number of studies as they are determinable by GC–MIP with TM_{010} cavities at sub-picogram per second detection limits. An example of an environmental analysis is shown in Figure 10.14, which compares lead- and carbon-specific detection for trialkyllead chlorides extracted from an industrial plant effluant and reacted with butyl Grignard reagent to form their trialkylbutyllead derivatives (48). The extent of chromatographic interference from the high level of carbon-containing compounds prevents any qualitative or quantitative determination of the trialkyllead compounds by GC–FID, GC–ECD, or GC–MS without extensive cleanup and loss of analyte. In their comparison study on reduced and atmospheric pressure MIP systems, Olsen et al. found for the latter system a 1-pg detection limit for mercury, with selectivity over carbon of 10,000 (38). GC–MIP of volatile elemental hydrides of germanium, selenium, and tin gave sub-nanogram detection (49), and there is considerable potential for the determination of these elements in environmental matrices.

The study of metal chelates having adequate volatility and thermal stability for GC has been considered over the past 20 years, with most emphasis being upon complexing ligands 2,4-pentanedione (acetylacetone) and its analogues; this area has recently been comprehensively reviewed (47). An example of the utility of GC–MIP detection was in a study of ligand redistribution and reaction kinetics of gallium, indium, and aluminum chelates (50). Figure 10.15 shows a capillary GC study of redistribution of trifluoroacetylacetone (TFA; 1,1,1-trifluoropentane-2,4-dione) and trifluoroisobutanoyl acetone (TIB); 1,1,1-trifluoro-6-methylheptane-2,4-dione) ligands on gallium and aluminum atom centers. In this study $Al(TFA)_3$ and $Ga(TIM)_3$ were reacted in chloroform

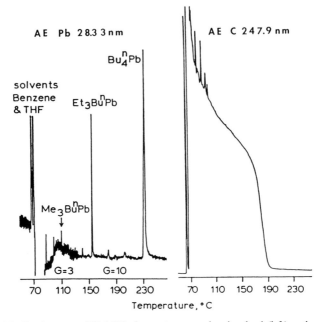

Figure 10.14. Simultaneous GC–MIP chromatograms showing lead (left) and carbon (right) detection of industrial plant water effluents derivatized as tributylalkylleads. (THF = tetrahydrofuran.) Reproduced by permission from S. A. Estes, P. C. Uden, and R. M. Barnes, *Anal. Chem.* **54**, 2402 (1982). Copyright 1982, American Chemical Society.

solution for 24 h at 25 °C and the products examined with carbon-, aluminum-, and gallium-specific GC detection. Individual peaks were identified as (1) $Al(TFA)_2(TIB)$; (2) $Ga(TFA)_2(TIB)$; (3) $Al(TFA)(TIB)_2$; (4) $Ga(TFA)(TIB)_2$; (5) $Al(TIB)_3$; and (6) $Ga(TIB)_3$. Kinetics and other physicochemical data can be obtained for these reactions, and the ability of GC–MIP to monitor such metals as Al, Ga, and In gives possibilities for determination of these elements in volatile organometallic compounds used in semiconductor and electronic materials production.

Many pi-bonded organometallics such as metallocenes are well behaved in capillary GC and are excellent model compounds for evaluation of GC–MIP detection for their constituent elements. Excellent detection of iron, cobalt, nickel, chromium, manganese, and vanadium in a series of cyclopentadienyl carbonyl/nitrosyl compounds was obtained, verifying elution of some previously unchromatographed compounds (51). Figure 10.16 presents chromatograms with carbon, chromium, and manganese element-specific detection. The identities of the eluted peaks in the universal carbon mode (left) are (a) $C_5H_5Mn(CO)_3$; (b) $CH_3C_5H_4Mn(CO)_3$; (c) $C_5H_5Cr(NO)(CO)_2(C_5H_5)_2Ni$

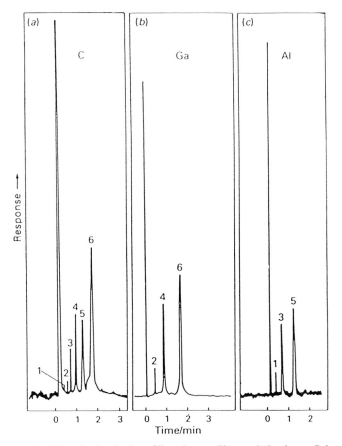

Figure 10.15. GC–AED of redistribution of ligands on gallium and aluminum. Column 5-m ×
0.2-mm-i.d. FSOT SE 30 capillary at 108 °C: (a) carbon detection at 247.9 nm; (b) gallium
detection at 294.3 nm; (c) aluminum detection at 396.1 nm. Reproduced by permission from
P. C. Uden and T. Wang, *J. Anal. At. Spectrom.* **3**, 919 (1988). Copyright 1988, Royal Society
of Chemistry.

(unresolved); (**d**) $C_5H_5V(CO)_4$; (**e**) $(C_5H_5)_2Fe$; and (**f**) $C_5H_55(CH_3)_5Co(CO)_2$.
On the right are shown manganese and chromium-specific chromatograms.
The slight responses noted for other compounds in these latter chromatograms
are due to inadequate selectivity among the different metals at the emission
lines measured with the low-resolution spectrometer used and are not indicative of the presence of these metals in other components of the mixture. Note
that the manganese organometallics detected have been measured as gasoline
additives by GC–AED using a DCP system (52), and GC–MIP could also be
used if solvent removal were carried out following sample injection.

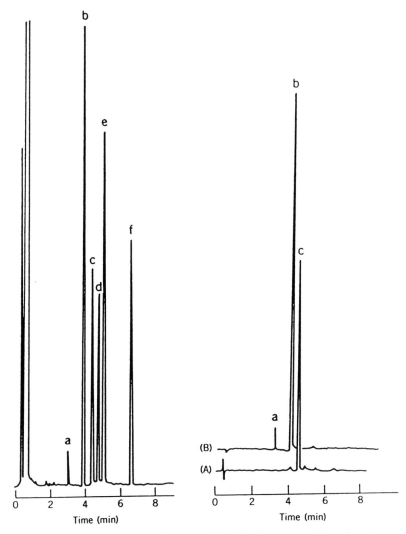

Figure 10.16. Atmospheric pressure microwave GC–AED of organometallics: left, carbon monitored at 247.9 nm; right (A) chromium monitored at 267.7 nm, and (B) manganese monitored at 257.6 nm. Reproduced by permission from S. A. Estes, P. C. Uden, M. D. Rausch, and R. M. Barnes, *HRC & CC, J. High. Resolut. Chromatogr. Chromatogr. Commun.* **3**, 471 (1980). Copyright 1980, Dr. Alfred Huethig Publishers.

An example of trace metal GC–MIP analysis, utilizing derivatization for organic functional group determination, involves the reaction of ferroceneboronic acid with diols (53), the chromatographic separation of pinacol and dicyclohexyl-1,1′-diol ferrocene boronates being monitored by iron- and boron-specific detection. Tailing noted on the boron peak is probably due to deposition of boron on the plasma tube walls, rather than chromatographic effects, since it does not appear in the iron trace.

4.2.2. GC–DCP Detection

The argon DCP system has been used to detect nonmetallic elements such as boron and silicon present in GC derivatizing groups. For silicon, the absence of interfering spectral response from the quartz discharge tube often used in the MIP is an added advantage. A selectivity of silicon over carbon of 2×10^6 with a detection limit of 25 pg/s has been reported (11). The 3-pg/s detection limit reported for boron was similar to that in the MIP. The nonconstrained design of the three-electrode DCP makes necessary a careful design of annular makeup argon flow to surround and entrain capillary GC effluant, to avoid undesirable postcolumn band broadening (11).

4.2.3. GC–ICP Detection

Despite its popularity as a spectroanalytical emission source, the ICP has received little attention as a GC detector, in contrast to its wide adoption in HPLC monitoring. In the first evaluation of GC–ICP performance, a packed column was interfaced to a demountable ICP torch through a T-shaped tube that enabled makeup argon to be added (54). A 0.35-m scanning monochromator and a 1.5-m, 0.02-nm-resolution multichannel direct reading spectrometer were used to determine Br, Cl, F, I, H, Si, and C. Detection limits for Si, C, and H were at the nanogram level, but limits for F, Cl, and Br were at or above the microgram level. Selectivities against carbon were 100 or lower, but in contrast to the reduced pressure MIP, oxygen or nitrogen did not have to be added to the plasma to reduce deposits. Oxygen-specific GC–ICP detection was investigated using near-IR emission. The major drawback to capillary GC–ICP lies in the necessity to introduce small capillary gas flows into ICP gas streams many orders of magnitude greater, thereby severely limiting detectability.

5. PLASMA EMISSION DETECTION FOR SUPERCRITICAL FLUID CHROMATOGRAPHY (SFC)

Although analytical SFC was demonstrated in the early 1960s, only in recent years has the availability of high-resolution packed and capillary SFC columns and instrumentation led to renewed interest. High-resolution SFC along with supercritical fluid extraction (SFE) may allow separations in areas where neither GC or HPLC may be possible. Adoption of detectors for SFC has proceeded in two main directions. Where methodology and instrumentation have derived from GC, the flame ionization detector (FID) has been favored. For development related more to HPLC, the UV–visible spectrophotometric detector has been adopted. Plasma emission is a natural development because of its use in GC and HPLC.

A Surfatron MIP sustained in helium was employed for SFC detection, giving sulfur-specific detection at 921.3 nm with a 25 pg/s limit for thiophene (55). Extensive spectral characterization was carried out in this system for two common SFC mobile phases: carbon dioxide and nitrous oxide (56). A RF plasma detector (RPD) has been evaluated for capillary SFC detection, measuring sulfur and chlorine at 921.3 and 837.6 nm, respectively, in the near-IR region. Minimal spectral interference was observed and detection limits were in the range of 50–300 pg/s. Figure 10.17 shows an RPD chlorine-specific chromatogram of chlordane and impurities (57). Since modification of plasma excitation by SFC solvents appears to be less troublesome than for typical organic HPLC solvents, it seems likely that as SFC becomes more widely adopted element-specific detection by atomic plasma emission will become a useful option.

6. LIQUID CHROMATOGRAPHIC APPLICATIONS

6.1. HPLC–MIP Detection

HPLC–plasma interfacing has developed most with the high-powered DCP and ICP argon plasmas, which are able to tolerate mobile phase solvents at the flow rates used in conventional packed column chromatographic procedures. The low-powered helium MIP, however, cannot be directly interfaced to standard packed HPLC columns, since the discharge will be quenched by continuously introduced milliliters per minute liquid flow streams. The much smaller liquid mobile phase flow in capillary HPLC suggests that MIP may have some potential for detection. The incentive for developing a viable capillary HPLC–helium MIP interface is considerable because of the possibility of monitoring nonmetallic elemental effluents; this is difficult or impossible with argon DCP and ICP systems. The direct injection nebulizer (DIN) (31) may provide one answer. Another involves the removal of the solvent, and it is possible that the Thermospray and Particle-beam approaches now used in HPLC–MS might be adapted; the interfacing problems for lower-powered MIPs are parallel to those in HPLC–MS and HPLC–FTIR. Another possibility is cryo-focusing, as used in the latter technique. Capillary HPLC columns with mobile phase flow rates of a few milliliters per minute provide an interesting possibility for helium MIP interfacing, but sample capacity may limit application for trace determinations.

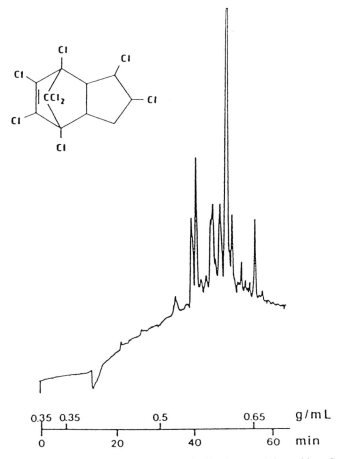

Figure 10.17. Supercritical fluid chromatogram of chlordane and impurities. Conditions: 2.5-m × 50-mm-i.d. biphenyl capillary column; CO_2 at $100\,°C$; chlorine detection at 837.6 nm. Reproduced by permission from R. J. Skelton, Jr., P. B. Farnsworth, K. E. Markides, and M. L. Lee, *Anal. Chem.* **61**, 1815 (1989). Copyright 1989, American Chemical Society.

7. FUTURE DIRECTIONS FOR AED–CHROMATOGRAPHIC DETECTION

The wider adoption of plasma spectral detection will depend on the introduction of standardized commercial instrumentation to permit interlaboratory comparisons of data and the development of recommended methods of analysis. Plasma chromatographic detection has already demonstrated a wide utility in academic, governmental, and industrial laboratories, and the recent commercial introduction of an integrated GC–MIP system suggests that the future of this technique is strong. Fully integrated units that circumvent the need for analysts to interface their own chromatograph, emission device, and spectrometer may become as familiar in the future as GC–MS and GC–FTIR systems are today. Integrated HPLC and SFC systems will be longer delayed, but their eventual adoption is probable in view of the broad scope of these analytical separation methods.

REFERENCES

1. P. C. Uden, in *Quantitative Analysis using Chromatographic Techniques* (E. Katz, ed.), p. 99. Chichester and New York, 1987.

2. L. Ebdon, S. Hill, and R. W. Ward, *Analyst* **111**, 1113 (1986).

3. L. Ebdon, S. Hill, and R. W. Ward, *Analyst* **112**, 1 (1987).

4. C. A. Bache and D. J. Lisk, *Anal. chem.* **39**, 786 (1967).

5. W. R. McLean, D. L. Stanton, and G. E. Penketh, *Analyst* **98**, 432 (1973).

6. T. H. Risby and Y. Talmi, *CRC Crit. Rev. Anal. Chem.* **14**, 231 (1983).

7. F. C. Fehsenfeld, K. M. Evenson, H. P. Broida, *Rev. Sci. Instrum.* **36**(3), 294 (1965).

8. C. I. M. Beenakker, *Spectrochim. Acta* **31B**, 483 (1976).

9. H. H. Abdellah, S. Coulombe, and J. M. Mermet, *Spectrochim. Acta* **37B**, 583 (1982).

10. R. J. Decker, *Spectrochim. Acta* **35B**, 19 (1980).

11. J. O. Beyer, Ph.D. dissertation, University of Massachusetts, Amherst, 1984.

12. R. M. Barnes, *CRC Crit. Rev. Anal. Chem.* **7**, 203 (1978).

13. R. B. Costanzo and E. F. Barry, *Anal. Chem.* **60**, 826 (1988).

14. B. M. Patel, E. Heithmar, and J. D. Winefordner, *Anal. Chem.* **59**, 2374 (1987).

15. D. C. Liang and M. W. Blades, *Anal. Chem.* **60**, 27 (1988).

16. D. Huang, D. C. Liang, and M. W. Blades, *J. Anal. At. Spectrom.* **4**, 789 (1989).

17. G. W. Rice, A. P. D'Silva, and V. A. Fassel, *Spectrochim. Acta* **40B**, 1573 (1985).

18. G. Knapp, E. Leitner, M. Michaelis, B. Platzer, and A. Schalk, *Int. J. Environ. Anal. Chem.* **38**, 369 (1990).

19. M. Zerezghi, K. J. Mulligan, and J. A. Caruso, *J. Chromatogr. Sci.* **22**, 348 (1984).

20. K. S. Brenner, *J. Chromatogr.* **167**, 365 (1978).

21. K. J. Slatkavitz, L. D. Hoey, P. C. Uden, and R. M. Barnes, *Anal. Chem.* **57**, 1846 (1985).

22. H. J. Perpall, P. C. Uden, and R. L. Deming, *Spectrochim. Acta* **42B**, 243 (1987).

23. Y. Takigawa, T. Hanai, and J. Hubert, *HRC & CC, J. High. Resolut. Chromatogr. Chromatogr. Commun.* **9**, 698 (1986).

24. J. J. Sullivan and B. D. Quimby, *HRC & CC J. High. Resolut. Chromatogr. Chromatogr. Commun.* **12**(5), 282 (1989).

25. R. L. Firor, *Am. Lab.* **21**(5), 40 (1989).

26. W. J. Hoskin, *The MPD 850 Organic Analyzer*, Appl. Chromatogr. Syst. Ltd., Luton, UK, 1977.

27. S. A. Estes, P. C. Uden, and R. M. Barnes, *Anal. Chem.* **53**, 1829 (1981).

28. C. Bradley and J. W. Carnahan, *Anal. Chem.* **60**, 858 (1988).

29. A. Bollo-Kamara and E. G. Codding, *Spectrochim. Acta* **36B**, 973 (1981).

30. S. R. Goode, B. Chambers, and N. P. Buddin, *Spectrochim. Acta* **40B**, 329 (1985).

31. K. E. LeFreniere, V. A. Fassel, and D. E. Eckel, *Anal. Chem.* **59**, 879 (1986).

32. T. H. Risby and Y. Talmi, *CRC Crot. Rev. Anal. Chem.* **14**, 250 (1983).

33. P. C. Uden, *Anal. Proc.* **18**, 189 (1981).

34. D. F. Hagen, J. S. Marhevka, and L. C. Haddad, *Spectrochim. Acta* **40B**, 335 (1985).

35. K. Zeng, Q. Gu, G. Wang, and W. Yu *Spectrochim. Acta* **40B**, 349 (1985).

36. D. B. Hooker and J. DeZwaan, *Anal. Chem.* **61**, 2207 (1989).

37. D. S. Sklarew, K. B. Olsen, and J. C. Evans, *Chromatographia* **27**, 44 (1989).

38. K. B. Olsen, D. S. Sklarew, and J. C. Evans, *Spectrochim. Acta* **40B**, 357 (1985).

39. S. A. Estes, P. C. Uden, and R. M. Barnes, *Anal. Chem.* **53**, 1336 (1981).

40. H. Muller and K. Cammann, *J. Anal. At. Spectrom.* **3**, 907 (1988).

41. B. D. Quimby, M. F. Delaney, P. C. Uden, and R. M. Barnes, *Anal. Chem.* **52**, 259 (1980).

42. P. C. Uden, K. J. Slatkavitz, R. M. Barnes, and R. L. Deming, *Anal. Chim. Acta.* **180**, 401 (1986).

43. C. Bradley and J. W. Carnahan, *Anal. Chem.* **60**, 858 (1988).

44. B. Riviere, J.-M. Mermet, and D. Derauz, *J. Anal. At. Spectrom.* **2**, 705 (1987).

45. C. Lauzon, K. C. Tran, and J. Hubert, *J. Anal. At. Spectrom.* **3**, 901 (1988).

46. R. J. Skelton, Jr., H.-C. K. Chang, P. B. Farnsworth, K. E. Markides, and M. L. Lee, *Anal. Chem.* **61**, 2292 (1989).

47. P. C. Uden, *J. Chromatogr.* **313**, 3 (1984).

48. S. A. Estes, P. C. Uden, and R. M. Barnes, *Anal. Chem.* **54**, 2402 (1982).

49. R. B. Robbins and J. A. Caruso, *J. Chromatogr. Sci.* **17**, 360 (1979).

50. P. C. Uden and T. Wang, *J. Anal. At. Spectrom.* **3**, 919 (1988).

51. S. A. Estes, P. C. Uden, M. D. Rausch, and R. M. Barnes, *HRC & CC, J. High Resolut. Chromatogr. Chromatogr. Commun.* **3**(9), 471 (1980).

52. P. C. Uden, R. M. Barnes, and F. P. DiSanzo, *Anal. Chem.* **50**, 852 (1978).

53. P. C. Uden, Y. Yoo, T. Wang, and Z. Cheng, *J. Chromatogr.* **468**, 319 (1989).

54. D. L. Windsor and M. B. Denton, *J. Chromatogr. Sci.* **17**, 492 (1979).

55. D. R. Luffer, L. J. Galante, P. A. David, M. Novotny, and G. M. Hieftje, *Anal. Chem.* **60**, 1365 (1988).

56. L. J. Galante, M. Selby, D. R. Luffer, G. M. Hieftje, and M. Novotny, *Anal. Chem.* **60**, 1370 (1988).

57. R. J. Skelton, Jr., P. B. Farnsworth, K. E. Markides, and M. L. Lee, *Anal. Chem.* **61**, 1815 (1989).

CHAPTER

11

THE FOURIER TRANSFORM INFRARED DETECTOR

DONALD F. GURKA

Environmental Monitoring Systems Laboratory
Office of Research & Development
U.S. Environmental Protection Agency
Las Vegas, Nevada

The evolution of gas chromatography (GC) may be viewed as a continuing drive to improve resolution, loading capacity, and detector sensitivity and selectivity. With its superior sensitivity and universal detectability, the mass spectrometer (MS) is sometimes considered the ultimate in GC detectors. However, the MS detector suffers from some serious shortcomings. Paramount among these are (*a*) its destructive nature, (*b*) its inability (with rare exceptions and in the absence of standards) to distinguish easily among isomers and conformers, and (*c*) its inability to provide significant functional group information.

Because infrared (IR) spectrometry is capable of overcoming these shortcomings, it would appear to be an ideal detector for GC effluents. However, traditional IR detectors possess gratings or prisms to disperse the polychromatic radiation from a heated source into its component frequencies (1). This process necessitates a slow translation of the dispersing device across the source beam. Furthermore, energy-wasting slits are required to direct the dispersed beam to the IR detector. The result is an IR technique too slow to scan GC effluents as they emerge from the GC column end and too insensitive for GC trace analysis.

Nevertheless, Welti demonstrated the feasibility of measuring grating-IR spectra of gases, albeit under static sampling conditions (2). He employed a heated cell of 25-mL volume and 9.2-cm length, in conjunction with a Perkin-Elmer Model 257 grating spectrophotometer, to measure the vapor-phase IR

Detectors for Capillary Chromatography, edited by Herbert H. Hill and Dennis G. McMinn.
Chemical Analysis Series, Vol. 121.
ISBN 0-471-50645-1 © 1992 John Wiley & Sons, Inc.

spectra of more than 300 compounds. Despite his success a more sensitive IR detector, with on-line detection capability, was needed for routine GC trace analysis.

1. THE MICHELSON INTERFEROMETER

If light from a source is divided by a suitable apparatus into two beams that are then superimposed, the intensity in the region of superimposition is found to vary from point to point between maxima that exceed the sum of the intensities in the beams and minima that may be zero (3). This phenomenon is called interference, and the apparatus producing this effect is called an interferometer. Although many types of interferometers have been proposed, the Michelson type has seen the greatest usage and will be the sole type discussed here. (See Reference 3 for a discussion of other interferometer types.)

Figure 11.1 shows the basic design (from a top view) of a Michelson interferometer. The system consists of a stationary and a moving mirror at right angles to each other. The mirrors are separated by a beam splitter at 45° to both mirrors, each of which is in the same plane. The splitter is coated with a partially reflective film causing the source beam to be partially reflected to the fixed mirror and partially transmitted to the moving mirror. The reflected and transmitted beams are again reflected and combined at the beam splitter, where the beams are split again with part going back to the source and part going on to the detector (ideally the final split should be 50:50). The path

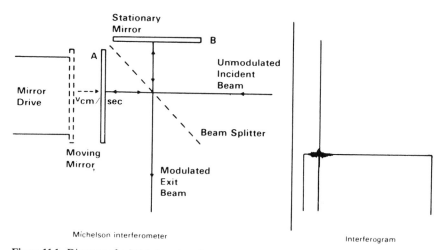

Figure 11.1. Diagram of a Michelson interferometer; right inset—an interferogram. Reprinted with permission from *American Laboratory* (September, 1974). Copyright 1974, American Laboratory.

difference between the fixed and moving mirrors is twice the difference between the mirror–beam splitter distances. This path difference is the retardation and is designated by the symbol δ. At equal mirror distance from the beam splitter (zero retardation) the combined beams are in phase and interfere constructively. A displacement of $\frac{1}{4}\lambda$ (wavelength) by the moving mirror results in a retardation of $\frac{1}{2}\lambda$, and the two beams are then out of phase. As the moving mirror translates, the beams are alternately in (constructive interference) and out (destructive interference) of phase. The output of this phenomenon is an interferogram (see Figure 11.1, right inset), and the maxima and minima grow more intense as the two path beams approach each other in length. Following the nomenclature of Griffiths and de Haseth (4) the intensity of the combined beam, at the detector, as a function of retardation ($I'[\lambda]$) for a monochromatic source is given by

$$I'(\delta) = 0.5I(\bar{v})\,(1 + \cos 2\pi\bar{v}\delta) \tag{1}$$

Because the $0.5I(\bar{v})$ term corresponds to a d.c. component that is unimportant in spectrophotometric measurements, Equation 1 may be reduced to the a.c. component and simplified to

$$I(\delta) = 0.5I(\bar{v})\cos 2\pi\bar{v}\delta \tag{2}$$

Because $I(\delta)$ at the detector is a function of beam splitter efficiency, $\bar{v}$, and the amplifier, Equation 2 should be modified to

$$I(\delta) = B(\bar{v})\cos \pi\delta \tag{3}$$

where $B(\bar{v}) = 0.5H(\bar{v})\,I(\bar{v})$ and $H(\bar{v})$ is a wavenumber-dependent correction factor.

The spectrum may then be obtained from the interferogram by calculating the cosine Fourier transform of $I(\delta)$.

Because IR sources are not monochromatic, Equation 3 must be further modified to account for polychromatic sources. Other factors affecting the final form of Equation 3 are resolution, phase errors, and beam divergence. (For a complete discussion of the effect of these parameters on the interferogram see Reference 4, which is Volume 83 of Wiley's Chemical Analysis monograph series.)

2. RELEVANCE OF THE INTERFEROMETER TO CAPILLARY CHROMATOGRAPHY

The advantages of using interferometry rather than grating IR as a detector for capillary GC are summarized in the following subsections.

2.1. Fellget's Advantage

Because the Fourier transform–infrared (FT–IR) sample beam is not dispersed, the detector receives all frequencies simultaneously. This provides the detection speed required to monitor GC effluents. The signal-to-noise ratio (S/N) of a spectrum measured on a FT–IR spectrometer will be greater than that measured on a grating spectrometer, by a factor equal to the square root of the number of resolution elements (if both spectrometers use the same source and detector, and operate at the same resolution, optical throughput, modulation frequency, optical efficiency, and measurement time). For a spectrum measured from 4000 to 700 cm^{-1}, at 8 cm^{-1} resolution (typical GC/FT–IR operating parameters), this corresponds to an S/N enhancement of

$$\left(\frac{4000 - 700}{8}\right)^{1/2} \quad \text{or} \quad 20.3$$

2.2. Jacquinot's Advantage

Jacquinot's advantage results from the superior optical throughput (especially at high and medium resolution) of interferometers relative to grating systems. Improved sensitivity is obtained, which is vitally needed for GC trace analysis. Following the nomenclature of Griffiths and de Haseth (5), the throughput advantage of FT–IR (ΘI) relative to that of grating (ΘG) IR is given by

$$\frac{\Theta I}{\Theta G} = \frac{2\pi A^I f a \bar{v}^2}{A^G h \bar{v}_{max}}$$

where A^I and A^G are respectively the area of the FT–IR mirror and grating being illuminated; f, a, and h correspond to focal length of the collimating mirror, grating constant, and slit height, respectively. Griffiths et al. have calculated the magnitude of this effect, as a function of frequency, for two commercially available spectrometers (6). At 4000 cm^{-1} the advantage exceeds 250, but drops off to about 10 at 750 cm^{-1}.

If an intermediate advantage value of 50 is used and the total effect of Fellget's and Jacquinot's advantage is assumed to be multiplicative, a total advantage under typical GC operating conditions would be about 1000. However, other factors such as detector and source performance (inter alia) act to lower the realized advantage.

2.3. Connes's Advantage

Because the IR spectral frequency is precisely referenced by a laser (usually helium–neon), the resulting spectra should be independent of any particular instrument. This is not true of grating IR spectra and should make GC/FT–IR spectral data uniform across different laboratories. The result may be the production of Class I IR spectra, which have been defined as "a physical constant of the compound in question" (7). This would provide a significant advantage to FT–IR, relative to quadrupole MS detectors, which require periodic tuning with a reference compound to provide spectra reproducible across different laboratories (8).

3. THE FT–IR SPECTROMETER SYSTEM

A block diagram of an integrated FT–IR spectrometer–data system is shown in Figure 11.2. The optical layout of this system is shown in Figure 11.3. The basic components are the spectrometer, computer, data storage disk, input and output devices, and analog–digital (A/D) converter. Computation speed is important because the amount of data collected during a GC/FT–IR run greatly exceeds that of other types of GC detectors. A typical 40-min run at a spectrometer scan speed of 4 scans/s would require computation of 9600

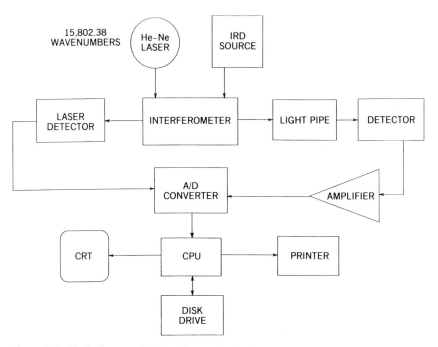

Figure 11.2. Block diagram of the Hewlett Packard infrared detector (IRD). Provided courtesy of Cyril Tang, Scientific Instruments Division, Hewlett Packard Corporation, Palo Alto, California.

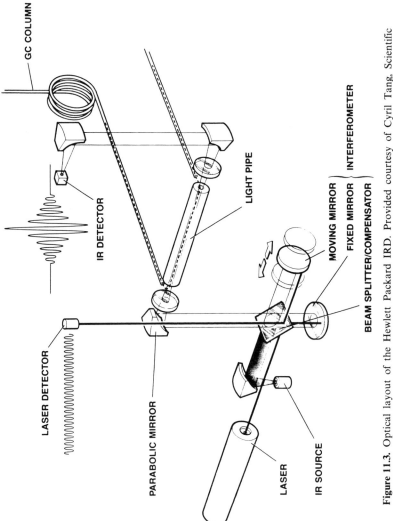

Figure 11.3. Optical layout of the Hewlett Packard IRD. Provided courtesy of Cyril Tang, Scientific Instruments Division, Hewlett Packard Corporation, Palo Alto, California.

256

scans (if all scans are computed). Furthermore, 2048 data points/scan are typically Fourier transformed. The problem is amplified if high-resolution spectra are required. This problem has been ameliorated to a great degree by the availability of array processors (9). The processor may Fourier transform the interferograms so fast that the spectra may be visualized on the display console, almost in real time, above the IR chromatographic peaks.

Computation speed is also critical for FT–IR quantification. That is because a narrow range of frequencies about the quantification band must be separated from the full range of frequencies collected during each interferometer scan.

The large amount of data collected during a GC/FT–IR run also stresses the disk storage capacity. Fortunately, hard storage disks with capacities up to 300 Mbytes are available. Even so, it may be necessary to format the FT–IR data to conserve disk space. Disk data access times may be significantly slower than computation time and can serve as the limiting factor in data treatment (10).

The dynamic range of the A/D converter should be capable of handling the S/N of the most intense region of the interferogram. Modern FT–IR A/D converters typically have 16-bit capacity, but 1–2 bits must be reserved for sampling noise. The A/D converter provides one limit to FT–IR signal throughput, but inadequate range may be compensated for by the technique of gain ranging. With this technique the electronic gain before the converter is altered to keep it within its dynamic range (11). After digitization, the interferogram is multiplied to restore it to its original form.

4. GC/FT–IR SOFTWARE

The true beginnings of modern GC/FT–IR may be dated from the availability of the Gas Chromatography/Fourier Transform Software (GIFTS). This software, prepared by J. de Haseth and D. A. Hanna for the U.S. Environmental Protection Agency (EPA), when used in conjunction with the EPA vapor-phase FT–IR spectral library, provided a routine analytical capability for GC/FT–IR similar to that already available for GC/MS. This capability included data collection and storage, generation of real-time chromatograms, and the ability to search reference spectral libraries.

The most popular chromatogram reconstruction technique is the Gram–Schmidt (G–S) vector orthogonalization. This approach was developed by de Haseth and Isenhour and generates a chromatogram directly from the interferometric data (12). A background vector is created when the light pipe (see Figure 11.3) is empty. Subsequently this vector is used to remove background signals from the vector that encodes the analyte information. A G–S chro-

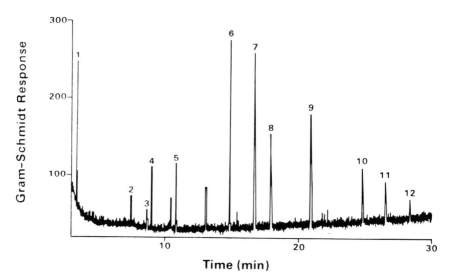

Figure 11.4. Gram–Schmidt chromatogram of 10 ng each of 12-component base-neutral extractables mixture. Reprinted with permission from D. F. Gurka and S. Pyle, *Environ. Sci. Technol.* **22**, 963 (1988). Copyright 1988, American Chemical Society.

matogram of 10 ng each of a 12-component mixture is shown in Figure 11.4. This chromatogram was generated on a 1987 vintage system and represents over an order of magnitude sensitivity improvement over 1982 vintage systems (13, 14). Chromatogram thresholds may be set during data acquisition, allowing the collection and storage of only the peaks of interest, thereby lowering computation times and conserving disk storage space.

A second form of chromatogram, which may be generated in real time or postrun, is the selected frequency range chromatogram (sometimes called the chemigram or multigram). Unlike the G–S, which utilizes data collected across all frequencies (multichannel advantage), this type of chromatogram displays only a preselected frequency region and may be used for functional group monitoring or quantification. Figure 11.5 shows a selected frequency chromatogram of a contaminated soil extract, collected over four spectral regions. The results indicate that several confirmatory functional group regions are required to avoid false positives. The results also suggest that slight spectral overlaps, which may not be a problem when two analytes are close to equal concentration, can be a problem when relative concentrations and/or spectral responses are significantly different. It is typical for analytes in multicomponent environmental and other complex sample types to span a wide concentration range (picogram to microgram is not unusual).

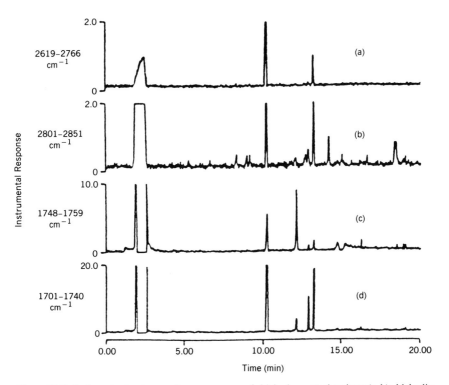

Figure 11.5. Sediment extract group frequency scans of aldehyde spectral regions: (a, b) aldehydic proton Fermi resonance vibrations; (c) aliphatic ester carbonyl stretch; (d) aromatic aldehyde carbonyl stretch. Reprinted with permission from D. F. Gurka, R. Titus, P. R. Griffiths, D. E. Henry, and A. Giorgetti, *Anal. Chem.* **59**, 2362 (1987). Copyright 1987, American Chemical Society.

The last component of GIFTS provides the analyst with library search capability. This GIFTS component reduces the library and unknown spectra to the same final form. Both are truncated and de-resolved to speed computation and reduce storage requirements. Dissimilarity or similarity between known and unknown is established by dot product or Euclidean distance approaches. A quality index is provided for each library hit that indicates the degree of spectral similarity between the library entry and the unknown. Table 11.1 shows some search results for environmental samples, and it has been suggested that the hit quality difference between consecutive library search hits is a measure of the search validity (14). Mass spectrometry confirmatory data was offered to substantiate this suggestion. The absolute magnitude of the quality index is proportional to the spectral signal noise.

Table 11.1. FSCC/GC/FT–IR Library Search Results for Contaminated Soil Extract

Retention Time (min)	GC/FT–IR Hit	GIFTS Hit Index
9.71	1,3-Diethylbenzene	0.4559
	1,2,3-Trimethylbenzene	0.5179
	1,2,3,4-Tetramethylbenzene	0.5216
9.84	1,3,5-Trimethylbenzene	0.4121
	1,3-Dimethylbenzene	0.8847
	1,2,4-Trimethylbenzene	0.6953
10.36	1,2,4-Trimethylbenzene	0.2125
	1,2,3,4-Tetramethylbenzene	0.2962
	1,2,4,5-Tetramethylbenzene	0.3761
13.01	Butylbenzene	0.5015
	1-Phenyldodecane	0.5069
	sec-Butylbenzene	0.5105
13.29	Octachlorocyclopentene	1.0043
	Hexachlorocyclohexane	1.0175
	2,5-Dichlorothiophene	1.0185
13.71	Naphthalene	0.4859
	1-Naphthalenesulfonic acid, dihydrate	0.6161
	1-Chloromethylnaphthalene	0.6566
14.22	2-Ethylbutanal	0.6872
	2-Ethylcaproaldehyde	0.6875
	Isovaleraldehyde	0.7166
14.79	2,4-Dibromophenol	0.5930
	4-Bromo-2-chlorophenol	0.6061
	2,4-Dichlorophenol	0.7125
15.40	2,6-Dimethylnaphthalene	0.6722
	Phenanthrene	0.6728
	1-Chloro-2-methylnaphthalene	0.6859
15.68	1-Methylnaphthalene	0.4240
	Naphthalene	0.5077
	1-Ethylnaphthalene	0.5243
15.80	N,N-Dipropylnicotinamide	0.5225
	Tetrabutylurea	0.5532
	2-Chloro-N,N-diisopropylacetamide	0.6252
16.20	Undecanenitrile	0.2768
	3-Decyne	0.2794
	1-Chlorodecane	0.2868

Table 11.1. *(Continued)*

Retention Time (min)	GC/FT–IR Hit	GIFTS Hit Index
17.05	1,2,3,4-Tetramethylbenzene	0.7490
	Hexamethylbenzene	0.7514
	Butylbenzene	0.7534
17.15	2,6-Dimethyl-4-heptanol	0.9264
	3,3-Dimethyl-1-butanol	0.9378
	2-Isopropyl-5-methyl-1-hexanol	0.9397
17.31	Benzene	0.0587
	Benzyl mercaptan	0.9139
	Bibenzyl	0.9206
17.35	1,2-Diphenylbenzene	1.1146
	α-Methylstyrene	1.1239
	Chlorodiphenylmethane	1.1255
17.74	α-Propylcyclohexanemethanol	0.3469
	Undecanenitrile	0.3523
	3-Decyne	0.3532
19.41	3-Decyne	0.4121
	Undecanenitrile	0.4199
	1-Chlorodecane	0.4218
19.58	1-Phenylhexane	0.4224
	Pentylbenzene	0.4422
	3-Decyne	0.4691
20.17	1-Methyl-2-pyrrolidinone	0.7265
	N-Methylformamide	0.7821
	Phenyl-2-propanone	0.8360
20.51	2-Ethylhexylamine	0.7139
	Heptane	0.7172
	Hexanethiol	0.7207
20.91	2-Methyl-2-butene	0.8428
	1-Isopropyl-4-methylbenzene	0.8441
	2-Methyl-2-pentene	0.8506
21.25	2-Chloro-5-nitrotrifluoromethylbenzene	0.7158
	2-Nitro(trifluoromethyl)benzene	0.7608
	5-Bromo-2-nitro-(trifluoromethyl)benzene	0.7958
21.52	3-Decyne	0.2630
	Myristonitrile	0.2693
	Undecane	0.2720

Table 11.1. *(Continued)*

Retention Time (min)	GC/FT–IR Hit	GIFTS Hit Index
21.66	2,4-Dimethylhexane	0.5036
	3,3-Dimethylhexane	0.5102
	2-Ethyl-4-methyl-1-pentanol	0.5125
23.33	2-Chloro-4-(ethylamino)-*s*-triazine	1.0581
	1,3-Dichloro-2-nitrobenzene	1.0754
	2,4-Bis(ethylamino)-*s*-triazine	1.0755
23.73	Myristonitrile	0.4490
	Tetradecane	0.4578
	1-Dodecene	0.4579
24.39	6-Methylpicoline aldehyde	0.6751
	Phosphoramidic acid, dibutyl ester	0.6999
	Ethyl phosphorothioate	0.7004
25.08	Ethylphosphonic acid, diethyl ester	0.5799
	α-Cyclopropylbenzyl alcohol	0.5876
	Vinylphosphoric acid	0.6141
26.14	1-Phenylhexane	0.5793
	Myristonitrile	0.5863
	3-Decyne	0.5875
27.24	*O,O*-Dimethylphosphorothioic acid	0.9332
	3,4-Dihydro-2-methoxy 2*H*-pyran	0.9430
	Methoxymethylphosphoric acid	0.9811
27.74	2-Aminoethanol	1.0338
	Hexyl phosphite	1.0586
	2-Amino-1-butanol	1.0663
28.00	1-*tert*-Butyl-4-dodecylbenzene	0.5849
	2-(4-*tert*-Butyl-phenoxy)ethanol	0.6120
	2,4-Di-*tert*-butylphenol	0.6396
28.39	2-Methyl-2-nitropropane	0.4694
	2,6-Dinitrotoluene	0.5432
	2-(Furfurylamino)ethane thiol	0.6080
28.72	1-Phenylhexane	0.3548
	Pentylbenzene	0.3621
	Undecanenitrile	0.3704
30.20	Ethyl phosphorothioate	0.6929
	Phosphonic acid, cyanomethyl ester	0.7439
	Vinylphosphonic acid, diethyl ester	0.7507

Table 11.1. *(Continued)*

Retention Time (min)	GC/FT–IR Hit	GIFTS Hit Index
30.39	Allyl phenyl ether	0.9306
	1,2-Diphenoxyethane	0.9369
	1,2-Diepoxy-3-phenoxyethane	0.9396
30.70	4,4′-Dichlorobenzophenone	0.6041
	4-Chlorobenzophenone	0.8059
	2,4′-Dichlorobenzophenone	0.8694
31.81	Methyl oleate	0.5361
	Butyl palmitate	0.5368
	Pentyl myristate	0.5384
33.31	1,2-Dichloroethane	1.1536
	1,1,1-Trichloroethane	1.1838
	1,1,2,2-Tetrachloroethane	1.2104
34.63	Benzylmercaptan	0.9591
	1,2-Dichloroethane	0.9546
	(2-Bromoethyl)benzene	0.9694
34.91	α-Chlorotoluene	0.8668
	1,1-Dichloroethane	0.9006
	Benzyl mercaptan	0.9113
34.93	Isobutyl chloride	0.9789
	1-Chloropentane	1.0032
	1-Chloro-3-methylbutane	1.0099
35.34	2-Methyl-2-propyl-1,3-propanediol	0.7823
	Isobutyl alcohol	0.7836
	2,2-Dimethyl-1,3-butanediol	0.8047
35.38	4-Methyl-2-phenylvaleronitrile	0.9982
	Pentylbenzene	0.9997
	Butylbenzene	1.0031
38.55	Ethyl phosphite	0.9693
	3-Bromo-1-propane	1.0133
	Ethyl 4-toluenesulfonate	1.0265

Source: Reprinted with permission from D. F. Gurka, M. Hiatt, and R. Titus, *Anal. Chem.* **56**, 1102 (1984). Copyright 1984, American Chemical Society.

5. THE GC/FT–IR INTERFACE

The key to optimum GC/FT–IR analysis is the interface. It is here that chromatographic and spectral requirements must be optimized and matched to provide the best analytical performance. Hardware, data acquisition, and chromatographic conditions must be adjusted to maximize sensitivity, selectivity, sample throughput, and data analysis.

5.1. Light Pipes

GC/FT–IR experiments were first performed by Low and Freeman employing an early Block Engineering interferometer (15). They performed trap and flow experiments by passing a packed column effluent through a copper cell of dimensions 5 cm in length × 4-mm i.d. Spectra were obtained for microliter quantities of neat, strong IR absorbers. It was necessary to trap the GC effluent and signal average to obtain sub-microliter detection limits. Obviously, trapping would lead to reduced sample throughputs and degraded chromatographic resolution if used on a routine analytical basis. Subsequently, Azarraga (16) demonstrated the feasibility of flow-through GC/FT–IR analysis with support coated open tubular (SCOT) columns. Because SCOT columns provide the superior chromatographic resolution required for environmental samples, Azarraga was then prompted to develop a new type of flow cell with superior sensitivity. Basically, this cell consists of a long narrow tube, internally coated with gold, capped with IR-transparent windows at both ends and

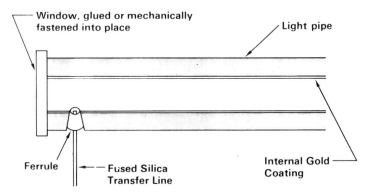

Figure 11.6. Cutaway diagram of a capillary light pipe (typical volume: ∼ 100 μL). The end of the fused silica capillary column (FSCC) is inserted directly to the pipe without blocking the source beam. An exit fused silica line vents the GC effluent from the pipe or directs it to a second detector. Reprinted with permission from P. R. Griffiths and J. A. de Haseth, *Fourier Transform Infrared Spectrometry*, p. 580. Wiley, New York, 1986. Copyright 1986, John Wiley and Sons, Inc.

possessing inlet and exit transfer lines for the GC effluent. He called it a light pipe because of its shape and function, and the gold coating provided a highly reflective surface that was inert to hot GC effluents. Figure 11.6 shows a typical GC/FT–IR light pipe. The light pipe input and exit beams should match the diameter of the light pipe and detector area, respectively. This ideal situation is termed "throughput matching."

Although the Azarraga method of light pipe gold coating produces high transmission, it is very difficult to reproducibly prepare the gold surface by this technique. For example, the present author purchased two light pipes from the same vendor in a 1-year period that differed by a factor of 3 in transmission. In 1984, Yang et al. investigated the conditions required to prepare light pipes reproducibly (17). They found that drawn capillary tubing had a much smoother surface than drilled tubing. They recommended washing the tubes with HF and $SiCl_2$ prior to gold coating, and a final annealing at 800 °C.

5.2.   Chromatography–Light Pipe Considerations

Griffiths has made a very careful evaluation of the interactive relationship between light pipe characteristics and chromatographic conditions (18). The evaluation considered both sampling factors and light pipe design.

The function of sampling experiments was to determine the precise relationship between light pipe volume and GC peak volume. He defined the following parameters:

Q_{tot} = total quantity of GC component;

$V_{1/2}$ = full width at half height (FWHH) of GC peak, mL;

C_{max} = maximum component concentration measured at GC detector;

V_R = GC peak retention volume;

F_c = carrier gas flow rate, mL/s;

V_{cell} = infrared cell volume;

q_{cell} = component quantity in cell;

Q_{cell} = maximum component quantity in cell;

C_{cell} = average cell concentration (q_{cell}/V_{cell});

V_L = transfer line volume between GC detector and cell entrance;

V = volume of carrier gas that has passed after component was first detected.

For a triangular GC peak and a nondestructive GC detector, if $V_{cell} \ll V_{1/2}$, then C_{cell} approximates the GC detector concentration profile and the cell contains only one GC-resolved peak. However, because of the detector–cell transfer line volume, there is a time delay between the detector maximum signal and C_{cell} given by

$$t'_d = \frac{V_L}{F_c}$$

However, although $C_{cell} \sim C_{max}$, Q_{cell} will be a small fraction of Q_{tot}:

$$Q_{cell} \simeq Q_{tot}(V_{cell}/V_{1/2})$$

If $V_{cell} \gg V_{1/2}$, then $C_{cell} < C_{max}$ and is given by

$$C_{cell} = Q_{tot}/V_{cell}$$

Under these conditions the cell may contain more than one GC-resolved component. Figure 11.7a shows how C_{cell} varies with $V_{cell}/V_{1/2}$ for a triangular peak when $V_L = 0$. Figure 11.7b shows how Q_{cell}/Q_{tot} varies with $V/V_{1/2}$. Both figures demonstrate an increasing time delay between GC detection and concentration, or quantity maximization, as $V/V_{1/2}$ increases. If C_{cell}/C_{max} and C_{cell} from Figures 11.7a, b are plotted against $V_{cell}/V_{1/2}$ in Figure 11.7c, the intersection of the two lines indicates the best compromise between chromatographic elution volume and cell volume. From the figure this is seen to be $V_{1/2} = $ cell volume. For light pipe GC/FT–IR work, the cell volume chosen is a function of the elution volume of a particular GC column type. For a typical packed column the pipe volume should be ~ 4 mL, and for a wide-bore fused silica capillary column (FSCC) the volume will be ~ 0.1 mL. It has been proposed that $V_{cell} \geqslant 4V_{1/2}$ is the optimum condition for the measurement of reference quality vapor-phase FT–IR spectra (P. R. Griffiths, private communication, 1989).

While the problem of matching light pipe and elution volumes may be simple for the situation of repetitive analysis of a few closely eluting compounds, multicomponent environmental analysis presents a different challenge. This type of sample can contain 50–100 components eluting over a wide range of retention times. For example, Gurka et al. have shown a variation in analyte GC elution volume of 4 for a still-bottom sample analyzed by packed column GC/FT–IR (19). They pointed out that it was not possible to simultaneously optimize the elution volume of every component of a multicomponent sample.

5.3. Cryogenic Detectors

The choice of IR detectors is another problem facing the GC/FT–IR analyst. At the 27th Pittsburgh Conference in 1976, L. V. Azarraga suggested that

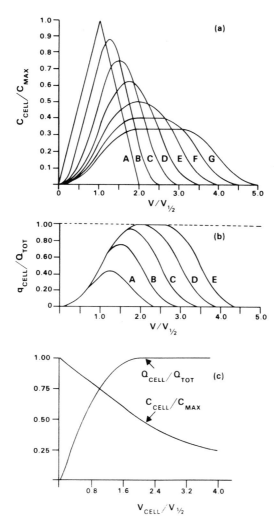

Figure 11.7. Light pipe volume–chromatography considerations. (a) Variations of C_{cell}/C_{max}, the ratio of the sample concentration in the cell to the maximum sample concentration at the GC detector, with $V/V_{1/2}$ (where V is the volume of carrier gas that has passed after the sample was first detected) for triangular GC peak, assuming $t'_d = 0$, for the following values of $V_{cell}/V_{1/2}$: A, 0.00; B, 0.50; C, 1.00; D, 1.50; E, 2.00; F, 2.50; G, 3.00. (b) Variation of Q_{cell}/Q_{tot}, the ratio of the quantity of sample in the cell to the total sample quantity, with $V/V_{1/2}$ for a triangular GC peak, assuming $t'_d = 0$, for the following values of $V_{cell}/V_{1/2}$: A, 0.50; B, 1.00; C, 1.50; D, 2.00; E, 2.50. (c) Variation of Q_{cell}/C_{tot} and C_{cell}/C_{tot} with $V_{cell}/V_{1/2}$; the values of Q_{cell} and C_{cell} were taken from plots of the type shown in parts a and b, respectively. It can be seen that the two curves intersect when $V_{cell}/V_{1/2} = 1$. Reprinted with permission from P. R. Griffiths, *Appl. Spectrosc.* **31**, 284 (1977). Copyright 1977, Society for Applied Spectroscopy.

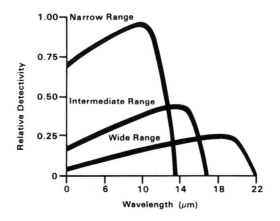

Figure 11.8. Variation of detectivity with wavenumber for narrow-range, medium-range, and wide-range MCT detectors. Reprinted with permission from P. R. Griffiths, J. A. de Haseth, and L. V. Azarraga, *Anal. Chem.* **55**, 1361A (1983). Copyright 1983, American Chemical Society.

cryogenically cooled mercury–cadmium–telluride (MCT) detectors would provide the best sensitivity for GC/FT–IR trace analysis. That suggestion was quickly accepted, and MCT detectors currently enjoy widespread usage. The MCT detector is photoconductive, and the detector element is composed of a mercury–tellurium and cadmium–tellurium semiconductor mixture. Varying the ratio of semiconductors alters the frequency at which the most intense response occurs (20). The capability to vary this frequency maximum leads to the availability of MCT detectors with three frequency detection ranges. These are called wide-band, mid-band, and narrow-band detectors. These detectors typically have low-frequency cutoffs of about 500, 700, and 800 cm^{-1}, respectively. There is a sensitivity difference of about 10-fold between the wide-band (least sensitive) and the narrow-band (most sensitive) MCT. Figure 11.8 shows the relationship between detector response and sensitivity (detectivity) for these detectors. Thus, the chromatographer is faced with a choice between selectivity and sensitivity when selecting the appropriate MCT for a GC/FT–IR run. Gurka and Pyle suggested that a 750 cm^{-1} MCT cutoff is a reasonable compromise, although they cautioned that small symmetrical molecules with few total fundamental vibrations but with one or more of these in the 750–650 cm^{-1} region might be difficult to identify with this detector (13). However, by analyzing 26 standards of various compound classes, they concluded that even when the standard's spectrum was not in the vapor-phase FT–IR library, a GC/FT–IR system containing a 750 cm^{-1} cutoff detector could still determine the compound class of the standard (see Table 11.2). Note that many of these standards were polynuclear hydrocarbons that typically

Table 11.2. Narrow-Band MCT Selectivity with Environmental Standards

Retention Time (min)	Compound	Amount Injected (ng)	IRD Library Search Hit[a]
3.31	N-Nitrosodimethylamine	5	1
7.44	Bis(2-chloroethyl) ether	25	1
7.90	1,4-Dichlorobenzene	25	1
8.64	Bis(2-chloroisopropyl) ether	25	(Di-n-propyl ether)
8.95	N-Nitrosodipropylamine	10	1
10.40	Bis(2-chloroethoxy)methane	25	1
10.72	1,2,4-Trichlorobenzene	25	1
10.85	Naphthalene	25	1
13.12	1,2,4,5-Tetrachlorobenzene	25	1
14.80	Acenaphthylene	25	1
14.80	Dimethyl phthalate	5	1
15.20	Acenaphthalene	25	2
16.58	Fluorene	25	(Phenanthrene)
16.60	Diethyl phthalate	5	1
16.76	4-Chlorophenyl phenyl ether	5	(4-Bromophenyl phenyl ether)
17.90	4-Bromophenyl phenyl ether	5	1
19.00	Phenanthrene	25	1
19.20	Anthracene	25	1
20.86	Di-n-butyl phthalate	5	1
22.10	Fluoranthene	50	(Naphthalene)
22.60	Pyrene	50	(3-Methylphenanthrene)
24.70	Butyl benzyl phthalate	25	(Di-n-propyl phthalate)
25.87	Benz[a]anthracene	250	(Phenanthrene)
26.02	Chrysene	250	(3-Methylbiphenyl)
26.40	Di-n-octyl phthalate	25	1
28.30	Bis(2-Ethylhexyl)phthalate	25	1
29.10	Benzo[b]fluoranthene	250	(Anthracene)

Source: Reprinted with permission from D. F. Gurka nd S. Pyle, Environ. Sci. Technol. 22, 963 (1988). Copyright 1988, American Chemical Society.

[a]Parentheses indicate compound not in spectral reference library. Best match is indicated.

exhibit few fundamental vibrations, at least one of which usually occurs in the $800–700 \text{ cm}^{-1}$ region (21).

Yang et al. (17) have examined various optical configurations that varied the MCT focal area and attempted to minimize the light pipe temperature effect. (Earlier it had been shown that packed column GC/FT–IR sensitivity declined by a factor of about 2/3 when the system light pipe was heated from 25 to 280 °C (19)). They concluded that a 0.2 mm × 0.2 mm MCT detector focal area was optimum for 1-mm-i.d. light pipes and that moving the interface collection optic away from the light pipe exit greatly reduced the light pipe temperature effect. The temperature effect was attributed to saturation of the detector or amplifier by unmodulated radiation emitted by the light pipe. The collection efficiency of this radiation drops off rapidly as the collection optic is moved away from the exit end of the light pipe. By chopping the unmodulated light pipe heat radiation, Brown et al. confirmed the conclusion of Griffiths and Yang, and they reduced the sensitivity drop by interposing a water-cooled aperture between the light pipe and the MCT detector (22).

In 1987, a prototype capillary GC/FT–IR interface, using a mid-band cutoff MCT ($700\,\text{cm}^{-1}$), was reported that reduced the sensitivity loss on heating from 25–280 °C to $\sim 20\%$ and also utilized a small focal area detector (23). Figure 11.9 shows a schematic of the prototype interface. Evaluation of

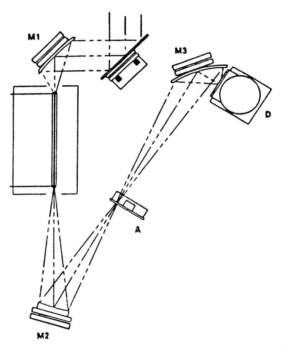

Figure 11.9. Optical configuration designed to discriminate against unmodulated emission from the light pipe. Toroidal mirror M2 (195-mm focal length) was used to collect the beam exiting a 1.06-mm-i.d. light pipe and refocus it onto an aperture (A, diameter = 1.06 mm i.d.). The apertured beam is then collected with a 90° off-axis ellipsoidal mirror (M3) with effective focal lengths of 250 and 42 mm and focused onto a 0.5-mm detector (D). Reprinted with permission from D. E. Henry, A. Giorgetti, A. M. Haefner, P. R. Griffiths, and D. F. Gurka, *Anal. Chem.* **59**, 2356 (1987). Copyright 1987, American Chemical Society.

Table 11.3. GC/FT–IR Quantification Parameters and Identification Limits

Compound	Concentration Range and Identification Limit (ng)[a]	Correlation Coefficient[b]	ν_{max} (cm^{-1})
N-Nitrosodimethylamine	10–100	0.993	1483
	10–250	0.989	
N-Nitrosodipropylamine	10–100	0.999	1482
	10–250	0.990	
Bis(2-chloroethyl) ether	10–100	0.993	1125
	10–250	0.966	
Bis(2-chloroisopropyl) ether	10–100	0.976	1092
	10–250	0.958	
Bis(2-chloroethoxy)methane	10–100	0.964	1085
	10–250	0.925	
Dimethyl phthalate	10–100	1.000	1281
	10–250	0.996	
Diethyl phthalate	10–50	0.996	1274
	10–50	0.998	
Di-n-butyl phthalate	10–100	0.994	1273
	10–250	0.989	
Butyl benzyl phthalate	10–100	0.990	1272
	10–250	0.970	
Ethyl hexyl phthalate	10–100	0.990	1271
	10–250	0.980	
Di-n-octyl phthalate	10–100	0.959	1273
	10–250	0.968	
4-Chlorophenyl phenyl ether	10–50	0.993	1244
	10–50	0.998	
4-Bromophenyl phenyl ether	10–100	0.989	1242
	10–250	0.993	
Pyrene	50–250	0.850	840
Naphthalene	25–250	0.979	781
Acenaphthylene	50–250	0.998	769
Acenaphthene	25–250	0.995	786
Fluorene	50–250	0.996	1451
Phenanthrene	50–250	0.888	808
	50–250	0.967	
Anthracene	50–250	0.910	3063
Fluoranthene	50–250	0.978	773
1,4-Dichlorobenzene	25–250	0.994	1091
1,2,4-Trichlorobenzene	25–250	0.993	1460
1,2,4,5-Tetrachlorobenzene	25–250	0.996	1441

Source: Reprinted with permission from D. F. Gurka and S. Pyle, *Environ. Sci. Technol.* **22**, 963 (1988). Copyright 1988, American Chemical Society.

[a] Lower end of range is at or near the identification limit. GS peak areas used for quantification.

[b] Determined at four concentration levels. Each level analyzed in duplicate.

271

this prototype with standards and samples revealed sensitivities in the range of 20–120 ng for strong and weak infrared absorbers, respectively (24). In 1988, a commercially available, GC/FT–IR system, using a narrow-band cutoff MCT (750 cm^{-1}), was evaluated that demonstrated sensitivity about three times better than that of this prototype (13). Table 11.3 shows the sensitivities achieved with typical environmental contaminants. Figure 11.10 shows spectra collected on this system near the analyte detection limit. At least part of the sensitivity enhancement is a result of using a narrower band MCT in this system (a 750 cm^{-1} cutoff vs. a 700 cm^{-1} cutoff for the prototype), but the commercial system also used a larger focal area detector (1 mm^2 vs. 0.25 mm^2). Both interfaces exhibited a ~ 20% sensitivity decline between a light pipe temperature of 25 °C and 280 °C. The larger detector area could result in a sensitivity loss of a factor of 2 for the commercial system relative to the prototype, whereas the narrow-band MCT could lead to a sensitivity enhancement of a factor of 3. Until now (late 1988), an improvement by over a factor of 40 has been achieved for commercially available capillary GC/FT–IR systems between 1982 and 1988 (13, 14). Without major breakthroughs in FT–IR hardware, remaining sensitivity improvements are expected to be less than a factor of 10, for light pipe, capillary GC/FT–IR.

6. MATRIX ISOLATION CAPILLARY GC/FT–IR

Capillary GC may be altered to a static sampling mode, from the dynamic flow-through mode of light pipe GC/FT–IR, by the technique of matrix isolation (MI) GC/FT–IR. This technique, pioneered by Reedy and co-workers, introduces a small amount of inert gas (typically argon) into the carrier gas stream (25). The GC-separated analytes are subsequently trapped in a frozen matrix of argon on a rotating gold disk that is refrigerated to ~ 12 K. Figure 11.11 shows the layout of the MI/GC/FT–IR system described by Borman (26). This approach is reported to yield sensitivities 10–100 times better than light pipe GC/FT–IR and close to that of quadrupole MS (27). This sensitivity enhancement is achieved in three ways:

1. Sharper IR bands are created by freezing out molecular rotations.
2. The analyte is deposited in a concentrated form, in an argon strip that is less than 0.1 mm wide.
3. The static sampling mode allows the improvements of spectral S/N via extensive signal averaging (S/N improves as the square root of the number of scans).

In practice, factor number one is important only for small molecules.

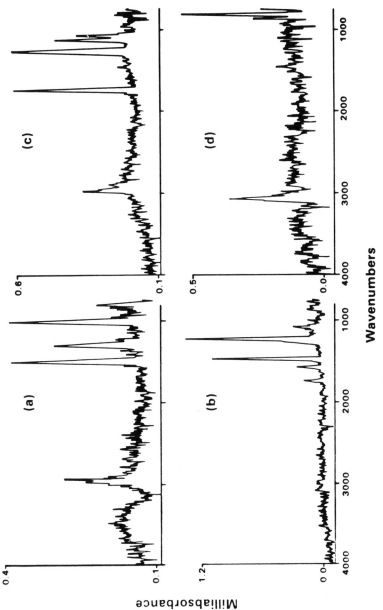

Wavenumbers

Figure 11.10. GC/FT–IR spectra of (a) 5 ng of *N*-nitrosodimethylamine; (b) 5 ng of 4-chlorophenyl phenyl ether; (c) 5 ng of di-*n*-butyl phthalate; and (d) 25 ng of phenanthrene. Reprinted with permission from D. F. Gurka and S. Pyle, *Environ. Sci. Technol.* **22**, 963 (1988). Copyright 1988, American Chemical Society.

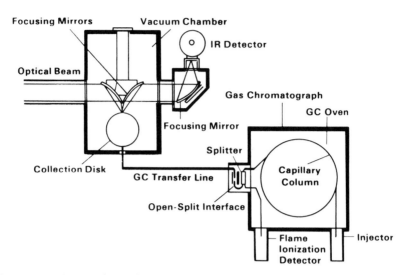

Figure 11.11. Diagram of a MI/GC/FT–IR interface. Note that the FT–IR spectrometer optics
are focused on the opposite side of the collection disk from the GC transfer line deposition
point, precluding the acquisition of real-time FT–IR data. Real-time data are provided by a
split to the flame ionization detector. Reprinted with permission from S. Borman, *Anal. Chem.*
56, 936A (1984). Copyright 1984, American Chemical Society.

Factor number 2 is important because the argon strip is more than 10 times
narrower than a typical light pipe i.d. Figure 11.12 shows the MI/GC/FT–IR
spectrum of 156 pg of of 1,2,3,4-tetrachlorodibenzodioxin (TCDD) taken after
30 minutes of signal averaging and the spectrum of 20 ng taken with 32 scans.
Because the signal averaging enhancement depends strongly on spectrometer
stability (see Reference 28 for a study of FT–IR spectrometer stability), it
remains to be seen how much sensitivity is to be gained by this approach.
No comprehensive study of FT–IR signal averaging has been published since
Cournoyer et al. obtained picogram-level sensitivities, by multihour signal
averaging, on a static sample (29). A second critical point is that real-time IR
data cannot be collected on current MI/GC/FT–IR systems and the analyst
must depend on an effluent split to a flame ionization detector to locate GC
peaks. Recently, a commercially available system utilizing a split to a MS has
been reported (30). Like the light pipe approach, MI produces spectra that
are dissimilar to condensed-phase IR spectra, and thus both require new
spectral data bases. Because the gold disk storage space is limited, sample
throughputs by the light pipe approach will be better. Although matrix
isolation FT–IR spectra initially were expected to be similar to vapor-phase
spectra, Coleman has reported that at sufficiently high analyte concentrations
nearest neighbor effects can change the MI spectra (31). This is especially true

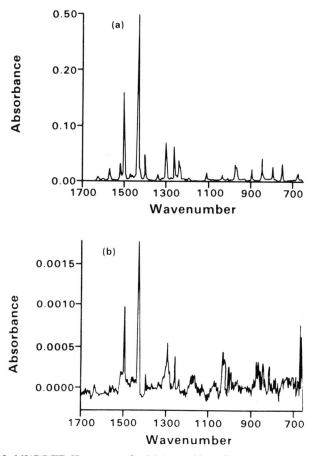

Figure 11.12. MI/GC/FT–IR spectra of 1,2,3,4-tetrachlorodibenzodioxin (TCDD): (a) 20 ng (32 scans); (b) 156 pg (5000 scans). Provided by Professor Charles Wurrey, University of Missouri–Kansas City on a Cooperative Agreement with the U.S. Environmental Protection Agency.

for protic compounds and can occur at masses as low as 5 ng for alcohols (32). This could be a serious problem for the analysis of total unknowns and would require a prior analysis, followed by reanalysis after appropriate sample dilution, to ensure the comparability of unknown and reference MI/GC/FT–IR spectra (assuming that reference MI spectra have been collected under truly matrix-isolated conditions). In spite of these shortcomings, the method is expected to find utility when its superior sensitivity is required (e.g., environmental monitoring of toxic isomeric compounds like TCDDs and furans, and polychloro and polybromo biphenyls).

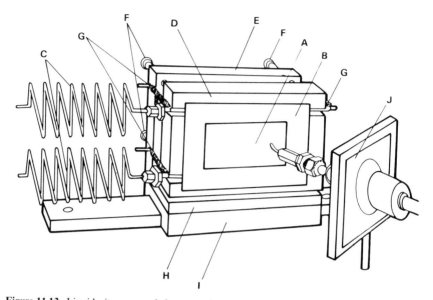

Figure 11.13. Liquid-nitrogen-cooled stage and transfer line positioner: (A) 50-mm × 25-mm × 2-mm ZnSe window; (B) brass cooling plate; (C) coiled thin-wall stainless steel tubing; (D) phenolic insulating plate; (E) vertical aluminum plate; (F) adjustment screws; (G) tension springs; (H) aluminum mounting plate; (I) translation stage; (J) fiber-optic positioner for transfer line. Reprinted with permission from A. M. Haefner, K. L. Norton, P. R. Griffiths, S. Bourne, and R. Curbelo, *Anal. Chem.* **60**, 2441 (1988). Copyright 1988, American Chemical Society.

A less expensive and less complicated "static" sampling approach called cryotrapping has been described by Griffiths (33). This approach traps capillary GC eluates on a liquid-nitrogen-cooled zinc selenide strip. This strip is translated across the GC column exit, and spectra may be collected postrun, or in "pseudo" real time. Sensitivity approaches that of the MI approach, and spectra are reported as similar to those produced by the KBr disk method. A diagram of the cryotrapping interface is shown in Figure 11.13, and commercial versions of this system have been described (34, 35).

7.   GC/FT–IR QUALITATIVE CONSIDERATIONS

7.1.   Sensitivity–Detectability

As pointed out earlier in this chapter, the sensitivity of the GC/FT–IR experiment requires matching the operational parameters of the GC and the FT–IR. The trade-offs between sensitivity, chromatographic resolution, and column

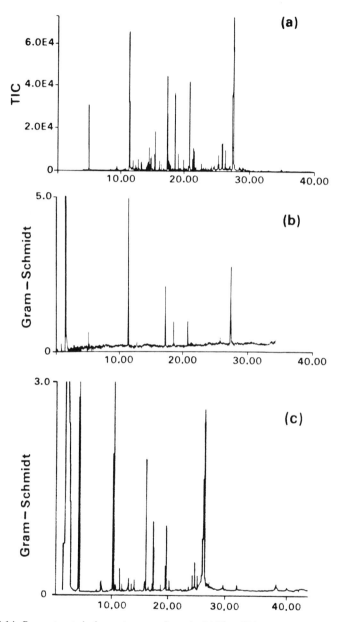

Figure 11.14. Reconstructed chromatograms from herbicide still-bottom extract analysis by (a) mass selective detector; (b) vintage 1982 commercial GC/FT–IR; and (c) new commercial GC/FT–IR (1987). Note that the new GC/FT–IR provides a comparable number of detections to that of the mass selective detector (used in full-scan mode). Reprinted with permission from D. F. Gurka and S. Pyle, *Environ. Sci. Technol.* **22**, 963 (1988). Copyright 1988, American Chemical Society.

loading must be considered prior to the analytical run. In the early 1980s typical packed column GC/FT–IR identification limits ranged from a few hundred nanograms for strong IR absorbers to a few micrograms for weak IR absorbers (19). Current (1992) light pipe identification limits are in the range of 5 to 25 ng (13) and thus are well within the loading capacity of thick-film wide and narrow-bore FSCCs. Figure 11.14 shows sample chromatograms demonstrating the progress achieved in improving light pipe GC/FT–IR sensitivity. It is seen that GC/FT–IR sensitivity now rivals that of full-scan GC/MS and it is estimated that MI/GC/FT–IR limits, without extensive signal averaging, are about a factor of 5–10 better than the light pipe approach.

One way to mitigate the loading–resolution–sensitivity trade-off is to utilize a twin GC column approach. Smith has described a dual-column approach for complex mulicomponent samples exhibiting a wide range of analyte concentrations (36). This type of sample can lead to detector "saturation" in the GC/MS experiment and may necessitate dilution of the concentrated analytes to within the MS dynamic range; however, low-concentration analytes may then be diluted below the detection limit. A schematic of the dual-oven approach is shown in Figure 11.15. Each GC column is housed in a separate oven, and a six-port microrotary valve can switch preselected analytes from column 1 to column 2. Column 1 can be a thick-film column to maximize loading, whereas the second column can be chosen to optimize resolution. Figure 11.16a shows a complex GC peak from column 1 that was diverted to yield its resolved components as shown in Figure 11.16b. This two-dimensional approach to FT–IR chromatography matching may come into more general use if its cost and complexity can be reduced.

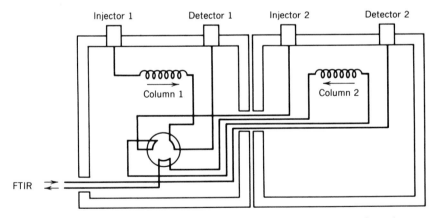

Figure 11.15. Schematic diagram of the dual-oven GC/FT–IR system. This configuration may be used for single-column or two-dimensional GC/FT–IR. Reprinted with permission from S. L. Smith, *Appl. Spectrosc.* **40**, 278 (1986). Copyright 1986, Society for Applied Spectroscopy.

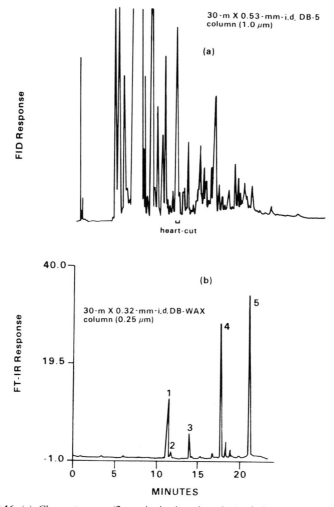

Figure 11.16. (a) Chromatogram (flame ionization detection) of the terpene mixture. The indicated section of the separation was heart-cut to a BD-WAX capillary column. (b) Infrared chromatogram (Gram–Schmidt) of the heart-cut section from part a. The DB-WAX column was temperature programmed from 35 °C at 20 °C/min and then to 150 °C at 5 °C/min. Reprinted with permission from S. L. Smith, *Appl. Spectrosc.* **40**, 278 (1986). Copyright 1986, Society for Applied Spectroscopy.

Detectability improvement may be defined as a means of increasing the number of detections, while sensitivity is defined as the slope of a plot of area counts vs. the amount of material injected (37). Typically the detectability is improved by increasing the amount, or the concentration, of material reaching the detector. Preconcentration via extract concentration is the easiest approach to increasing detectability. Unfortunately, this approach concentrates all analytes simultaneously and at some point must lead to column overloading and coelution. The two-dimensional GC/FT–IR approach of Smith allows only preselected analytes on the first column to be prefocused via repetitive cryotrapping. Kollar et al. have reported that purgeable volatiles may be preconcentrated by headspace analysis leading to greater detectability in subsequent GC/FT–IR analyses (38). Finally, Grob has demonstrated that detectability may be improved by the injection of large extract volumes (80–100 μL) onto the GC column (39). Retention gaps of up to 50 m must be used to prefocus these large volumes and prevent flooding of the analytical column. The applicability of large volume injection to environmental analysis has been described (40).

7.2. Molecular Responses

It is now well known that the magnitude of the IR response is a function of molecular structure (i.e., strong and weak IR absorbers). Nevertheless, the frequent quoting of low-nanogram sensitivities for the FT–IR industry standard (isobutyl methacrylate), during the late 1970s to the early 1980s deluded many into believing that other compounds would exhibit similar behavior. In 1984, it was demonstrated that, like quadrupole GC/MS, GC/FT–IR response factors exhibited over an order of magnitude in range (14). Furthermore, the response factor scales for these two detectors are essentially inverted, with GC/FT–IR being least sensitive to hydrocarbons and most sensitive to oxygen-containing molecules. The reverse is often true for electron-impact, quadrupole GC/MS. A 1988 commercial light pipe GC/FT–IR system was reported to exhibit similar sensitivity to that of full-scan quadrupole GC/MS for strong IR absorbers and about one-tenth the sensitivity for weak absorbers (13).

Recently, a directly linked GC/FT–IR/MS mode has been used to compare the two types of detectors for their relative responses in various data acquisition modes (41). Data were acquired by MS in the total-ion and single-ion chromatogram (TIC, SIC) approaches, and FT–IR data were collected in the G–S and integrated and maximum absorbance approaches. Relative MS and FT–IR responses, as determined by the differences in relative slopes of detector response vs. nanograms injected plots, decreased in the order G–S > integrated absorbance = SIC > maximum absorbance > TIC. Because the G–S and TIC measure total molecular effects, the G–S is indicated as being more sensitive than the TIC to overall molecular structure. The order of sensitivity to molecular substructure, was integrated absorbance = SIC > maximum absorbance. Because the reported parameters were measured on only 15 environmentally important compounds, it remains to be seen whether or not these response orders will be generally applicable.

7.3. Structure Assignment

Tentative identification or compound class assignment are the two types of structural assignment obtainable from GC/FT–IR analysis. Identification within a reasonable time frame is possible only if a reference spectrum or authentic standard is obtainable. Because gas-phase FT–IR spectra are only qualitatively similar to condensed-phase spectra, the emergence of the GC/FT–IR method required new spectral data bases. The EPA's 3500-member data base became available in the early 1980s, and current commercially available vapor-phase data bases contain some 9000 spectra. Because the Chemical Abstracts Service (CAS) chemical registry list contains more than 9×10^6 chemicals, a typical chemist analyzing a total unknown will generally not have a reference spectrum to work with. Under these conditions the best that can be done is to ascertain the compound class or functionality of the unknown. This capability is an intrinsic strength of the infrared technique, and the functionality information may lead to useful generalizations about the unknown. This functionality information may be displayed by the FT–IR in real time or be determined after the GC run.

8. QUANTIFICATION

While the qualitative aspects of GC/FT–IR have been explored in depth, little has thus far been reported on quantification. To a large extent, this has been a result of the insensitivity of the FT–IR technique. Prior to about 1984, the method was too insensitive to quantify over any appreciable dynamic range on open tubular columns.

As early as 1982, Sparks et al. showed that the G–S reconstruction method was linear over the range of 0.3–1.0 μg (42). Cooper and Taylor reported absorbance calibration curves for some strongly absorbing analytes, on a wide-bore capillary column, at a light pipe temperature of 230 °C (43). In 1981, Hembree et al. demonstrated MI/GC/FT–IR linearity for pyrene, with Beer's Law being followed over the range of 700 ng to 8 μg (44).

An exhaustive study of light pipe GC/FT–IR quantification was reported in 1989 (41). A directly linked FSCC/GC/FT–IR/MS system was employed. allowing the on-line confirmation of FT–IR data by the MS detector.

Table 11.4. Comparison of MS and IR Quantification Methods by Regression Analysis[a]

Compound	MS Detector[b]				IR Detector					
	Single Ion		Total Ion		Integrated Absorbance Slope[c]	r^2	Maximum Absorbance Slope[d]	r^2	Gram–Schmidt Slope[e]	r^2
	Slope ×10⁴	r^2	Slope ×10⁴	r^2						
Acenaphthene	4.359	0.9983	19.65	0.9983	0.4473	0.9995	0.04901	0.9985	51.75	0.9047
Acenapthylene	7.445	0.9986	17.28	0.9996	0.4072	0.9959	0.04716	0.9985	34.18	0.9093
Anthracene	8.552	0.9850	23.24	0.9999	0.4104	0.9969	0.03927	0.9971	35.96	0.9653
Benz[a]anthracene	9.115	0.9979	19.35	0.9995	0.5155	0.9918	0.02372	0.9921	24.33	0.9410
Benzoic acid	2.070	0.9944	8.05	0.9925	1.9649	0.9864	0.05289	0.9892	302.8	0.9929
Benzo[a]pyrene	10.936	0.9894	21.77	0.9632	0.2592	0.9966	0.0094	0.9074		
Bis(2-chloroethoxy)methane	4.700	0.9988	13.14	0.9993	0.61020	0.9992	0.07707	0.9991	65.68	0.9966
Bis(2-chloroethyl) ether	3.884	0.9997	10.49	0.9993	0.6462	0.9955	0.09352	0.9992	77.79	0.9959
Bis(2-chloroisopropyl) ether	1.014	0.9990	4.74	0.9994	0.5874	0.9981	0.05475	0.9998	68.18	0.9980
4-Bromophenyl phenyl ether	4.668	0.9983	24.09	0.9995	2.2566	0.9995	0.2148	0.9996	375.7	0.9982
Butyl benzyl phthalate	5.600	0.9988	25.04	0.9978	2.4327	0.9999	0.1477	0.9994	526.9	0.9988
4-Chloroaniline	3.231	0.9973	11.44	0.9989	0.9128	0.9991	0.0992	0.9965	187.9	0.9960
4-Chloro-m-cresol	2.053	0.9934	13.90	0.9489	0.6648	0.9975	0.0722	0.9946	134.9	0.9872
2-Chloronaphthalene	2.905	0.9982	15.02	0.9995	0.01450	0.9897	0.00290	0.9988	40.84	0.8804
2-Chlorophenol	2.908	0.9971	11.41	0.9989	0.6427	0.9976	0.08249	0.9965	130.6	0.9896
4-Chlorophenyl phenyl ether	4.345	0.9963	21.40	0.9999	2.8761	0.9999	0.2830	0.9997	431.6	0.9986
Chrysene	12.044	0.9962	21.77	0.9632	0.7304	0.9985	0.04472	0.9984	24.18	0.9644
o-Cresol	2.798	0.9983	14.04	0.9992	0.4024	0.9972	0.04999	0.9964	80.66	0.9906
p-Cresol	3.048	0.9937	14.20	0.9980	1.0361	0.9972	0.1216	0.9959	124.7	0.9964
Dibenzofuran	9.472	0.9996	13.04	0.9995	0.01297	0.9697	0.03469	0.8579	101.8	0.9725
Di-n-butylphthalate	13.312	0.9993	23.81	0.9997	2.4062	0.9998	0.2153	0.9996	540.5	0.9984
1,2-Dichlorobenzene	3.578	0.9973	19.98	0.9998	0.01301	0.9937	0.03034	0.9947	60.24	0.9881

Compound										
1,3-Dichlorobenzene	3.457	0.9967	9.50	0.9998	0.02037	0.9985	0.05296	0.9950	65.91	0.9892
1,4-Dichlorobenzene	3.911	0.9990	10.21	0.9998	0.02460	0.9994	0.06158	0.9994	81.77	0.9893
2,4-Dichlorophenol	2.537	0.9991	14.22	0.9977	1.2167	0.9964	0.1459	0.9969	200.5	0.9904
Dimethyl phthalate	7.831	0.9995	18.25	0.9997	2.7816	0.9998	0.2833	0.9996	595.5	0.9983
Diethyl phthalate	7.201	0.9991	19.53	0.9996	—	0.9998	—	0.9997	582.5	0.9986
4,6-Dinitro-o-cresol	1.705	0.9631	9.70	0.9872	1.8689	0.9936	0.1635	0.9967	276.9	0.9984
2,4-Dinitrophenol	1.288	0.9582	6.21	0.9706	1.6330	0.9920	0.1506	0.9916	291.8	0.9924
2,4-Dinitrotoluene	2.469	0.9987	9.67	0.9998	0.06511	0.9966	0.1023	0.9928	286.4	0.9839
2,6-Dinitrotoluene	1.814	0.9989	10.22	0.9989	0.06785	0.9947	0.1077	0.9966	305.3	0.9763
Di-n-octylphthalate	17.291	0.9963	33.43	0.9970	0.06255	0.9983	0.06255	0.9991	386.7	0.8362
Ethylhexyl phthalate	9.109	0.9989	35.43	0.9987	1.9710	0.9991	0.09021	0.9993	432.8	0.9974
Fluoranthene	11.387	0.9969	24.51	0.9989	1.8774	0.9983	0.07181	0.9966	65.55	0.9022
Fluorene	5.646	0.9985	18.88	0.9980	0.8562	0.9987	0.09232	0.9989	44.89	0.9172
Hexachlorobenzene	6.282	0.9994	17.67	0.9981	0.04836	0.9981	0.06835	0.9995	127.7	0.9979
1,3-Hexachlorocyclobutadiene	2.315	0.9965	10.18	0.9992	0.01960	0.9960	0.03580	0.9979	35.37	0.8575
Hexachlorocyclopentadiene	2.905	0.9982	9.70	0.9996	0.01857	0.9862	0.03333	0.9845	46.51	0.8875
Hexachloroethane	1.379	0.9972	7.42	0.9991	0.05959	0.9986	0.1416	0.9992	49.74	0.6081
Isophorone	9.270	0.9996	14.38	0.9997	0.03971	0.9990	0.08679	0.9984	144.9	0.9899
2-Methylnaphthalene	6.213	0.9991	14.75	0.9990	0.01391	0.9950	0.02403	0.9981	33.07	0.9907
Naphthalene	7.966	0.9977	17.64	0.9994	0.8441	0.9954	0.08342	0.9956	39.09	0.9781
2-Nitroaniline	2.371	0.9973	10.05	0.9979	1.1820	0.9994	0.1114	0.9996	220.3	0.9936
3-Nitroaniline	2.974	0.9962	10.22	0.9968	1.4008	0.9990	0.1220	0.9985	239.9	0.9928
4-Nitroaniline	2.559	0.9915	8.86	0.9895	1.8692	0.9992	0.9715	0.9936	277.4	0.9954
Nitrobenzene	1.932	0.9991	9.74	0.9997	0.04047	0.9979	0.9221	0.9997	199.8	0.9765
4-Nitrophenol	1.465	0.9953	9.89	0.9955	1.6075	0.9953	0.1124	0.9951	246.2	0.9918
N-Nitrosodimethylamine	2.420	0.9991	2.47	0.9954	0.9889	0.9993	0.1198	0.9982	272.5	0.9987
N-Nitrosodiphenylamine	2.708	0.9997	10.36	0.9979	0.8995	0.9971	0.0468	0.9994	117.8	0.9885
N-Nitroso-di-n-propylamine	3.154	0.9991	6.32	0.9986	0.8665	0.9995	0.0988	0.9991	196.5	0.9988

Table 11.4. (*Continued*)

| Compound | MS Detector[b] | | | | IR Detector | | | | | |
| | Single Ion | | Total Ion | | Integrated Absorbance Slope[c] | r^2 | Maximum Absorbance Slope[d] | r^2 | Gram–Schmidt Slope[e] | r^2 |
	Slope $\times 10^4$	r^2	Slope $\times 10^4$	r^2						
Pentachlorophenol	4.044	0.9685	19.01	0.9824	0.7966	0.9883	0.06790	0.9859	132.7	0.9882
Phenanthrene	7.5356	0.9975	19.47	0.9998	0.5537	0.9989	0.05169	0.9941	34.18	0.9827
Phenol	4.3279	0.9991	10.61	0.9995	0.7462	0.9966	0.09232	0.9978	116.5	0.9825
Pyrene	10.9600	0.9821	24.99	0.9973	0.8724	0.9977	0.07105	0.9971	45.97	0.9011
1,2,4-Trichlorobenzene	3.1692	0.9987	10.49	0.9998	0.02551	0.9979	0.05258	0.9969	63.98	0.9970
2,4,5-Trichlorophenol	2.3832	0.9759	14.75	0.9974	0.9709	0.9966	0.1083	0.9952	164.8	0.9924
2,4,6-Trichlorophenol	2.1624	0.9945	14.32	0.9965	1.0598	0.9965	0.1135	0.9969	143.2	0.9901
Mean		0.9950		0.9965		0.9929		0.9962		0.9695

[a] Based on duplicate 3-μL injections of 25, 50, 100, and 250 ng.
[b] Based on area counts of most abundant ion.
[c] Integrated absorbance across multiple peak scans.
[d] FT–IR scan with highest absorbance.
[e] Based on area counts of IR chromatogram peak.

284

Regression analysis was applied to G–S, TIC, SIC, and the integrated and maximum absorbance approaches. Table 11.4 shows the regression analysis results on 15 compounds for the GC/FT–IR/MS data. Correlation coefficients indicated that the degree of plot scatter increased in the order SIC < maximum absorbance = integrated absorbance ≪ G–S. Correlation coefficients for both absorbance approaches exceed 0.9950 over the range of 25–250 ng, indicating sufficient precision for routine analysis. As shown in Figure 11.17, G–S plots, unlike those of the other approaches, generally did not pass through the graph origin. Thus, where possible, the G–S approach should not be used for IR quantification.

9. APPLICATIONS OF CAPILLARY GC/FT–IR

Applications of the GC/FT–IR technique have been reviewed by Smith (45) and White (46). The technique has been applied successfully to environmental monitoring, fossil fuels (energy), flavors and fragrances, natural products, and industrial samples. Selected applications of the technique for significant reports after 1984, along with their references (47–60), are listed in Table 11.5. The cited applications utilize preconcentration, MS confirmation of FT–IR data, sampling by on-the-fly or trapping, and two-dimensional GC (inter alia). Because, like GC/MS, GC/FT–IR is a universal detection system, the range of applications is seriously limited only by analyte volatility (47), and super-critical fluid (SCF) FT–IR or high-performance liquid chromatography (HPLC) FT–IR may be used directly on nonvolatiles (48). At first glance it may seem surprising that so many of the cited applications utilize GC/MS in conjunction with GC/FT–IR (either as independent or directly linked systems). However, Figures 11.14 and 11.16 show the complexity of the chromatograms obtained from some of these applications. It is immediately suspected that either spectral technique, used alone, would be insufficient to characterize more than a small percentage of the detected analytes. Indeed, it has been shown that the data collected from a linked GC/FT–IR/MS system is superior to that of GC/MS and provides confirmed spectral information on 41% of the jointly detected analytes in complex environmental samples (49).

10. FUTURE TRENDS

10.1. Hyphenated FT–IR Approaches

Hyphenated techniques, a term coined by Hirschfeld (61), can be as simple as a separation device linked to a detector (for example, GC/FT–IR) or as

Table 11.5. Selected Capillary GC/FT–IR Applications[a]

Category	Sample Type	Analyte Type	GC Column	Comment	Reference
Energy	Coal combustion extract	Hydrocarbon isomers	$1.5\,m \times 0.32\,mm$, $1\,\mu m$ DB-5	Characterized isomers when GC/MS failed	50
Energy	Coal gasification and coal oxidation extracts	Hydrocarbons, acids, esters	$30\,m \times 0.32\,mm$, $1.0\,\mu m$ DB-5	Linked FT–IR to ion trap	51
Energy	C_8 petroleum feedstock	Hydrocarbons	$50\,m \times 0.25\,mm$, OV-101 or Carbowax 20 m	Correlated GC/MS molecular weight with FT–IR data	52
Energy	Jet fuel	Hydrocarbons	$60\,m \times 0.33\,mm$, $1\,\mu m$ DB-5	Studied effect of data acquisition parameters	53
Environmental	Sediment, oil	Chlorinated aromatics, hydrocarbons, phosphorus pesticides	$30\,m \times 0.32\,mm$, $1.0\,\mu m$ DB-5	Compared light-pipe FT–IR and MS results	54
Environmental	Soil	Chlorinated hydrocarbons	$30\,m \times 0.32\,mm$, $1.0\,\mu m$ DB-5	Compared light-pipe FT–IR and MS results	14
Environmental	Still bottom	Chlorinated phenols	$60\,m \times 0.25\,mm$, $0.25\,\mu m$ DB-5	Employed cryotrapping FT–IR	55
Environmental	Soil	Dioxins	$50\,m \times 0.2\,mm$, $0.25\,\mu m$ DB-5	Employed MI/GC/ FT–IR	56

286

Environmental	Sediment	Chlorinated compounds	30 m × 0.33 mm, 1 μm DB-5	Reported quantitative results	13
Environmental	Crude oil	Hydrocarbon	30 mm × 0.33 mm, 1 μm DB-5	Studied effect of extract background on detectability	41
Flavors and fragrances	Chocolate	Polymethyl pyrazines	60 m × 0.32 mm, 1 μm DB-1	Used injector preconcentrator	57
Flavors and fragrances	Essential oil	Terpenes	30 m × 0.53 mm, 1 μm DB-5 30 m × 0.32 mm, 0.25 μm DB-wax	Used two-dimensional GC	36
Flavors and fragrances	Citrus extract	Ketones, aldehydes, esters		Compared nasal and FT–IR results	58
Industrial	Fermentation by products	Carbohydrate degradation products	30 m × 0.33 mm, 1 μm DB-1	Used pyrolysis GC	45
Industrial	Polymer adhesive	Acrylic compounds	50 m × 0.33 mm, DB-5	Studied aging process	59
Industrial	Hydrogenation mixture	Polyfluoroethylenes	1 m × 3 mm, porosil C	Study of hydrogenation products	60
Natural products	Grass oil extract	Terpenes	30 m × 0.33 mm, 1.0 μm DB-1	Determined functionality	45
Natural products	Sunflower oil extract	Diterpenoids	30 m × 0.33 mm, 1.0 μm DB-5	Compared FT–IR and MS data	47

[a] Selected from 1984 or later references.

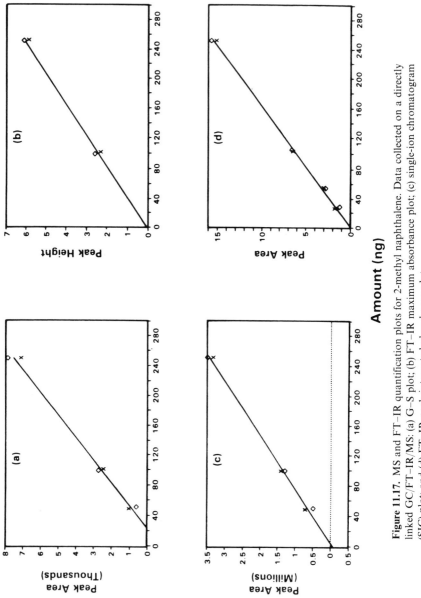

Figure 11.17. MS and FT–IR quantification plots for 2-methyl naphthalene. Data collected on a directly linked GC/FT–IR/MS: (a) G–S plot; (b) FT–IR maximum absorbance plot; (c) single-ion chromatogram (SIC) plot; and (d) FT–IR peak integrated absorbance plot.

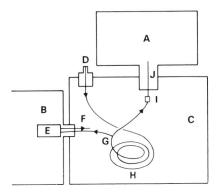

Figure 11.18. GC/FT–IR/MS interface: (A) MSD; (B) FT–IR interface; (C) GC oven; (D) oncolumn injector; (E) light pipe assembly; (F) effluent vent; (G) splitter; (H) 30-m × 0.32-nm DB-5 column; (I) 0.3-m × 0.23-mm DB-5 coated line; (J) MSD interface. Reprinted with permission from D. F. Gurka and R. Titus, *Anal. Chem.* **58**, 2189 (1986). Copyright 1986, American Chemical Society.

complex as a separation device linked to several detectors (for example, GC/FT–IR/MS). Because many samples are too complex to be characterized by a single detector, a dual-detector approach has been postulated as an alternative (14, 49). Williams et al. have defined the problems associated with GC/FT–IR/MS and noted that such a system may operate in a parallel or a serial (tandem) mode (62). The parallel split mode has the advantage that the MS chromatographic resolution is not degraded by the light pipe dead volume and the disadvantage that neither detector receives 100% of the analyte. Figure 11.18 shows a schematic of the parallel approach. In either a parallel or a serial mode, the effluent split to the MS detector must be chosen such that the flow rate is compatible with the MS vacuum pumping capability. The serial configuration directs all of the GC effluent through the nondestructive FT–IR detector and, with a well-chosen light pipe volume, can minimize chromatographic resolution problems at the MS. However, the transfer line between the detectors can lead to longer retention times at the MS than the FT–IR, and a computerized solution to this problem has been proposed (63).

Commercial systems utilizing both GC/FT–IR/MS and MI/GC/FT–IR/MS were exhibited at the 1987 Pittsburgh Conference (30). Wilkins has reviewed the state of the art of this hyphenated technique to 1987 (64). Environmental applications of hyphenated FT–IR/MS to volatiles and nonvolatiles have also been reported (65, 66).

10.2. Supercritical Fluid/FT–IR

While many analytes in their derivatized or underivatized forms are volatile enough for GC, many must be separated by HPLC or other techniques appropriate for nonvolatiles. Removal of the HPLC mobile phase presents unique problems for both FT–IR (48) and MS detectors (67). These problems may

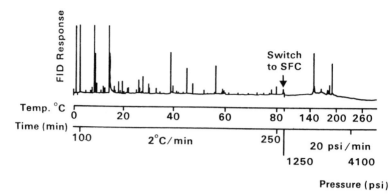

Figure 11.19. Combined GC/SCF chromatogram of a still-bottom residue extract. Mobile phase was switched from helium at 250 °C to supercritical carbon dioxide at 120 °C and 1250 psi at the time indicated by the arrow. Reprinted with permission from S. L. Pentoney, Jr., A. Giorgetti, and P. R. Griffiths, *J. Chromatogr. Sci.* **25**, 93 (1987). Copyright 1987, Preston Publications, a Division of Preston Industries, Inc.

be obviated by using SCF mobile phases, which are gases at atmospheric pressure. SCF possess unique properties at or near their critical state of pressure or temperature. Of particular use to the separation sciences are the lower mobile phase viscosities (as much as a factor of 10 better than the liquid state) and the increased solute diffusivities (68). The utilization of these advantages offers the possibility of increasing chromatographic resolution by decreasing column inside diameter or by reducing packing-particle diameter (48). Griffiths and co-workers [see Haefner et al. (33) and Fuoco et al. (69)] have reported a cryotrapping FT–IR interface (see Figure 11.13) that provides real-time data in the low-nanogram range. Their demonstration that GC and SFC chromatograms can be generated from the same injection (55) presents the possibility of sequential analysis of volatiles and nonvolatiles. Figure 11.19 shows the realization of this goal on an environmental still-bottom sample. Successful application of SFC/FT–IR to agricultural compounds (70) and to coal tar pitch (71) indicates an important role for this technique.

Because most spectral interfaces must solve the problem of mobile phase removal, Griffiths has postulated a universal FT–IR interface (72,73). Such an interface could accommodate GC, SFC, and HPLC mobile phases.

10.3. Interpretative and Data Acquisition Software

While spectral detectors offer higher information content and selectivity relative to that of detectors more commonly in use (flame ionization, electron

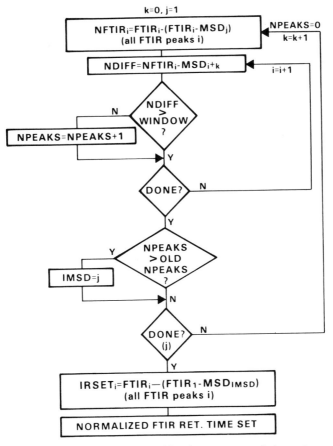

Figure 11.20. Flow chart of correlation algorithm for reconstructed chromatogram overlap: DONE?, is (i) = the total number of IR peaks detected?; NPEAKS = number of FTIR peaks coincident with MSD peaks; IMSD = Nth MSD peak number giving largest NPEAKS; FT–IR$_i$ and MSD$_i$ = respective component peak apex time into run; DONE? is (j) = (the total number of IR peaks detected)/3. Reprinted with permission from J. R. Cooper, I. C. Bowater, and C. L. Wilkins, *Anal. Chem.* **58**, 2791 (1986). Copyright 1986, American Chemical Society.

capture, photoionization), they place an increased interpretative demand upon the analyst. Nevertheless, the increased detector selectivity is required for complex samples such as those that are environmentally derived. Demand for this type of sample analysis rose to 1×10^5 samples during 1987 in the EPA's Superfund Program (74).

To balance the problems of increased interpretative demand and high sample loads with a need for more sample characterization than can be provided by a single detector, an integrated computerized GC/FT–IR/MS system is needed. Although the hardware for such a system is now available, the integrated software for data acquisition, data processing, and spectral interpretation is not. A software routine capable of cross-correlating and maximizing the library search routines for data derived from both detectors has been reported (63). The software algorithm (see Figure 11.20) cross-correlates corresponding peaks from each detector even when retention times and the number of detections is different for MS and FT–IR. Transfer-line-induced time lags, as well as differences in detector sensitivity and selectivity, all can lead to peak assignment problems.

Because the efficacy of the library search approach depends on the availability of reference spectra, it is not expected that such an approach will be useful for all analytical applications. This is especially true for environmental analysis, which typically deals with total unknowns, in a world where the number of known chemicals is increasing exponentially (75). Ultimately, the capability to generate several types of spectra in real time may catalyze efforts to predict molecular structure from spectral data. Several groups have already reported success along these lines, but computational times were excessive (76). The biggest drawback to structural prediction from spectral data may be the necessity for more than MS and FT–IR spectral data. Efforts reported thus far usually employ nuclear magnetic resonance (NMR) data in conjunction with those obtained from MS and FT–IR (77, 78). Current trends in FT–IR and MS miniaturization, along with FT–IR sensitivity improvements, have encouraged their hyphenation. No such trends in NMR improvement have yet emerged. Until such improvement occurs, approaches that maximize the amount, type, and interpretation of data available from MS and FT–IR should be pursued. The reported use of simultaneous pulsed mode real-time electron impact (EI) and chemical ionization (CI) for Fourier transform MS, in conjunction with FT–IR, is one such approach (79). Faster parallel processing computers (80) and optical storage devices (81) should alleviate the computational speed and data storage problems inherent to this data-intense approach. Reference 75 reports that a 2400-ft reel of optical storage tape, of the same dimensions as a standard $10\frac{1}{2}$-in magnetic tape, can store 600,000 Mbytes, as opposed to 320 Mbytes for its magnetic counterpart. Similar progress in the development of spectral-data-based molecular structure prediction software could eventually provide capillary GC detector systems exhibiting a decreased reliance on standards and reference spectra. This decreased reliance is an essential response to the explosive annual increase in the number of known chemicals. The annual increase has been reported to be 4×10^5 for Chemical Abstracts Service registration, with about a thousand new chemicals entering into commercial use yearly and thus eventually into the environment (82).

NOTICE

Although the research described in this paper has been funded wholly or in part by the U.S. Environmental Protection Agency, it has not been subjected to EPA review and, therefore, does not necessarily reflect the views of the EPA, and no official endorsement should be inferred. Mention of trade names or commercial products does not constitute endorsement or recommendation for use.

REFERENCES

1. N. L. Alpert, W. E. Keiser, and H. A. Szymanski, *IR-Theory and Practice of Infrared Spectroscopy*, pp. 30–42. Plenum/Rosetta, New York, 1970.

2. D. Welti, *Infrared Vapour Spectra*, Heyden/Sadtler, New York, 1970.

3. M. Born and E. Wolf, *Principles of Optics* 6th ed., p. 256. Pergamon, New York, 1980.

4. P. R. Griffiths and J. A. de Haseth, *Fourier Transform Infrared Spectrometry*, Chapter 1. Wiley, New York, 1986.

5. P. R. Griffiths and J. A. de Haseth, *Fourier Transform Infrared Spectrometry*, pp. 274–275. Wiley, New York, 1986.

6. P. R. Griffiths, H. J. Sloane, and R. W. Hannah, *Appl. Spectrosc.* **31**, 485 (1977).

7. T. Hirschfeld, in *Fourier Transform Infrared Spectroscopy: Applications to Chemical Systems* (J. R. Ferraro and L. J. Basile, eds.), Vol. 2, Chapter 6, p. 193. Academic Press, New York, 1979.

8. G. M. Message, *Practical Aspects of Gas Chromatography/Mass Spectrometry*, pp. 196–198. Wiley (Interscience), New York, 1984.

9. J. N. Berube and H. L. Buijs, *ACS Symp. Ser.*, **57**, 106 (1977).

10. P. R. Griffiths and J. A. de Haseth, *Fourier Transform Infrared Spectrometry*, p. 229. Wiley, New York, 1986.

11. P. R. Griffiths and J. A. de Haseth, *Fourier Transform Infrared Spectrometry*, pp. 73–74. Wiley, New York, 1986.

12. J. A. de Haseth and T. L. Isenhour, *Anal. Chem.* **49**, 1977 (1977).

13. D. F. Gurka and S. Pyle, *Environ. Sci. Technol.* **22**, 963 (1988).

14. D. F. Gurka, M. Hiatt, and R. Titus, *Anal. Chem.* **56**, 1102 (1984).

15. M. J. D. Low and S. K. Freeman, *Anal. Chem.* **39**, 194 (1967).

16. L. V. Azarraga, *Appl. Spectrosc.* **34**, 224 (1980).

17. P. W. J. Yang, E. L. Ethridge, J. L. Lane, and P. R. Griffiths, *Appl. Spectrosc.* **38**, 613 (1984).

18. P. R. Griffiths, *Appl. Spectrosc.* **31**, 284 (1977).

19. D. F. Gurka, P. R. Laska, and R. J. Titus, *J. Chromatogr. Sci.* **20**, 145 (1982).

20. P. R. Griffiths and J. A. de Haseth, *Fourier Transform Infrared Spectrometry*, p. 215. Wiley, New York, 1986.

21. G. Mamantov, A. A. Garrison, and E. L. Wehry, *Appl. Spectrosc.*, **36**, 339 (1982).

22. R. S. Brown, J. R. Cooper, and C. L. Wilkins, *Anal. Chem.* **57**, 2275 (1985).

23. D. E. Henry, A. Giorgetti, A. M. Haefner, P. R. Griffiths, and D. F. Gurka, *Anal. Chem.* **59**, 2356 (1987).

24. D. F. Gurka, R. Titus, P. R. Griffiths, D. E. Henry, and A. Giorgetti, *Anal. Chem.* **59**, 2362 (1987).

25. G. T. Reedy, S. Bourne, and P. T. Cunningham, *Anal. Chem.* **51**, 1535 (1979).

26. S. Borman, *Anal. Chem.* **56**, 936A (1984).

27. Advertisement in *Anal. Chem.* **59**, 489A (1987).

28. J. A. de Haseth, *Appl. Spectrosc.* **36**, 544 (1982).

29. R. Cournoyer, J. C. Shearer, and D. H. Anderson, *Anal. Chem.* **49**, 2275 (1977).

30. S. Borman, *Anal. Chem.* **59**, 769A (1987).

31. W. M. Coleman, III and B. M. Gordon, *Appl. Spectrosc.* **41**, 1163 (1987).

32. W. M. Coleman, III and B. M. Gordon, *Appl. Spectrosc.* **42**, 671 (1988).

33. A. M. Haefner, K. L. Norton, P. R. Griffiths, S. Bourne, and R. Curbelo, *Anal. Chem.* **60**, 2441 (1988).

34. S. Bourne, A. M. Haefner, K. L. Norton, and P. R. Griffiths, *Anal. Chem.* **62**, 2448 (1990).

35. T. Visser and M. J. Vredenbregt, *Vibrat. Spectrosc.* **1**, 205 (1990).

36. S. L. Smith, *Appl. Spectrosc.* **40**, 278 (1986).

37. J. V. Johnson and R. A. Yost, *Anal. Chem.* **57**, 758A (1985).

38. R. G. Kollar, R. Citerin, M. Markelov, and K. L. Gallaher, *Proc. SPIE—Int. Soc. Opt. Eng.* **553**, 192 (1985).

39. K. Grob and B. Schilling, *J. Chromatogr.* **299**, 415 (1984).

40. D. F. Gurka, S. M. Pyle, I. Farnham, and R. Titus, *J. Chromatogr. Sci.* **29**, 339 (1991).

41. D. F. Gurka, I. Farnham, B. B. Potter, S. Pyle, R. Titus, and W. Duncan, *Anal. Chem.* **61**, 1584 (1989).

42. D. T. Sparks, R. B. Lam, and T. L. Isenhour, *Anal. Chem.* **54**, 1922 (1982).

43. J. R. Cooper and L. T. Taylor, *Anal. Chem.* **56**, 1989 (1984).

44. D. M. Hembree, A. A. Garrison, R. A. Crocombe, E. L. Wehry, and G. Mamantov, *Anal. Chem.* **53**, 1783 (1981).

45. J. L. Smith, in *Advances in Capillary Chromatography* (J. G. Nikelly, ed.), Chapter 3. Huethig, New York, 1986.

46. R. L. White, *Appl. Spectrosc. Rev.* **23**, 1 (1987).

47. S. L. Smith, *J. Chromatogr. Sci.* **22**, 143 (1984).

48. P. R. Griffiths, S. L. Pentoney, Jr., A. Giorgetti, and K. H. Shafer, *Anal. Chem.* **58**, 1349A (1986).

49. D. F. Gurka and R. Titus, *Anal. Chem.* **58**, 2189 (1986).

50. K. S. Chiu, K. Biemann, K. Krishnan, and S. L. Hill, *Anal. Chem.* **56**, 1610 (1984).

51. E. S. Olson and J. W. Diehl, *Anal. Chem.* **59**, 443 (1987).

52. V. N. Garg, B. D. Bhatt, V. K. Kaushik, and K. R. Murthy, *J. Chromatogr. Sci.* **25**, 237 (1987).

53. J. R. Cooper and L. T. Taylor, *Appl. Spectrosc.* **38**, 366 (1984).

54. K. H. Shafer, T. L. Hayes, J. W. Brasch, and R. J. Jakobsen, *Anal. Chem.* **56**, 237 (1984).

55. S. L. Pentoney, Jr., A. Giorgetti, and P. R. Griffiths, *J. Chromatogr. Sci.* **25**, 93 (1987).

56. D. F. Gurka, J. W. Brasch, R. H. Barnes, C. J. Riggle, and S. Bourne, *Appl. Spectrosc.* **40**, 978 (1986).

57. A. J. Fehl and C. Marcott, *Anal. Chem.* **58**, 2578 (1978).

58. G. Fischbock, W. Pfannhauser, and R. Kellner, *Mikrochim. Acta*, **1**, 27 (1988).

59. G. A. Luoma and R. D. Rowland, *J. Chromatogr. Sci.* **24**, 210 (1986).

60. J. M. Chalmers, M. W. MacKenzie, and H. A. Willis, *Appl. Spectrosc.* **38**, 763 (1984).

61. T. Hirschfeld, *Anal. Chem.* **52**, 297 (1980).

62. S. S. Williams, R. B. Lam, D. T. Sparks, and T. L. Isenhour, *Anal. Chim. Acta* **138**, 1 (1982).

63. J. R. Cooper, I. C. Bowater, and C. L. Wilkins, *Anal. Chem.* **58**, 2791 (1986).

64. C. L. Wilkins, *Anal. Chem.* **59**, 571A (1987).

65. D. F. Gurka, L. D. Betowski, T. A. Hinners, E. M. Heithmar, R. Titus, and J. M. Henshaw, *Anal. Chem.* **60**, 454A (1988).

66. D. F. Gurka, L. D. Betowski, T. L. Jones, S. M. Pyle, R. Titus, J. M. Ballard, Y. Tondeur, and W. Niederhut, *J. Chromatogr. Sci.* **26**, 301 (1988).

67. T. R. Covey, E. D. Lee, A. P. Bruins, and J. D. Henion, *Anal. Chem.* **58**, 1451A (1986).

68. M. Novotny, S. R. Springston, P. A. Peaden, J. C. Fjeldsted, and M. L. Lee, *Anal. Chem.* **53**, 407A (1981).

69. R. Fuoco, K. H. Shafer, and P. R. Griffiths, *Anal. Chem.* **58**, 3249 (1986).

70. S. Shah and L. T. Taylor, *HRC & CC, J. High Resolut. Chromatogr. Chromatogr. Commun.* **12**, 599 (1989).

71. M. W. Raynor, I. L. Davies, K. D. Bartle, A. A. Clifford, A. Williams, J. M. Chalmers, and B. W. Cook, *HRC & CC, J. High Resolut. Chromatogr. Chromatogr. Commun.* **11**, 766 (1988).

72. P. R. Griffiths, S. L. Pentoney, Jr., G. L. Pariente, and K. L. Norton, *Mikrochim. Acta* **1**, 47 (1988).

73. P. R. Griffiths, A. M. Haefner, K. L. Norton, D. J. J. Fraser, D. Pyo, and H. Makishima, *HRC & CC, J. High Resolut. Chromatogr. Chromatogr. Commun.* **12**, 119 (1989).

74. W. Worthy, *Chem. Eng. News* **65** (36), 33 (1987).

75. *Opportunities in Chemistry*. National Academy of Sciences, Washington, DC, 1985.

76. Z. Hipp, in *Computer Supported Spectroscopic Databases* (J. Zupan, ed.), Chapter 8. Wiley, New York, 1986.

77. M. E. Munk, M. Farkas, A. H. Lipkis, and B. D. Christie, *Mikrochim. Acta* **2**, 199 (1986).

78. D. A. Laude, Jr. and C. L. Wilkins, *Anal. Chem.* **58**, 2820 (1986).

79. D. A. Laude, Jr., G. M. Brissey, C. F. Ijames, R. S. Brown, and C. L. Wilkins, *Anal. Chem.* **56**, 1163 (1984).

80. R. W. Hockney and C. R. Jesshope, *Parallel Computers.* Adam Hilger, Bristol, 1983.

81. D. O'Sullivan, *Chem. Eng. News* **66**(8), 6 (1988).

82. J. C. Arcos, *Environ. Sci. Technol.* **21**, 743 (1987).

CHAPTER

12

THE ION MOBILITY DETECTOR

HERBERT H. HILL

Department of Chemistry
Washington State University
Pullman, Washington

DENNIS G. McMINN

Department of Chemistry
Gonzaga University
Spokane, Washington

The use of ambient pressure ionization for the detection of compounds after capillary chromatography provides the chromatographer with a varied selection of methods from which to choose. For the nonselective detection of organic compounds, flame ionization detection has evolved as the method of choice. For halogenated compounds and nitro compounds, electron capture detection is preferred. For nitrogen- and phosphorous-containing compounds, the alkali-doped detection methods provide enhanced sensitivity. And for metal- and silicon-containing compounds, the hydrogen atmosphere flame ionization detector (HAFID) has proven useful.

In each case, these detectors rely on an ionization process for analytical information. By combining ionization chemistry with ion mobility measurements, the ion mobility detector (IMD) provides a flexibility for chromatographic detection that is not available from other methods. Yet it maintains the simplicity and low cost of ambient pressure ionization detection methods in general. In addition, the IMD is the only ambient pressure ionization detector that has been successfully coupled to all three forms of chromatography: gas (GC), supercritical fluid (SFC), and liquid chromatography (LC).

Detectors for Capillary Chromatography, edited by Herbert H. Hill and Dennis G. McMinn.
Chemical Analysis Series, Vol. 121.
ISBN 0-471-50645-1 © 1992 John Wiley & Sons, Inc.

1. BASIC PRINCIPLES OF ION MOBILITY SPECTROMETRY

Figure 12.1 provides a schematic diagram of an ion mobility spectrometer. The spectrometer can be divided into five regions: (1) the ionization region; (2) the ion–molecule reaction region; (3) the ion gate region; (4) the ion drift region; and (5) the ion collection region.

1.1. The Ionization Region

In theory, any ambient pressure ionization source can be used as the primary ion source of an ion mobility spectrometer. Although flame ionization has not been investigated for ion mobility spectrometry (IMS), other methods such as radioactive ionization (1), photoionization (2), and electrified spray ionization (3) have been employed successfully to create ions in IMD after chromatography.

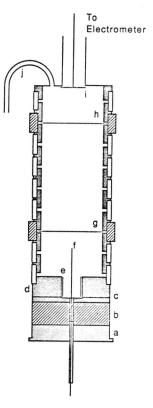

Figure 12.1. Schematic drawing of an ion mobility spectrometer: (a) detector base; (b) macro insulator; (c) drift gas outlet; (d) high-voltage supply; (e) ^{63}Ni foil; (f) capillary column; (g) entrance gate; (h) exit gate; (i) collector; and (j) drift gas inlet. Reprinted with permission from R. H. St. Louis, W. F. Siems, and H. H. Hill, Jr., *LC-GC* **9**, 810 (1988). Copyright 1988, Aster Publications.

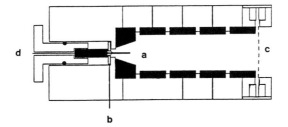

Figure 12.2. Schematic drawing of coronaspray ionization source: (a) electrospray needle; (b) high-voltage supply; (c) ion entrance gate; and (d) LC column. Adapted from C. B. Shumate, Ph.D. thesis, Washington State University, Pullman, 1989.

In electrified spray ionization, a corona discharge is established from the orifice of an electropolished needle into a neutral gas of nitrogen or air. Electrons from the discharge create a plasma of positive and negative ions, with concentration maxima near the point of the orifice. Figure 12.2 provides a detailed schematic of the coronaspray ionization source. Although this source can be used for GC and SFC, its primary application has been for the nebulization and ionization of compounds separated by liquid chromatography.

Figure 12.3 provides a schematic diagram of a photoionization source that has been used for IMD after capillary GC. In this approach, a 10.2-eV (123.6-nm) krypton lamp is positioned such that the effluent from a capillary gas chromatograph enters the detector in front of the lamp's magnesium fluoride window. With photoionization, the inert gases of the detector and the mobile phase of the chromatograph have ionization potentials too large to be ionization by the photons of the lamp. When organic compounds enter the detector, photoionization of these compounds occur, producing unfragmented ions with characteristic mobilities.

By far the most common ionization source used with IMD after chromatography is the ^{63}Ni radioactive source. This source is similar to that used in most electron capture detectors (ECDs). Typically, the ^{63}Ni foil is positioned inside the ion mobility spectrometer and the beta emission from the foil ionizes the nitrogen drift gas to produce positive reactant ions and thermal electrons.

1.2. The Ion–Molecule Reaction Region

One advantage of photoionization is that secondary ion–molecule reactions appear to be minimized and the mobility of the primary ions produced from the photoionization process is directly measured in the detector. For coronaspray and ^{63}Ni ionization, however, gas-phase ion–molecule chemistry is

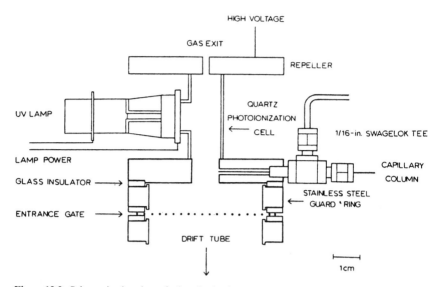

Figure 12.3. Schematic drawing of photoionization source. Reprinted with permission from M. A. Baim, R. L. Eatherton, and H. H. Hill, Jr., *Anal. Chem.* **55**, 1761 (1983). Copyright 1983, American Chemical Society.

of primary importance. Once primary ions have been formed in the detector, a variety of ion–molecule reactions occur to produce stable response ions on which mobility measurements are made. Although the energy for primary ionization occurring with coronaspray is derived from the electric field whereas that from the ^{63}Ni comes from the radioactive emission of beta particles, the secondary ionization processes in the coronaspray source may be similar to those of the ^{63}Ni source.

After initial ionization of the nitrogen gas, the primary N_2^+ undergoes a series of ion–molecule reactions with trace amounts of H_2O, NH_3, and NO to produce stable secondary ion clusters called *reactant ions*. For the ^{63}Ni source, these positive reactant ions have been identified as $(H_2O)_nNH_4^+$, $(H_2O)_nNO^+$, and $(H_2O)_nH^+$. Similarly, thermal electrons produced during the primary ionization process can undergo capture reactions with oxygen, water, and carbon dioxide to form negative reactant ion clusters that have been identified as $(H_2O)_nO_2^-$ and $(H_2O)_n(CO_2)_mO_2^-$.

With the ^{63}Ni source, IMD response to analytes introduced into the detector from the chromatograph occurs through ion–molecule interactions of the analytes with these positive and negative reactant ions. Stable ions produced from this interaction are called *product ions*, and their characteristic mobilities are measured to provide qualitative information about the analyte while their concentration is measured to provide quantitative information.

1.3. The Ion Gate Region

The entire ionization and ion–molecule reaction region is placed under an electric field so that ions formed are directed down the field until they are stopped by an ion gate. Called a Bradbury and Neilsen gate [see Bradbury and Neilsen (4)], this consists of a single grid of finely spaced wires. Alternate wires are biased above and below the drift field potential. Thus an orthogonal electric field is produced that is some three times the drift field potential, directing the ions into the wires where they are neutralized. To open the gate and allow the ions to enter or exit the drift region, the gate wires are referenced to the drift potential.

1.4. The Ion Drift Region

An ion entering the drift region of the spectrometer is accelerated by an electric field until it collides with a neutral molecule, accelerates again until it has another collision, and so forth. This chaotic sequence of accelerations and collisions at the molecular level translates into a constant ion velocity over macroscopic distances. This constant ion velocity (v) is directly proportional to the electric field strength (E), and the proportionality constant is called the ion mobility constant (K):

$$v = KE$$

According to the Mason–Schamp equation (5), this ion mobility constant is determined by the charge on the ion (e), the mass of the ion (m), the mass of the neutral drift gass (M), the temperature of the drift gas (T), the number density of the drift gas (N), and the cross section of the ion–molecular interaction (Ω_0):

$$K = \frac{3}{16} \frac{e}{N} \left(\frac{1}{m} + \frac{1}{M} \right)^{1/2} \left(\frac{2\pi}{kT} \right) \left(\frac{1}{\Omega_0} \right)$$

For conditions of constant temperature and pressure, ion mobility is directly proportional to the charge on the ion (Z) and inversely proportional to the collision cross section (Ω_0) and the square root of the reduced mass (μ):

$$K = ZC/\mu^{1/2}\Omega_0$$

where C is a constant, and $\mu = mM/(m + M)$.

Because different ions have different velocities in a constant electric field, the time it takes for an ion to traverse the drift region becomes the analytical measurement of its mobility. Thus, for a constant electric field and drift length, the drift time (t) of an ion through the drift region depends directly on the collision cross section of the ion–drift gas interaction (Ω_0) and the square root of the reduced mass (μ) and inversely on the ionic charge (Z):

$$t = C' \frac{\mu^{1/2}\Omega_0}{Z}$$

where C' is a constant.

1.5. The Ion Collection Region

The design of the ion collection region depends upon the manner in which ion mobility spectra are obtained. The most common way to record spectra is by following arrival time spectra with a fast electrometer and averaging a number of these spectra to increase the signal-to-noise ratio (S/N). A second mode of operation that increases resolution and sensitivity and that permits the use of slower electronics is a time-dispersive interferometric method (6). With this mode of operation a second ion gate is positioned in the spectrometer such that the drift region is located between the first and the second ion gates. Both ion gates are opened and closed at the same frequency. By scanning this gating frequency (for example, from 10 to 2000 Hz), ions come in and out of phase with the gating and form an interferogram at the collecting electrode. Fourier transformation of this interferogram produces a traditional ion mobility spectrum in the time domain (7). For more detailed information on data collection methods in IMS, the reader is referred to St. Louis and Hill (8).

2. DEVELOPMENT OF IMS AS A CHROMATOGRAPHIC DETECTOR

2.1. Early Developments

Although IMS was first introduced as an analytical instrument that was capable of "stand-alone" operation, it was also recognized for its potential as a versatile detection method for GC. As early as 1970, M. J. Cohen and F. W. Karasek numerically evaluated the potential of IMS to respond to samples eluting from both packed and capillary column gas chromatographs (9). The first examples of IMS after GC were from packed columns (1). After separation of musk ambrette in benzene on a 6-ft × $\frac{1}{8}$-in.-o.d. stainless steel column packed with 6% SE-130 on 180/100 mesh Anachrom ABS, the effluent

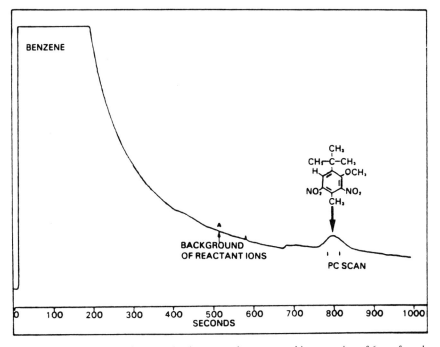

Figure 12.4. FID tracing of a packed column gas chromatographic separation of 6 ng of musk ambrette in benzene. Reprinted with permission from F. W. Karasek and R. A. Keller, *J. Chromatogr. Sci.* **10**, 626 (1972). Copyright 1972, Preston Publications, a division of Preston Industries.

was split 2:1 between an FID and IMS. Figure 12.4 shows the chromatographic separation with FID detection of 6 ng of the musk ambrette, while Figure 12.5 shows the IMS scan of 3 ng of musk ambrette as the compound eluted from the column. Although chromatographic resolution is poor and the ion mobility spectrum is complicated by column bleed, a comparison of the two figures shows that IMS is clearly superior to FID detection of this compound.

In a similar early investigation of IMS as a detection method for gas chromatography, Cram and Chesler separated a series of freons and obtained characteristic IMS scans of each compound as it eluted from the column (10).

In addition to IMS–GC interface systems that relied on splitting the effluent after separation, a valved interface of a gas chromatograph to an ion mobility spectrometer was developed by Karasek and co-workers (11). A schematic of this interface is shown in Figure 12.6.

With this approach, a 0.2-s portion of the effluent from a chromatographic separation could be electrically switched into the spectrometer for ion mobility

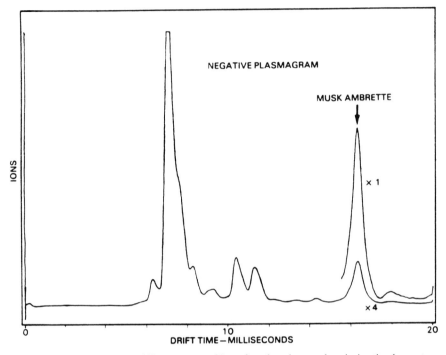

Figure 12.5. Negative ion mobility spectrum of 3 ng of musk ambrette taken during the chromato-
graphic run shown in Figure 12.4. Reprinted with permission from F. W. Karasek and R. A.
Keller, *J. Chromatogr. Sci.* **10**, 626 (1972). Copyright 1972, Preston Publications, a division of
Preston Industries.

analysis. A continuous chromatographic tracing was obtained by placing a
thermal conductivity detector in line between the chromatographic column
and the interface valve. In a similar approach, effluent was split between an
FID and the valve (12). Advantages of the valved interface over the continuous
split introduction method were reduced contamination of the spectrometer
from large quantities of eluant and column bleed while only small quantities
of a highly purified compound of interest were injected into the IMS.

The aformentioned investigations were directed toward operating IMS as
a spectrometer after GC. However, by continuously monitoring individual
ions or ion clusters at specific mobilities or mobility windows, a continuous
chromatographic tracing can be achieved. In this continuous mobility monitoring mode, the IMS behaves more as a traditional chromatographic detector.
In this mode of operation, the IMS is referred to as an ion mobility detector
(IMD). Thus IMS is a method for obtaining spectra, and the IMD is used for
obtaining chromatograms.

The first example of a chromatogram obtained with continuous ion mobility

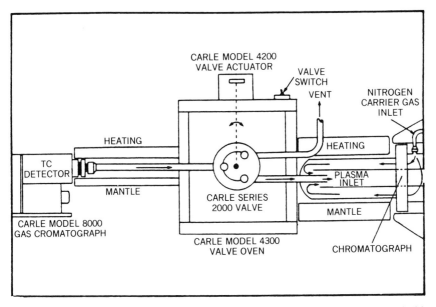

Figure 12.6. Schematic diagram of a valved gas chromatograph–IMS interface. Reprinted with permission from F. W. Karasek, O. S. Tatone, and D. W. Denney, *J. Chromatogr.* **87**, 137 (1973). Copyright 1973, Elsevier Science Publishers, Physical Sciences & Engineering Division.

detection was provided by Franklin GNO Corp. (now PCP Inc.) in an early application note. In this example, 0.125 ng of resmethrin [(5-benzyl-3-furyl)-methyl-2,2-dimethyl-3-(2-methylpropenyl)cyclopropanecarboxylate] in a peanut oil extract was detected after separation on an OV-17 packed column with a 1:1 effluent split (13).

Karasek and colleagues expanded this approach to demonstrate the versatility of response that could be achieved with ion mobility detection (14). Still using packed columns, they illustrated selective and nonselective detection modes for both positive and negative ions. Figure 12.7 provides an example of these detection modes for negative ions: tracing a is a selective detection of iodobenzene; tracing b is a selective detection of chlorobenzene; and tracing c is a nonselective electron capture detection of both compounds. Table 12.1 (15) provides a chronological listing of the early developments in IMS with references to the original literature.

2.2. Developments Since 1980

Early investigations encountered several obstacles to successfully interfacing IMS with chromatographic systems. First, the clearance times of stand-alone

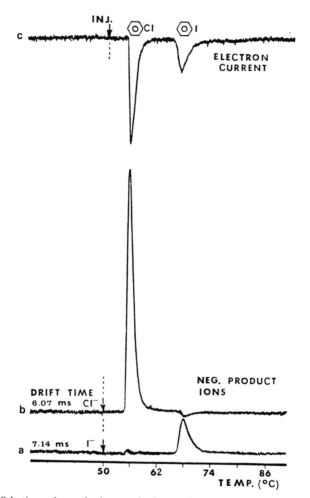

Figure 12.7. Selective and nonselective negative ion continuous detection mode of an IMD after gas chromatography. Reprinted with permission from F. W. Karasek, H. H. Hill, S. H. Kim, and S. Rokushika, *J. Chromatogr.* **135**, 329 (1977).

spectrometers were often on the order of several minutes. Thus much of the resolution gained from the chromatographic separation step could be lost in the detection step. Secondly, the high sensitivity of IMS meant that the instrument was easily contaminated. Column bleed, which was continuously introduced into the detector, would slowly degrade the analytical response and complicate the ion mobility spectrum. Thirdly, the opposing drift flow

Table 12.1. Early Developments in Chromatography–IMS (1970–1979)

Year	Development	Reference
1970	Original concept	9
1972	Splitter interface to GC	1
1973	Multiple scans from single chromatograms	10
	Valved interface to GC	10,11
1975	Continuous monitoring mode	13
1977	Multiple detection modes	14
1978	Column bleed investigations	15

design permitted the introduction of solvent into the drift region of the spectrometer where ion–molecule reactions perturbed mobility measurements.

For interfacing IMS with capillary GC, it is especially important to consider the problems of clearance time and spectrometer cleanliness. Figure 12.1 shows a schematic of an IMS system designed specifically for capillary GC. In this design: (1) drift gas and reactant gas flow patterns were altered so that a unidirectional flow was established throughout the detector; (2) the drift tube was completely sealed to more efficiently flush neutral solvent and sample molecules from the reaction and drift regions of the spectrometer; (3) the ionization region of the detector was reduced in size to lower sample residence time and maintain the integrity of the gas chromatographic resolution; and (4) the effluent from the column was introduced between the ionization region and the entrance gate to the drift region. By introduction of the sample at this point, the neutral molecules are swept away from the drift region with the full velocity of the drift gas and through the ionization region toward the exit. As the neutral molecules pass through the ionization region, ions that are formed in the ionization zone are extracted from the ionizer and directed toward the entrance gate of the spectrometer. Neutrals are swept out of the rear of the instrument.

Because of the efficiency of this design, even samples separated by supercritical fluids can be detected by IMS (16, 17). When modifiers are used with supercritical fluids, it is especially important to introduce the effluent in such a way that it is efficiently swept through the ionization region and out of the detector. When large flow volumes of effluent from packed SFC columns were introduced into the detector, interferences from the modifier were observed (18) and, when the sample was introduced at the base of the detector, reduced sensitivity was reported (19). While more investigations of IMD after SFC appear to be warranted, especially with respect to the ion–molecule chemistry and the effect on response of modifiers, initial results with pure CO_2 as the

Table 12.2. Developments in Chromatography–IMS (1980–1990)

Year	Development	Reference
1982	Microprocessor-controlled IMS	20
	Unidirectional flow IMS	21
	Capillary GC inerface	21
1983	Photoionization source	2
1985	Fourier transform IMS	7
1986	Capillary SFC interface	16, 17
1989	Electrified spray ionization source	3
1990	LC interface	22

supercritical mobile phase have demonstrated ion mobility as a versatile detection method.

Investigations of alternate ionization processes have led to the development of two new ion sources for IMS: photoionization (2) that provides a wider dynamic range than ^{63}Ni sources, and coronaspray ionization (3), which permits liquid samples to be efficiently introduced into the spectrometer. Table 12.2 (20–22) chronologically lists those developments in IMS as a detection method after chromatography that have occurred since 1980.

3.   ANALYTICAL FIGURES OF MERIT

3.1.   Qualitative Analysis

The primary figure of merit with respect to qualitative analysis is the separation efficiency of the ion mobility process and the reproducibility of the data. Separation efficiency is quantified by taking the ratio of the spatial variances of the ion zone at the end of the separation (σ^2) to the length of the drift region (L). This provides a theoretical plate height (H) similar to that in chromatography:

$$H_{\text{IMS}} = \frac{\sigma^2}{L}$$

Separation efficiency is commonly reported in terms of the total number of theoretical plates (N):

$$N = \frac{L}{H} = 5.54 \left(\frac{t}{W_h} \right)^2$$

where t is the measured drift time of the ion, and W_h is the temporal width (at half peak maximum) of the ion zone as it leaves the drift region.

A maximal number of theoretical plates possible for IMS can be determined by considering only diffusion broadening of the ion zone during migration in the drift region:

$$N_{max} = \frac{qV}{2\eta kT}$$

where q is the charge on the ion; V is the voltage across the drift region; T is the temperature in kelvins; and η is the ratio of the mean agitation energy of the ion to the mean thermal energy of the molecules in the drift tube (23). Here η is called the Townsend energy factor and accounts for the energy provided to the ions from the electric field, giving them not only a net velocity but also greater random kinetic energy than the surrounding neutral molecules.

With typical IMS operating conditions of $V = 3500$ V, $T = 150\,°C$, and $\eta = 2.7$, N_{max} is on the order of 18,000 for singly charged ions. For current IMS designs, N is only about 5000, so it would appear that considerable improvement in IMS efficiency is still possible. Note that separation efficiency increases as drift field energy (qV) increases relative to random thermal energy (kT). Thus efficiency should increase with increased voltage across the drift region and for multiply charged ions.

To aid in qualitative applications of IMS, a compilation of ion mobility constants has been published for more than 725 ions, with values ranging from 0.53 to 3.37 cm^2/V·s (24). Because mass identification was not obtained for many of these ions, it is possible that some of them were misidentified. Generally, however, it is believed that reduced mobility (K_0) values vary by only a few percent when different instruments and conditions are used. When instrumental parameters and the environment of the drift region are carefully controlled, reproducibility within 2% can be expected. To correct for minor differences in drift conditions, it has been suggested that standards be used to calibrate ion mobility experiments (25).

For identification of compounds by IMS, ions must be separated such that unit resolution can be achieved. Resolution (R) is defined as

$$R = \frac{t_{d2} - t_{d1}}{2(\sigma_{t1} + \sigma_{t2})}$$

where t_{d2} and t_{d1} are drift times of two ions, and σ_{t1} and σ_{t2} are the temporal standard deviations of the respective ion peaks. From the expression for the number of theoretical plates, $N = (t_d/\sigma_t)^2$, a useful expression relating resolution to separation efficiency and mobility constants can be derived (26):

$$R = \frac{|\Delta K|}{K} \frac{N^{1/2}}{4}$$

where $|\Delta K|$ is the absolute difference in the mobility, and K is the average mobility of the two ions. Thus, for a typical separation efficiency of $N = 5000$, ions must have a mobility difference of 6% to be separated with unit resolution:

$$\frac{|\Delta K|}{K} = \frac{4}{(5000)^{1/2}}(100) = 6\%$$

By using either the standard signal averaging or the Fourier transform mode of operation, ion mobility spectra can be obtained from GC, SFC, or LC peaks. For most compounds at atmospheric pressure, with either ^{63}Ni, photoionization or coronaspray ionization sources, simple spectra are produced. Figure 12.8 presents typical IMS spectra of several analyses taken after LC

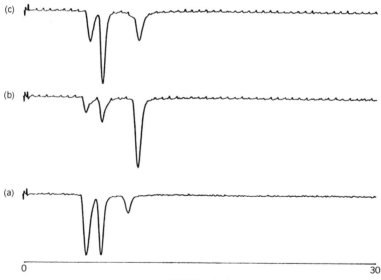

Drift Time (ms)

Figure 12.8. Coronaspray (signal-averaged) ion mobility spectra of (a) acetaminophen, (b) caffeine, and (c) phenacetin. Mobile phase was methanol/water (1:1); $N \simeq 3000$. Reprinted with permission from D. G. McMinn, J. A. Kinzer, C. B. Shumate, W. F. Siems, and H. H. Hill, *J. Microcolumn Sep.* **2**, 188 (1990). Copyright 1990, Aster Publication Corp.

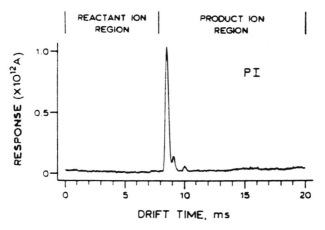

Figure 12.9. Photoionization (signal-averaged) ion mobility spectrum of naphthalene in nitrogen gas; $N \simeq 4200$. Reprinted with permission from M. A. Baim, R. L. Eatherton, and H. H. Hill, *Anal. Chem.* **55**, 1761 (1983). Copyright 1983, American Chemical Society.

with coronaspray nebulization and ionization. The separation efficiency of these spectra is about $N = 3000$. Similarly IMS spectra from a photoionization source after GC (Figure 12.9) and ^{63}Ni sources after SFC (Figure 12.10) provide separation efficiencies on the order of $N = 4200$ and $N = 5100$, respectively. These separation values are equivalent to a good packed GC or LC column.

Because the ion mobility separation process is based on physical and chemical properties different from those of the chromatographic separation process, the use of IMS after chromatography appears to add a second dimension of qualitative data. We have plotted several series of ion mobility data (K_0) with respect to chromatographic retention time data (k') for a variety of columns. Although some correlation between the two separation methods may exist, it is not obvious, and these K_0 vs. k' plots appear to be random. By matching $(K_0^2 + k'^2)^{1/2}$ values with those of standards, compound identification may be possible in many cases (27).

3.2. Quantitative Analysis

For the evaluation of IMS as a quantitative detector, it is important to consider the dynamic range and minimum detectability. As with any secondary ionization source, the ^{63}Ni source for IMS exhibits a nonlinear response. In the region near the detection limits where the reactant ions are more abundant than the analyte, a quasi-linear dynamic range is observed. Analysts often dilute samples in order to work within this range.

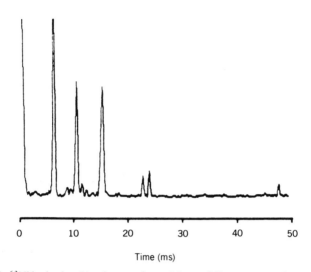

Figure 12.10. ^{63}Ni ionization (Fourier-transformed) ion mobility spectrum of 2,4-dichlorophen-oxyacetic acid. Mobile phase was supercritical CO_2; $N \simeq 5100$. Reprinted with permission from M. A. Morrissey and H. H. Hill, *J. Chromatogr. Sci.* **27**, 529 (1989). Copyright 1989, Preston Publications, a division of Preston Industries.

The ionization process can be represented as

$$R^+ + A \rightarrow P^+ + N$$

where P^+ represents the product ions; A, the analyte; R^+, the reactant ions; and N, neutral reaction products. As a rule of thumb, good quantitative data can be obtained if the ratio of the reactant ions to the product ions is greater than 0.5.

To increase the operating range of the IMS, a general logarithmic relationship can be used (28):

$$\ln R = S \ln W + \text{constant}$$

where R is the IMS response; S, the logarithmic sensitivity of the instrument; and W, the amount of the analyte. Comparing to a known standard, the quantitative relationship becomes

$$W_x = W_s \left(\frac{R_x}{R_s} \right)^{1/S}$$

where W_x is the amount of the unknown; W_s, the amount of the known standard; R_x, the IMS response to the unknown analyte; R_s, the IMS response to the known standard; and S, the slope of the logarithmic calibration curve.

In general, IMS is a sensitive analytical detector. Almost every publication on IMS mentions its tremendous sensitivity toward organic compounds, yet few document these claims with more than cursory experiments. From early work, it is clear that the detection limit of the instrument lies well below the nanogram region. Detection limits of 40 pg for p-chlorotoluene (29) and 23 pg for dodecane (21) were reported.

In a recent report on IMS (30), minimum detectability (D_{min}) was defined as the mass flow rate required to produce an analytical signal three times the root mean square noise (N_{rms}) of the system

$$D_{min} = 3N_{rms}/\varepsilon F$$

where F is Faraday's constant (96,500 C/mol) and ε is the overall detection efficiency measured as the ratio of the flow rate of ions detected (moles per second) to the flow rate of molecules in the sampled stream (moles per second).

For IMS, detection efficiency can be divided into several components:

$$\varepsilon = (\varepsilon_n)(\varepsilon_i)(\varepsilon_t)(\varepsilon_d)$$

where ε_n is the fraction of neutrals transferred from the sampling stream to the spectrometer; ε_i is the efficiency of the source; ε_t is the efficiency of ion transport through the drift tube; and ε_d accounts for intensity loss from diffusion, gate depletion, and electronic time constants. Typical values for IMS are $\varepsilon_n = 1.0$, $\varepsilon_i = 0.001$, $\varepsilon_t = 0.5$, and $\varepsilon_d = 0.3$. For a signal averaging experiment of 1.0-s duration, N_{rms} is approximately 1.5×10^{-13} A. If we use these values, a reasonable minimum detectability should be on the order of

$$D_{min} = 3.1 \times 10^{-14} \, \text{mol/s}$$

The minimum detectable amount (MDA) for this analyte can be calculated from the equation for the area (A) under a normal distribution curve:

$$A = (2\pi)^{1/2}\sigma Y_0$$

where σ is the standard distribution, and Y_0 is the maximum ordinate.

For a chromatographic peak with normal distribution, $D_{min} = Y_0$ and the width at half height is $W_h = 2.3\sigma$. Thus

$$\text{MDA} = (2\pi)^{1/2}\left(\frac{W_h}{2.3}\right)(D_{min}) = 1.1 W_h D_{min}$$

If the width at half height of a chromatographic peak is 5 s, then the MDA is 1.7×10^{-13} mol; so for an analyte with a molecular weight of 200, the minimum detectable mass (MDM) is

$$MDM = (MDA)(MW) = (11.7 \times 10^{-13})(200) = 3.4 \times 10^{-11} \, g$$

This amount is in good agreement with detection limits reported earlier for IMS.

Lower detection limits can be achieved if the analyte has a higher ionization efficiency or if the noise can be reduced by using a larger time constant. With noise reduced to 5×10^{-15} A and compounds with large ionization efficiencies, a minimum detectable weight of about 0.2 pg for tributylamine, phenyl sulfide, and tributyl phosphate has been reported after capillary GC (31).

4. APPLICATIONS

In the past, the IMD has been used primarily as a stand-alone instrument for the detection of explosives (32–34), drugs (35–38), chemical warfare agents (39–42), and industrial atmospheric toxins (40, 43–45). However, difficulties have been encountered when investigators have attempted to quantify analytes in complex and dynamic mixtures (46). Changing background matrix or sample overload generally causes serious problems when IMS is used as a stand-alone detection method unless the monitoring conditions are carefully controlled. Matrix and overload problems generally can be eliminated by using chromatographic separation prior to sample introduction.

Microcolumn separation methods are especially well matched for use with IMS detection. Capillary GC, SFC, and LC have been used successfully as an introduction step for IMS detection. Unfortunately commercial IMS detectors have not been available in the past for direct interfacing with chromatographic systems. This lack of easy access to instrumentation has limited the number of applications that have been demonstrated. Nevertheless, a few examples of the types of applications for which IMS is well suited are available in the literature.

Table 12.3 provides a chronological listing of compounds detected by IMS after GC. Before 1982 packed column GC was used. After 1982 all applications were performed using capillary columns.

Figures 12.11–12.14 are examples of the different modes of operation of the IMS when used as a chromatographic detector. Figure 12.11 shows nonselective positive ion mobility detection of silylated glycoprotein sugars of fucose, xylose, mannose, galactose, glucose, *N*-acetylglucosamine, *N*-acetylgalactosamine, and myoinositol (47). These monosaccharides were cleaved

Table 12.3. GC–IMS Applications

Year	Application
1972	Musk ambrette
1973	Halogenated alkanes
	Freons
1974	Benzoic and phthalic acids
1975	Resmethrin in peanut oil
1977	Halogenated benzenes
1982	Naphthalene and methylnaphthalenes in gasoline
	α-Pinene and limonene in orange extract
1983	Chlorinated pesticides: lindane, heptachlor, Aldrin, Dieldrin, and DDT
	Aromatic hydrocarbons: Toluene, ethylbenzene, o-xylene, cumene, mesitylene, tert-butylbenzene, p-cymene, durene, naphthalene, phenylcyclohexane
	2,4-Dichlorophenoxyacetic acid (2,4-D) in soils
1986	Barbiturates: aprobarbital, butabarbital, amobarbital, pentobarbital, secobarbital
1988	Steroids: estorone, progesterone, testosterone
	Opiates: hydromorphone, codeine, morphine tri-n-butylamine
	Methyl esters in cabbage seed oil extract
	Peppermint oil
1989	Di-n-hexyl ether
	Light petroleum
1990	Phenyl sulfide
	Fragrance separation
	Tributyl phosphate
	Polychlorinated biphenyls (PCBs)
	Brominated compounds in PCBs

from Sertoli cell glycoprotein and then silylated prior to GC separation. Identifications were assigned based on chromatographic retention time and ion mobility spectra. The trimethylsilyl derivatives form positive product ions efficiently and can be detected below the 100-pg level.

Most organic compounds undergo ion–molecule reactions with positive reactant ions to form positive product ions either through proton transfer, positive ion attachment, or hydride abstraction reactions. Thus, in the nonselective positive ion monitoring mode the response of the IMD resembles that of an FID. For most organic compounds the IMD is more sensitive than the FID. The FID, however, provides a longer linear dynamic range and

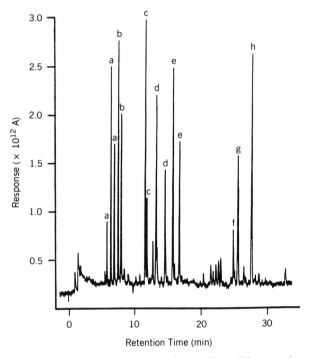

Figure 12.11. Nonselective positive ion detection after capillary GC separation of silylated mono-saccharides cleaved from Sertoli cell glycoprotein: (a) fucose; (b) xylose; (c) mannose; (d) galactose; (e) glucose; (f) *N*-acetylglucosamine; (g) *N*-acetylalactosamine; and (h) myoinositol. Adapted from R. L. Eatherton, Ph.D. thesis, Washington State University, Pullman, 1987.

a more consistent response factor between classes of organic compounds. Response factors for the IMD vary depending upon the affinity of the analyte for positive charges. A rule of thumb is that the analytes with larger proton affinities produce larger response factors. Thus, compounds such as amines, organophosphates, and sulfides provide the best response in the positive ion mode of the IMD.

By choosing to monitor a narrow mobility window, selective detection of compounds forming positive ions can be achieved. For example, Figure 12.12 shows the selective detection of phenylpropanolamine in a base-neutral urine extract (47). Although no further sample cleanup was performed after extraction, the phenylpropanolamine was easily detected in the complex matrix.

Negative ions are formed in the IMD through either free-electron attachment or by electron transfer from the negative reactant ions. Nonselective detection of negative product ions provides responses similar to those obtained

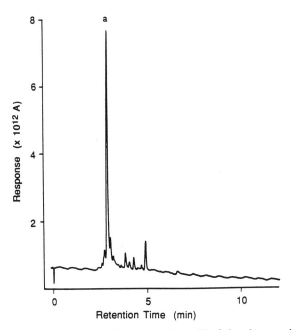

Figure 12.12. Selective positive ion detection after capillary GC of phenylpropanolamine in urine extract. Adapted from R. L. Eatherton, Ph.D. thesis, Washington State University, Pullman, 1987.

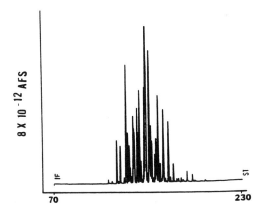

Figure 12.13. Nonselective negative ion detection after capillary GC of Aroclor 1248 (AFS = amps full scale). Reprinted with permission from R. H. St. Louis and H. H. Hill, *J. High Resolut. Chromatogr.* **13**, 628 (1990). Copyright 1990, Alfred Huethig Publishing.

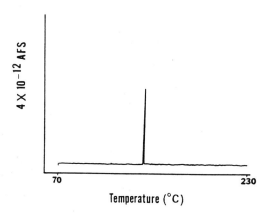

Figure 12.14. Selective negative ion detection after capillary GC of 4,4′-dibromobiphenyl in Aroclor 1248. Reprinted with permission from R. H. St. Louis and H. H. Hill, *J. High Resolut. Chromatogr.* **13**, 628 (1990). Copyright 1990, Alfred Huethig Publishing.

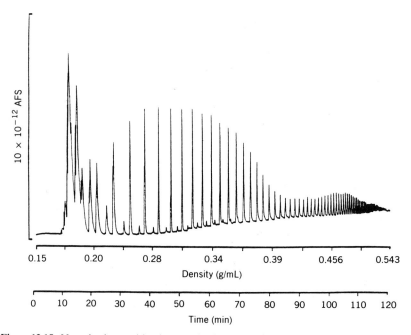

Figure 12.15. Nonselective positive ion monitoring mode for the detection of polydimethyl-silicone oligomers after SFC with CO_2 as the mobile phase. Reprinted with permission from M. A. Morrissey, W. F. Siems, and H. H. Hill, *J. Chromatogr.* **505**, 215 (1990). Copyright 1990, Elsevier Science Publishers, Physical Sciences & Engineering Division.

by ECD (48). Figure 12.13 depicts a GC separation with nonselective negative ion detection of Aroclor 1248. Rather spectacular selectivity can be achieved by selective ion monitoring, as in Figure 12.14, where 4,4′-dibromobiphenyl was detected in the Aroclor sample shown in the previous figure.

Similar modes of detection are available for SFC. Figure 12.15 demonstrates the nonselective positive ion detection of polydimethylsilicone oligomers following SFC separation (49). Selective positive ion mobility detection of cholesterol in a commercial fish oil extract is shown in Figure 12.16A. The amount of cholesterol in this sample was determined to be 2.2 mg/mL, or about 2 mg in each capsule. This value is comparable to 0.5 mg/mL found in bacon grease (Figure 12.16B).

With nonselective negative ion monitoring, the IMD can be used to provide ECD-type information after SFC. Figure 12.17 is an example of nonselective negative ion detection following SFC of a solid extract. In this study, soil samples were obtained by random grab sampling of a tilled field near Pullman, Washington. After Soxhlet extraction of the soil with hexane and acetone, the sample was injected into the SFC without derivatization or sample cleanup

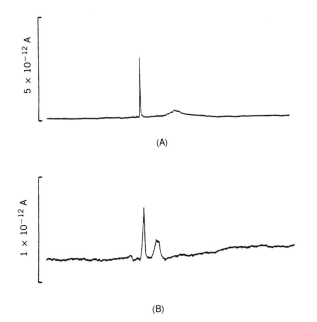

(A)

(B)

Figure 12.16. (A) Selective positive ion mobility detection of cholesterol in a commercial fish oil after SFC separation with CO_2 mobile phase. (B) Selective positive ion mobility detection of cholesterol in bacon grease after SFC separation with CO_2 mobile phase. Adapted from M. A. Morrissey, Ph.D. thesis, Washington State University, Pullman, 1988.

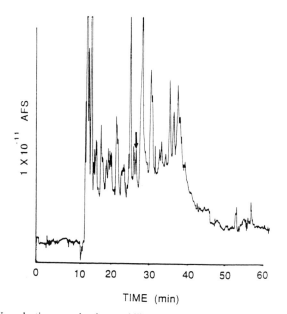

Figure 12.17. Nonselective negative ion mobility detection of underivatized soil extract after separation by SFC with CO_2 as mobile phase. The arrow denotes where underivatized 2,4-dichlorophenoxyacetic acid elutes. Reprinted with permission from M. A. Morrissey and H. H. Hill, *J. Chromatogr. Sci.* **27**, 529 (1989). Copyright 1989, Preston Publications, a division of Preston Industries.

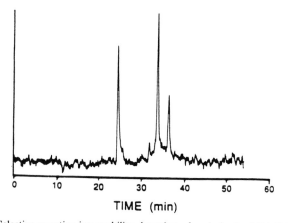

Figure 12.18. Selective negative ion mobility detection of underivatized 2,4-dichlorophenoxyacetic acid in the soil extract shown in Figure 12.17 after SFC separation with CO_2 as the mobile phase. Reprinted with permission from M. A. Morrissey and H. H. Hill, *J. Chromatogr. Sci.* **27**, 529 (1989). Copyright 1989, Preston Publications, a division of Preston Industries.

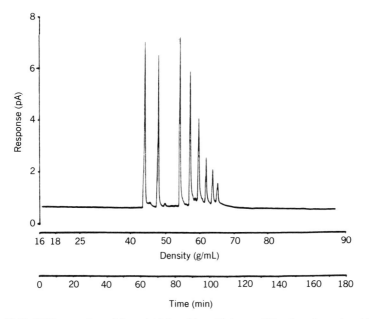

Figure 12.19. SFC separation of di- and triglycerides with ion mobility detection using chlorodifluoromethane as the mobile phase. Adapted from M. A. Morrissey, Ph.D. thesis, Washington State University, Pullman, 1988.

Table 12.4. SFC–IMD Applications

Year	Application	Reference
1986	Triton X-100	16
	Polystyrene 800	16
	Benzoates	17
	Fatty acid methyl esters	17
1988	Steroids: estrone, progesterone, testosterone	51
	Opiates: hydromorphone, codeine, morphine	51
	Benzodiazopinones: diazepam, nitrazepam, fluorazepam	51
1989	2,4-Dichlorophenoxyacetic acid in soil	52
1990	Polydimethylsilicone	53
1991	Lipids from onion seeds: palmitic acid, linoleic acid, α-toopherol, γ-sitosterol	54
	Benzoquinone	18
	Polycyclic aromatic hydrocarbons	19

procedures. Although the ECD type chromatogram shown in Figure 12.17 was extremely complex, selective negative ion monitoring for the 2,4-dichloropheonoxyacetic acid product ion (shown in Figure 2.18) produced a relatively simple chromatogram in which 2,4-D could be quantified to be present in the soil at the 0.5-ppm level. This was consistent with levels of 2,4-D commonly found in wheat fields in eastern Washington.

Another potential advantage of ion mobility detection after SFC is that it may be used with mobile phases other than pure CO_2. Figure 12.19 shows an SFC separation of di- and triglycerides in supercritical fluid chlorodifluoromethane with nonselective positive ion mobility detection (50). Other mobile phases including CO_2 + methanol and CO_2 + acetronitrile have been investi-

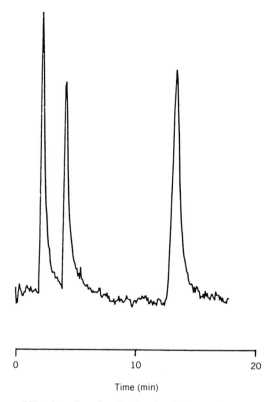

Time (min)

Figure 12.20. Ion mobility detection of analgesics after LC separation. From left to right, peaks are acetaminophen (0.88 ng), caffeine (0.97 ng), and phenacetin (1.6 ng). Mobile phase was methanol in water (20:80). Reprinted with permission from D. G. McMinn, J. A. Kinzer, C. B. Shumate, W. F. Siems, and H. H. Hill, *J. Microcolumn Sep.* **2**, 188 (1990). Copyright 1990, Aster Publishing Corp.

Table 12.5. LC–IMD Applications

Year	Application	Reference
1990	p-Hydroxybenzoic acids: Methyl Propyl Butyl	22
	Analgesics: Acetaminophen Caffeine Phenacetin	22
	Nitroanilines: *ortho* *meta* *para*	22

gated (18, 19). Additional applications of IMD for SFC detection are given in Table 12.4 (51–54).

Ion mobility detection after LC requires a change of ionization source from the radioactive ^{63}Ni source to an electrified spray source. Figure 12.20 shows the first example of direct ion mobility detection after LC, and Table 12.5 lists the only examples in the literature using this technique. Electrified spray nebulization and ionization have also been investigated for capillary electrophoretic (55) and flow injection analysis (56) methods.

Although the lack of commercially available instruments has limited applications of the IMD as a chromatographic detector, those analyses to which it has been applied demonstrate its potential for a wide range of analytical determinations for which it can provide unique data. Commercial chromatographic detectors are being developed, and as they become available more applications will inevitably be demonstrated. For GC, IMD's sensitivity and multimode operation provide a more sensitive and versatile detector than is currently available. Its potential compatibility with mixed mobile phases and non-CO_2 phases coupled with its ability to provide electron-capture-type responses will drive further development for its use after SFC. In LC, IMD's ability to respond to compounds that do not contain chromophores will ensure its development as a detection method to complement UV–visual detection.

REFERENCES

1. F. W. Karasek and R. A. Keller, *J. Chromatogr. Sci.* **10**, 626 (1972).
2. M. A. Baim, R. L. Eatherton, and H. H. Hill, *Anal. Chem.* **55**, 1761 (1983).

3. C. B. Shumate and H. H. Hill, *Anal. Chem.* **61**, 601 (1989).

4. N. E. Bradbury and R. A. Neilsen, *Phys. Rev.* **49**, 388 (1936).

5. E. A. Mason and H. W. Schamp, *Am. Phys.* (N. Y.) **4**, 233 (1958).

6. F. J. Knorr, R. L. Eatherton, W. F. Siems, and H. H. Hill, U.S. Patent 4,633,083 (1986).

7. F. J. Knorr, R. L. Eatherton, W. F. Siems, and H. H. Hill, *Anal. Chem.* **57**, 402 (1985).

8. R. H. St. Louis and H. H. Hill, *CRC Crit. Rev. Anal. Chem.* **21**(5), 321 (1990).

9. M. J. Cohen and F. W. Karasek, *J. Chromatogr. Sci.* **8**, 330 (1970).

10. S. P. Cram and S. N. Chesler, *J. Chromatogr. Sci.* **11**, 391 (1973).

11. F. W. Karasek, O. S. Tatone, and D. W. Denney, *J. Chromatogr.* **87**, 137 (1973).

12. F. W. Karasek and S. H. Kim, *J. Chromatogr.* **99**, 257 (1974).

13. R. F. Wernlund and M. J. Cohen, "Identification of Pesticide Residues in Picogram Quantities," Application Note No. 2, Franklin GNO Corp. (now PCP Inc.), West Palm Beach, FL 33402.

14. F. W. Karasek, H. H. Hill, S. H. Kim, and S. Rokushika, *J. Chromatogr.* **135**, 329 (1977).

15. T. Ramstad, T. J. Nestrick, and J. C. Tou, *J. Chromatogr. Sci.* **16**, 240 (1978).

16. R. L. Eatherton, M. A. Morrissey, and H. H. Hill, *HRC & CC, J. High Resolut. Chromatogr. Chromatogr. Commun.* **9**, 154 (1986).

17. S. Rokushika, H. Hatano, and H. H. Hill, *Anal. Chem.* **59**, 8 (1986).

18. M. A. Morrissey and H. M. Widmer, *J. Chromatogr.* **552**, 531 (1991).

19. M. X. Huang, K. E. Markides, and M. L. Lee, *Chromatographia* **314**, 163 (1991).

20. M. A. Baim, F. J. Schuetze, J. M. Frame, and H. H. Hill, *Am. Lab.* Feb., p. 59 (1982).

21. M. A. Baim and H. H. Hill, *Anal. Chem.* **54**, 38 (1982).

22. D. G. McMinn, J. A. Kinzer, C. B. Shumate, W. F. Siems, and H. H Hill, *J. Microcolumn Sep.* **2**, 188 (1990).

23. E. W. McDaniel, *Collision Phenomena in Ionized Gases,* Chapter 10. Wiley, New York, 1964.

24. C. Shumate, R. H. St. Louis, and H. H. Hill, *J. Chromatogr.* **373**, 141 (1986).

25. Z. Karpas, *Anal. Chem.* **61**, 684 (1989).

26. G. E. Spangler, K. N. Vara, and J. P. Carrico, *J. Phys. E,* **19**, 191 (1986).

27. D. Atkinson and H. H. Hill, Federation of Analytical Chemistry and Spectroscopy Societies. *Conf. Anaheim,* CA, 1990.

28. J. C. Tou and G. U. Boggs, *Anal. Chem.* **48**, 1351 (1976).

29. F. W. Karasak and G. E. Spangler, in *Electron Capture-Theory and Practice in Chromatography* (A. Zlatkis and C. F. Poole, eds.), Elsevier, Amsterdam, 1981.

30. H. H. Hill, W. F. Siems, R. H. St. Louis, and D. G. McMinn, *Anal. Chem.* **62**, 1201A (1990).

31. R. H. St. Louis, W. F. Siems, and H. H. Hill, *J. Microcolumn Sep.* **2**, 138 (1990).

32. F. W. Karasek, *Res./Dev.*, **25**, 32 (1974).

33. F. W. Karasek, *J. Chromatogr.* **93**, 141 (1974).

34. A. H. Lawrence and P. Neudorfl, *Anal. Chem.* **60**, 104 (1988).

35. F. W. Karasek, D. E. Karasek, and S. H. Kim, *J. Chromatogr.* **105**, 345 (1975).

36. F. W. Karasek, H. H. Hill, and S. H. Kim, *J. Chromatogr.* **117**, 327 (1976).

37. A. H. Lawrence, *Anal. Chem.* **58**, 1269 (1986).

38. D. S. Ithakissions, *J. Chromatogr. Sci.* **18**, 88 (1980).

39. J. M. Preston, F. W. Karasek, and S. H. Kim, *Anal. Chem.* **49**, 1746 (1977).

40. C. S. Leasure and G. A. Eiceman, *Anal. Chem.* **57**, 1890 (1985).

41. J. P. Carrico, A. W. Davis, D. N. Campbell, J. E. Roehl, G. R. Sima, G. E. Spangler, K. N. Vora, and R. H. White, *Am. Lab, Feb.*, p. 152 (1986).

42. J. E. Roehl, *Opt. Eng.* **24**, 985 (1985).

43. W. M. Watson and C. F. Kohler, *Environ. Sci. Technol.* **13**, 1241 (1979).

44. R. J. Dam, In *Plasma Chromatography* (T. W. Carr, ed.). Plenum, New York, 1984.

45. R. F. Wernlund and M. J. Cohen, *Res./Dev.* July, p, 32 (1975).

46. G. A. Eiceman, C. S. Leasure, and V. J. Vandiver, *Anal. Chem.* **58**, 76 (1986).

47. R. L. Eatherton, Ph.D. thesis, Washington State University, Pullman, 1987.

48. R. H. St. Louis and H. H. Hill, *J. High Resolut. Chromatogr.* **13**, 628 (1990).

49. M. A. Morrissey, W. F. Siems, and H. H. Hill, *J. Chromatogr.* **505**, 215 (1990).

50. M. A. Morrissey, Ph.D. thesis, Washington State University, Pullman, 1988.

51. R. L. Eatherton, M. A. Morrissey, and H. H. Hill, *Anal. Chem.* **60**, 2240 (1988).

52. M. A. Morrissey and H. H. Hill, *J. Chromatogr. Sci.* **27**, 529 (1989).

53. M. A. Morrissey, W. F. Siems, and H. H. Hill, *J. Chromatogr.* **505**, 215 (1990).

54. R. M. Hannan and H. H. Hill, *J. Chromatogr.* **547**, 393 (1991).

55. R. W. Hallen, C. B. Shumate, W. F. Siems, T. Tsuda, and H. H. Hill, *J. Chromatogr.* **480**, 233 (1989).

56. C. B. Shumate, Ph.D. Thesis, Washington State University, Pullman, 1989.

CHAPTER

13

MASS SPECTROMETRIC DETECTORS

RAYMOND E. CLEMENT AND ERIC J. REINER

Ontario Ministry of Environment LSB
Rexdale, Ontario, Canada

The mass spectrometer as an analytical tool has been used since about 1900. Initially, the principal use of this device was to study nuclides composing the various elements, for which F. W. Aston was awarded a Nobel Prize in 1922. The first quantitative analysis of mixtures by mass spectrometry was performed in 1927, when gaseous organic compounds from reaction mixtures were investigated. Coupling of a mass spectrometer with a gas chromatograph was first reported by R. S. Gohlke in 1957.

One of the principal challenges was to devise a means of enriching the gas chromatographic effluent by selective removal of carrier gas before entering the mass spectrometer. This was accomplished by an interface device. Since gas chromatography (GC) and mass spectrometry (MS) are each techniques in their own right, their combination (GC–MS) is one of many so-called hyphenated techniques and is capable of providing far greater information about complex mixtures of compounds than either GC or MS separately.

Recent developments have greatly increased the applicability and ease of use of GC–MS. One of the more important was the introduction of fused silica wall coated open tubular (capillary) GC columns. Owing to the reduced flow of capillary columns compared to packed columns, capillary columns can be connected directly to mass spectrometers without the need of a special interface. The flexibility of fused silica capillary columns simplifies their installation in GC–MS systems so even relatively unskilled chromatographers can quickly achieve excellent results. Direct connection of the GC column to the mass spectrometer and the low-bleed characteristics of bonded-phase

Detectors for Capillary Chromatography, edited by Herbert H. Hill and Dennis G. McMinn.
Chemical Analysis Series, Vol. 121.
ISBN 0-471-50645-1   © 1992 John Wiley & Sons, Inc.

capillary columns have allowed GC–MS systems to achieve sub-picogram detection of some analytes. The introduction of relatively inexpensive bench-top GC–MS systems (1) was an important advance because this afforded many laboratories the opportunity to acquire GC–MS instrumentation, which in turn expanded the application of GC–MS to many more complex analytical problems.

Current GC–MS systems are also equipped with computer hardware and software that allow the user to perform complex analyses and difficult, laborious data interpretation automatically. Modern GC–MS systems have advanced to the stage where the mass spectrometer can be treated as a "black box" detection system when used in routine applications.

Several excellent books are available that describe GC–MS (2–4). It is not the intent of this chapter to describe in detail GC–MS techniques as they are treated in such books. The emphasis here will be on the mass spectrometer as a capillary column GC detector. Although it is necessary to know some of the basic principles of mass spectrometry to optimize application of GC–MS to the study of complex mixtures, detailed understanding of theory is not needed. It is more important for the chromatographer to know the characteristics of the mass spectrometric detector compared to other GC detectors and to understand the types of data obtainable and their use. Examples of the use of the mass spectrometric detector for environmental applications of GC are described to illustrate the advantages of GC–MS.

1. INSTRUMENTAL COMPONENTS

A gas chromatograph, mass spectrometer, and data system (GC–MS–DS) are the three components required for the separation and mass spectrometric analysis of complex mixtures. Each of these components are divided further as shown in Figure 13.1. The gas chromatograph consists of an injector (which may have an autosampler), an oven that houses the GC column, and a transfer line that allows the GC effluent to enter the mass spectrometer. The mass spectrometer consists of an ionization chamber (the source), the mass analyzer, and an ion detector, all of which are under vacuum. The vacuum, 10^{-6}–10^{-8} Torr, is achieved using diffusion or turbomolecular pumps backed by rotary pumps. A data system may be used to completely automate the entire system from the injection of samples using an autosampler to a final formatted analytical report. The complex software used by the data system allows a computer to control components such as the autosampler, the GC temperature programs, heated zones such as the transfer line and ion source, and the detector. In some instruments, mass spectrometric calibration and data reduction are also automated.

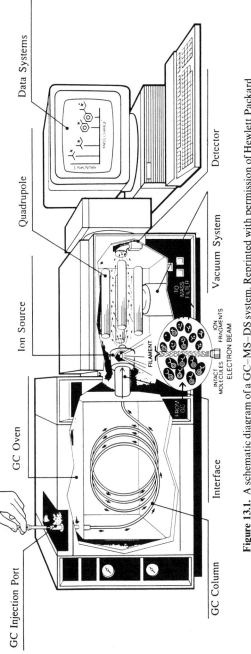

Figure 13.1. A schematic diagram of a GC–MS–DS system. Reprinted with permission of Hewlett Packard Corporation.

Data Systems

Quadrupole

Ion Source

GC Oven

GC Injection Port

Detector

Vacuum System

Interface

GC Column

ION
FRAGMENTS
ELECTRON BEAM

FILAMENT

INTACT
MOLECULES

TO
MASS
FILTER

FROM
GC

329

2. THEORY OF OPERATION

As sample molecules exit from the gas chromatograph into the mass spectrometer ion source, some of them (a small fraction) become ionized. The ionization technique imparts a distribution of energies into the sample molecule, M, giving rise to an excited molecular ion (or parent ion), $M^{\ddot{+}}$, under kinetic control (5–8). At the low ion source pressures employed, $(10^{-4}-10^{-6}\,\text{Torr})$, the mean free path for collisions between ions or molecules is large. Therefore, the molecular ions behave as isolated systems in which the energy absorbed during the ionization process is distributed rapidly with respect to the time required for fragmentation to take place by vibrational relaxation, vibronic relaxation, and internal conversion over all energy states. When a sufficient amount of energy accumulates in a particular bond or combination of bonds, the molecular ion will dissociate. The energetically excited parent ions decompose unimolecularly through a series of reactions to form the progeny ions that eventually are recorded to give the mass spectrum.

The ions formed in the source are separated from each other according to their mass-to-charge ratios (m/z) by an analyzer. The analyzer operates either by making the trajectory of ions through the analyzer stable or unstable or by dispersing the ions according to m/z somewhat like a prism diverges a beam of light. Quadrupole mass analyzers operate by making ions stable/unstable depending on their m/z, whereas magnetic sector analyzers disperse ions according to their m/z as they pass through a magnetic field. In some of the newer mass spectrometric techniques, such as ion traps and Fourier transform mass spectrometry (FTMS), ionization and mass analysis occur in the same region.

3. IONIZATION METHODS

A large number of ionization methods are available, but only a few are suitable for GC–MS analysis. Electron ionization (EI) is the oldest and most widely used ($>90\%$). Chemical ionization (CI) is also used in GC–MS; however, it is a more specialized technique not available on all commercial systems. Although most GC–MS work has been performed by monitoring the positive ions formed in the ion source, either positive or negative ion spectra may be obtained.

3.1. Electron Ionization

The sample molecules entering the ion source from the GC are ionized by electrons that are emitted from a rhenium or tungsten filament (9, 10). As these electrons approach the sample molecules, interaction of the molecule with the electron waves causes excitation and ionization of the sample molecule into a series of different electronic energy states. The subsequent energy randomization forms a series of parent ions with a distribution of internal energies. The distribution of internal energies—and consequently the appearance of the mass spectrum—is strongly dependent on the nominal electron beam energy (E_{el}). The standard E_{el} is set at 70 eV for a number of reasons: (a) for most compounds this is the electron energy at which maximum ion formation occurs; (b) the appearance of the mass spectrum changes little with changes in E_{el}; (c) a relatively intense molecular ion as well as a large number of fragment ions are formed, giving both molecular weight and structural information; and (d) the internal energy distribution of the ions formed should be the same from instrument to instrument, giving rise to instrument-independent spectra. The ability to produce instrument-independent mass spectra allows such spectra to be compiled in libraries (11, 12) and unknown spectra to be statistically compared with library spectra (2, 13).

3.2. Chemical Ionization

Chemical ionization is the technique of ionizing molecules through gas-phase ion–molecule reactions (14–19). The major advantage of CI compared to EI is that usually the distribution of energies of the parent ions is much narrower. By selecting the proper conditions, much less fragmentation of molecules

Table 13.1. Chemical Ionization Processes

A. *Positive ions:*

(1) Proton transfer	$XH^+ + M \rightarrow MH^+ + X$	
(2) Hydride abstraction	$XH^+ + M \rightarrow (M-H)^+ + X + H_2$	
(3) Charge exchange	$X^{\ddot{+}} + M \rightarrow X + M^{\ddot{+}}$	
(4) Adduct formation	$X^+ + M \rightarrow MX^+$	

B. *Negative ions:*

(1) Resonance electron capture	$MN + e^- \rightarrow MN^{\dot{-}}$	
(2) Dissociative resonance capture	$MN + e^- \rightarrow M^- + N\cdot$	
(3) Ion pair production	$MN + e^- \rightarrow M^+ + N^- + e^-$	
(4) Proton abstraction	$MH + X^- \rightarrow M^- + XH$	
(5) Adduction formation	$M^- + X \rightarrow MX^-$	

and thus more abundant molecular ions can be obtained by using CI. This is useful for determining the molecular weights of some compounds that do not give measurable molecular ion peaks under EI conditions. For this reason, CI is often called a "soft" ionization technique. Both positive and negative ions can be formed through a variety of CI processes. These are summarized in Table 13.1.

Another advantage of CI is selectivity. Reagent gases can be chosen such that they react with only specific components in a sample. For example, if a reagent gas such as ammonia is used, only amines and other very basic compounds are ionized and detected.

4. CHARACTERISTICS OF MASS SPECTROMETRIC DETECTORS

4.1. Basic Principles

The main characteristics of mass spectrometers are mass range, mass resolution, scan speed, linear dynamic range, selectivity, sensitivity, and minimum detectable quantity (MDQ). The mass range of a mass spectrometric detector is the range of ion masses (m/z) that can be formed in the ion source, passed through the analyzer region, and detected by the electron multiplier.

Resolution (R) or resolving power (RP) is a numerical measure of the ability of a mass spectrometer to separate two ions of different m/z. The resolution required to separate two mass peaks just barely resolved is given by $R = M/\Delta M$, where M is the nominal mass of the ion of interest and ΔM is the difference in mass of the two peaks to be resolved. In MS, "resolved" usually means peaks that overlap so that the valley between them is 10% of their peak height (10% valley definition). Sometimes a 5% valley definition is employed. Figure 13.2 illustrates the definition of mass resolution. The resolution with respect to mass range is instrument dependent. In a quadrupole instrument ΔM remains constant. Therefore resolution is mass dependent. The resolution is set to resolve adjacent unit masses and is therefore called unit mass resolution. In a magnetic sector instrument the peak width does change with mass and resolution remains constant over the entire mass range. Resolving powers of up to 10^5 can be achieved by using magnetic sector analyzers.

The scan speed of an instrument is the speed at which a mass analyzer can scan over a decade of mass (1 order of magnitude; for example, from m/z 600 to m/z 60). Fast scan speeds ($\leqslant 1$ s/decade) are necessary in capillary column GC–MS applications. As capillary column GC peak widths become increasingly narrow, GC–MS scan speeds must be faster so that data from at least 3–5 full scans can be obtained across the GC peak.

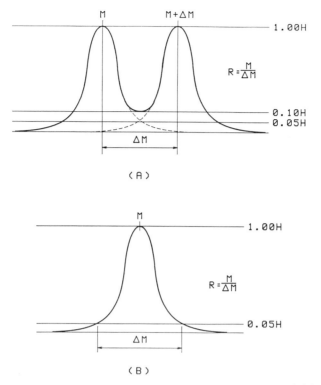

Figure 13.2. Resolution defined by (A) the 10% valley definition and (B) the 5% height definition.

The selectivity of the mass spectrometric detector is much greater than the selectivity of conventional GC detectors, because mass spectra provide more specific information. By increasing mass resolution, selectivity can be further increased. High resolving power can be used to identify the elemental composition of ions or to differentiate between ions of similar nominal mass. For example, all ions listed in Table 13.2 have similar nominal masses but different exact masses. Of these four ions the first two are the closest in mass,

Table 13.2. Resolution of Ions with Similar Nominal Mass

Ion	Nominal Mass	Exact Mass
$C_5H_{11}O_2$	103	103.07595
$C_4H_{11}N_2O$	103	103.08718
$C_6H_{15}O$	103	103.11236
$C_5H_{15}N_2$	103	103.12359

with a ΔM of 0.01123 daltons (Da). In order to resolve all four of these ions, a resolving power of 9172 would be required.

The sensitivity and MDQs of mass spectrometers make this type of GC detector essential for use in ultra-trace organic analysis applications. Some high-resolution mass spectrometers can detect absolute quantities of compounds that are less than $10\,\mathrm{fg}(1 \times 10^{-14}\,\mathrm{g})$. For routine work, GC–MS systems can achieve detection limits at least as low as other sensitive GC detectors. However, the performance of the GC–MS system is much superior to other GC detectors when factors such as the ability to operate in either universal or selective mode and the ability to identify GC peaks are taken into account. The linear dynamic range of GC–MS systems varies from 2 to 5 orders of magnitude.

4.2. Quadrupole Mass Analyzer

The quadrupole mass analyzer (filter) consists of a set of four parallel rods in a square array (Figure 13.3) (20, 21). Mass selection occurs by applying d.c. (U) and RF[$V\cos(wt)$] voltages to the opposite pairs of these rods. Characteristic voltages (V and U), frequencies (w), and rod separations (r) determine which ions of a specific mass range have stable trajectories within the bounds of the rods. In practice, this mass range is made narrow enough to transmit only ions of the same nominal m/z. The mass spectrum can be scanned by varying V and U such that the ratio of U/V remains constant.

The quadrupole mass spectrometer is the principal type of mass spectrometric detector used in GC–MS analysis because it is relatively inexpensive, compact, and requires less experience to operate and maintain than does a magnetic sector instrument. Its design allows it to scan at fast speeds ($< 1\,\mathrm{s/decade}$). In recent years data system control has been accomplished by personal computers (PCs), which has significantly reduced the cost of the instrument. In addition, the ability to switch scan modes rapidly enables the detector to record positive and negative ions in sequential scans. A number of disadvantages are also associated with quadrupole instruments. They show mass discrimination and are less sensitive than magnetic sector instruments, especially at m/z values greater than 500 Da. The upper mass range of most quadrupole instruments is between m/z 1000 and m/z 2000 Da. This is not critical in GC–MS applications since most compounds that are amenable to gas chromatography do not have molecular weights above these values.

4.3. Magnetic Sector Instruments

Double-focusing magnetic sector instruments (Figure 13.4) are composed of a magnetic sector that disperses the ion beam made of ions of differing

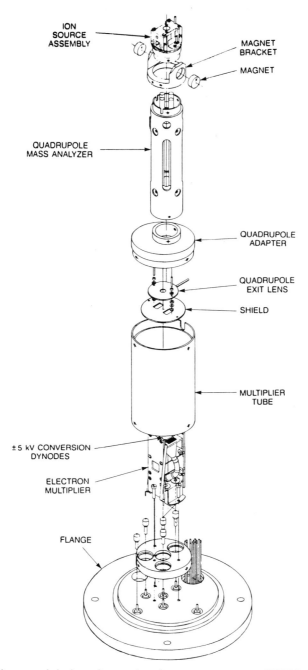

ION
SOURCE
ASSEMBLY

MAGNET
BRACKET

MAGNET

QUADRUPOLE
MASS ANALYZER

QUADRUPOLE
ADAPTER

QUADRUPOLE
EXIT LENS

SHIELD

MULTIPLIER
TUBE

±5 kV CONVERSION
DYNODES

ELECTRON
MULTIPLIER

FLANGE

Figure 13.3. An expanded view of a quadrupole mass spectrometer (INCOS 50) and its components Reprinted with permission of Finnigan MAT.

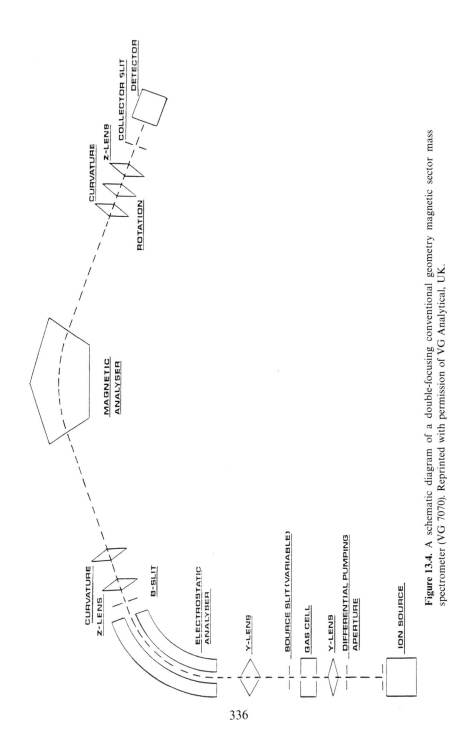

Figure 13.4. A schematic diagram of a double-focusing conventional geometry magnetic sector mass spectrometer (VG 7070). Reprinted with permission of VG Analytical, UK.

momentum (m/z ratios) and an electrostatic analyzer (ESA) that focuses ions of differing kinetic energy (22–24). This double-focusing effect enables magnetic sector instruments to obtain much higher resolving powers than quadrupole instruments. The ions that exit the ion source are subjected to high acceleration potentials, typically ranging between 3000 and 10,000 V. Ions that pass through these high potential gradients have a considerable spread in kinetic energies that makes both momentum (magnetic sector) and kinetic energy (ESA) analysis necessary for obtaining high resolving powers.

By convention, the ion beam is usually passed through the ESA first. However, in recent years, instruments of reversed geometry (the magnet preceding the ESA) have become available commercially. A distinct advantage of reversed geometry instruments is that an ion of interest (parent ion) may be selectively transmitted through the magnetic sector and all fragment (daughter) ions of that particular parent ion may be recorded by scanning the ESA. This technique is known as mass-analyzed ion kinetic energy spectrometry (MIKES). Modern magnetic sector instruments have resolving powers greater than 100,000 and can analyze ions of m/z that exceed 20,000 Da. They are much more sensitive than quadrupole instruments at lower (unit mass) resolution and at least as sensitive at resolving powers of 10,000. At these high resolving powers, exact mass and elemental composition information can be obtained from the mass spectrum. The scan rates of modern instruments are sufficient to yield GC–MS data even from narrow capillary column peaks.

Magnetic sector instruments are some two to five times more expensive than quadrupole instruments. Therefore they are most often used in applications that require high sensitivity, high resolution, or a greater mass range than is available from quadrupole instruments. Magnetic sector instruments are usually much larger, require more maintenance, are more difficult to tune and calibrate, and require more operator training.

4.4. Mass Spectrometry/Mass Spectrometry (MS/MS)

An even greater degree of selectivity and structural information can be obtained by using MS/MS. The sequential analysis of ions that fragment outside the ion source enables fragmentation reactions and not just ions to be monitored. A collision gas such as helium or argon is used in order to induce fragmentation [collision-induced dissociation (CID) or collisionally activated dissociation (CAD)] and enhance the fragmentation reaction(s) to be monitored. There are three different types of MS/MS scans that can be obtained:

- *Daughter scan (D)*. This scan gives all daughter (fragment) ions of a preselected (parent) ion in the mass spectrum.

- *Parent ion scan (P).* This scan gives all parent ions that fragment to give the preselected daughter ion in the mass spectrum.

- *Neutral loss scan (NL).* This scan gives all ions that fragment by the loss of a preselected neutral; it can be used to monitor compounds that contain specific functional groups such as chlorinated compounds (M^+-35) or carboxylic acids (M^+-45).

There are several ways of obtaining MS/MS spectra. Combinations of analyzers such as magnetic (B), electrostatic (E), and quadrupole (Q) are tandem-in-space instruments, whereas ion traps and FTMS instruments are tandem-in-time instruments.

A major advantage of MS/MS is its ability to reduce chemical noise and therefore increase the signal-to-noise ratio (S/N) for detected peaks. This results in increased sensitivity as well as better selectivity compared to those of conventional mass spectrometry. The increased selectivity is a result of monitoring reactions rather than just ions. For example, two ions may have the same combination of atoms but a different structure. The masses of the ions will be identical but, because of their different structures, parent ions may decompose differently in the collision cell. Therefore, the second stage of mass selection can be used to differentiate between the two possible structures. The increase in selectivity by monitoring reactions in MS/MS is analogous to that achieved by using multidimensional chromatography.

4.5.   Fourier Transform Mass Spectrometry

In Fourier transform mass spectrometry (FTMS) (25–28), ions are trapped in a cylindrical or cubic cell in a high-vacuum ($< 10^{-6}$ Torr) chamber centered in a homogeneous magnetic field. The ions move in a cyclic motion within the cell with a frequency W, where $W = zeB/m$. By applying a short broad-band RF (radio-frequency) signal, the ions become excited and move into a higher coherent orbit. The packet of ions that is formed induces a small alternating current on the receiver plates. A Fourier transformation is then used to deconvolute this signal and determine the m/z values for the ions present in the cell. The major advantage in using FTMS is that resolving powers greater than 10^8 are possible (25). When FTMS is coupled to GC, a differentially pumped dual cell (27) or external source (28) is required to achieve these high resolving powers. The unambiguous determination of the elemental composition of the ions in a mass spectrum can be achieved at such resolving powers. Other advantages of FTMS include instrument simplicity, simultaneous detection of all ions, the ability to trap ions, and the ability to do MS/MS.

FTMS instruments are used principally when high resolving powers are required, and in that respect they are superior to magnetic sector instruments. Although their cost is similar to sector instruments, they currently cannot handle higher source pressures and therefore are much less sensitive. In addition, a large computer is needed to perform the many complex calculations. The development of FTMS for complex mixture analysis is at an early stage. However, the superior performance possible from this technique eventually may make conventional GC–MS obsolete.

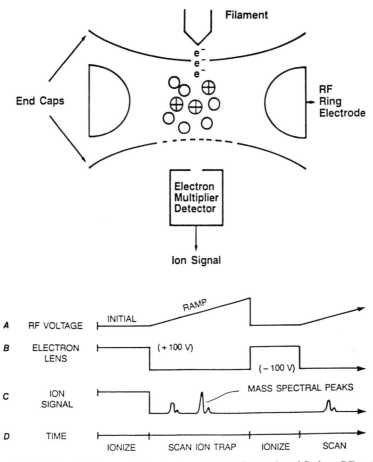

Figure 13.5. A schematic diagram of the ion trap detector. Traces A and B show RF and d.c. voltages required to produce a mass spectrum. Trace C and Trace D show the various times involved.

4.6. Ion Trap Mass Spectrometers

With an ion trap mass spectrometer, ions are both formed and mass analyzed in the same region of the quadrupole ion storage trap (29, 30), or quistor (Figure 13.5). All ions in the ion trap with m/z values greater than a minimum value will have stable trajectories in the ion cell. This minimum value is proportional to an RF voltage applied to the ring electrode. As the RF voltage is scanned upward, ions of increasing m/z values will become unstable and be ejected through the end cap, where they are detected by an electron multiplier. The principal advantage of the ion trap is its versatility. It requires no hardware changes to do CI or MS/MS analysis. It is also sensitive, inexpensive, and much simpler than other mass spectrometers, therefore requiring little experience to operate and maintain.

5. TYPES OF DATA OBTAINED

As a GC detector, the mass spectrometer is unique because it is both a universal and tunable selective detector. When operated in the universal detection mode, its response vs. retention time resembles that of a flame ionization detector (FID). Unlike the FID or other conventional GC detectors, data from the mass spectrometer are much more suitable for the identification of organic compounds. When it is operated a selective detector, detection limits about 100 times lower than those obtained in the universal mode are achieved. For halogenated organic compounds, detection limits of the mass spectrometer operated in the selective mode are comparable to those achievable by using an electron capture detector (ECD). To effectively use the full capabilities of the mass spectrometer it is necessary to understand how the data are generated in various mass spectrometer modes of operation and how these data are used for qualitative and quantitative determinations.

5.1. Mass Spectrum

Compound ionization and fragmentation were discussed previously. For GC–MS work, this occurs continuously as the GC effluent enters the mass spectrometer ion source. The mass spectrometer can be used as a mass filter to pass ions of any m/z within its operating limits. By setting this filter to pass initially ions of highest m/z and then rapidly varying the setting to pass ions of successively lower m/z until the lower limit is reached (or vice versa), the information needed to form the mass spectrum is obtained. Because of the narrow peak widths of capillary GC columns this scan must be rapid, typically $\leqslant 1.0$ s. An example of the formation of the mass spectrum is shown in Figure 13.6. The molecule ABCD enters the mass spectrometer ion source, where removal of an electron has formed the ion ABCD^{+} (The superscript "plus" notation indicates that this is a radical cation.) If it is stable, ABCD^{+} will survive through the analyzer and be detected. However, sufficient excess energy may be imparted by the electron ionization process to cause the rupture of bonds and formation of additional ions after the loss of neutral fragments. Thus, Figure 13.6 contains the additional ions A^{+}, BC^{+}, ABC^{+}, and BCD^{+}. The number, relative abundance, and structure of the ions formed depends upon the ionizing energy, the molecular structure, and the ion source temperature and pressure. Rearrangement ions may form within the ion source to give ions at m/z values that would not be predicted by the simple fragmentation scheme described above.

The general pattern of the mass spectrum is like a fingerprint that can be used to identify the original structure of a molecule. An experienced mass spectrometrist can interpret the individual mass fragments and mass losses to deduce the original structure, or can compare the mass spectral pattern of the unknown to the known patterns of many thousands of compounds by a computerized searching process. A compound's mass spectrum will be identical regardless of the mass spectrometer used, providing conditions in the ion source are the same. Some ions may be able to support two charges. These ions will appear at one-half the mass (m/z; $z = 2$) of the parent ion in the mass spectrum. (Abundance is plotted according to the m/z of ions, not the mass.)

Figure 13.7 illustrates how a mass spectrum is interpreted for a simple structure, trichloromethane ($CHCl_3$). From the molecular ion at m/z 118, a loss of mass 35 is observed to give a peak at m/z 83 (indicative of the Cl atom). Loss of a second Cl atom from the structure gives a peak at m/z 48. Other peaks are observed that cannot be accounted for by combinations of C, H, and Cl atoms. These are isotope peaks from the presence of natural, stable isotopes of chlorine (^{37}Cl) and carbon (^{13}C). Such isotope peaks provide important information in the interpretation of mass spectra. For example, Figure 13.8 is an expanded plot of the portion of the mass spectrum surrounding the base peak (m/z 83). Peaks at m/z 83, 85, and 87 are all

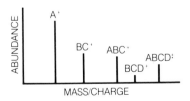

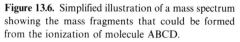

Figure 13.6. Simplified illustration of a mass spectrum showing the mass fragments that could be formed from the ionization of molecule ABCD.

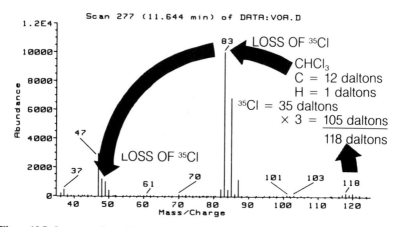

Figure 13.7. Interpretation of the mass fragments observed in the mass spectrum of $CHCl_3$.

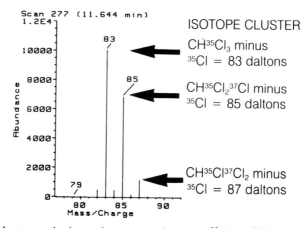

Figure 13.8. Isotope peaks due to the presence of atoms of ^{37}Cl and ^{35}Cl in a mass fragment.

identified as $CHCl_2$, but having zero, one, or two ^{37}Cl atoms present, respectively. The small peaks at m/z 84, 86, and 88 are ones that also contain contributions from the ^{13}C isotope. The number and relative abundance of isotope peaks depends upon the number of specific heteroatoms present, the masses of their stable isotopes, and the natural abudances of the isotopes. The isotopes of common elements and their natural abundances are shown in Table 13.3.

Table 13.3 Isotopic Abundances for Some Common Elements

Element	Mass (m)	Mass (m + 1)	Percentage of m	Mass (m + 2)	Percentage of m
H	1	2	0.015	—	—
C	12	13	1.1	—	—
N	14	15	0.37	—	—
O	16	17	0.04	18	0.20
F[a]	19	—	—	—	—
Si	28	29	5.1	30	3.4
P[a]	31	—	—		
S	32	33	0.80	34	4.4
Cl	35	—	—	37	32.5
Br	79	—	—	81	98.0
I[a]	127	—	—	—	—

[a] No other isotope.

5.2. Reconstructed Gas Chromatogram

In modern GC–MS analysis, the mass spectrometric detector is set to scan repetitively across a preset mass range as the GC eluant enters the MS ion source. Typically, scan speeds are 0.5–1.0 s for capillary GC–MS work. For each scan, a mass spectrum is generated and stored on a computer disk drive. To monitor the progress of the analysis, these mass spectra can be used to produce a reconstructed gas chromatogram (RGC).

The RGC, which is also called the total ion chromatogram (TIC), is produced by summing the measured ion abundances of all ions detected in a mass spectrum and plotting this value as a function of the mass spectrum scan number or retention time. Figure 13.9 shows an RGC from the analysis of volatile organic compounds. In appearance, it is similar to the trace that would be obtained by using an FID. If a single mass spectrum is obtained every second, then it would require about 1000 mass spectra to generate the RGC shown in Figure 13.9. Modern GC–MS systems can be operated so that the RGC trace is shown on a computer terminal as each mass spectrum is obtained. Of the 1000 mass spectra used to generate the RGC in Figure 13.9, only about 20—those taken at the center of eluting GC peaks—are important to the analyst. Mass spectra taken from the valleys between peaks may also be useful to provide data for the subtraction of background ions from the analyte mass spectra.

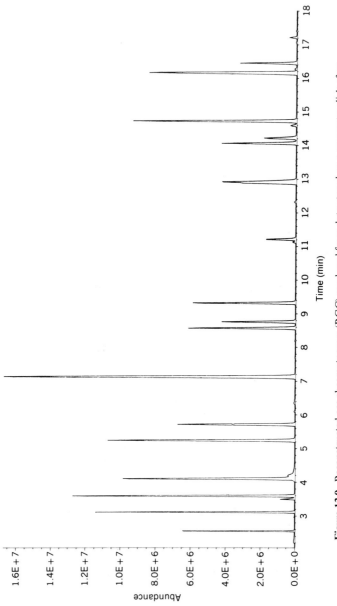

Figure 13.9. Reconstructed gas chromatogram (RGC) produced from data stored on a computer disk after the GC–MS analysis of volatile organic compounds.

5.3. Mass Chromatogram

Instead of plotting the sum of all ions in each mass spectrum vs. time, an increase in selectivity can be achieved by plotting the abundance of only one specific ion. The resulting trace is called a mass chromatogram. The increased selectivity of a mass chromatogram compared to the RGC is obvious, since only eluting GC peaks that produce a mass spectrum containing the selected ion will produce a corresponding peak in the mass chromatogram output. Of course, the degree of selectivity depends upon the specific ion chosen—it should be specific to the compound(s) of interest (compared to known interferences) and should be a prominent ion in the mass spectrum in order to provide the best detection limits. Also, greater selectivity generally is achieved by chosing high-mass ions.

Figure 13.10 shows a mass chromatogram for m/z 130 of the standard analysis in Figure 13.9. Only two peaks are observed in Figure 13.10, since the other peaks shown in Figure 13.4 are from compounds whose mass spectra do not contain this ion. By plotting two or more characteristic ions, selectivity is further improved. By using the technique of mass chromatograms, it is possible to detect minor sample components that coelute with other compounds that are at much greater concentrations. Since mass chromatograms are generated from mass spectra stored on computer disks, they can be generated as needed for any m/z value within the chosen scanning range of the mass spectrometer.

5.4. Selected Ion Monitoring

Selected ion monitoring (SIM) is one of the most important techniques for the ultra-trace determination of organic compounds. In a 10 g soil sample, it is possible to obtain low-ppt $(10^{-12}\,\text{g/g})$ detection of selected organic compounds by using GC–MS operated in the SIM mode. SIM is similar to the mass chromatogram technique described in the previous subsection, except that for the SIM experiment only those ions chosen prior to sample analysis are monitored; all others are ignored by the detector. Therefore, if improper ions are chosen before sample injection, the sample will have to be reanalyzed. However, by choosing only a few selected ions that are characteristic of the analyte(s), much lower detection limits (~ 100 times lower) are obtained with SIM.

Improved detection limits are the result of the increased time spent by the detector on the chosen ions. For example, instead of scanning the mass spectrometer over a 100-Da range for each mass spectrum in 1 s, the entire 1-s time frame can be used to monitor a single characteristic ion. The improvement in S/N for monitoring the selected ion in the foregoing example would

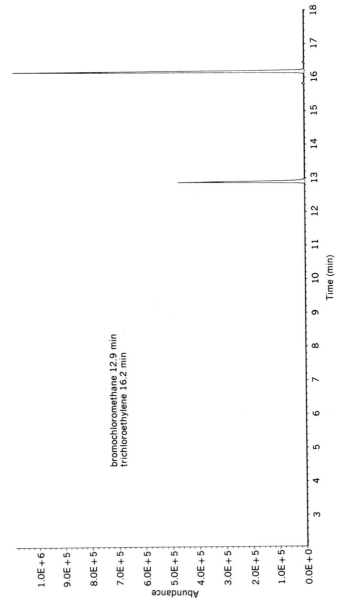

Figure 13.10. Mass chromatogram of m/z 130 from the GC–MS analysis of volatile organic compounds shown in Figure 13.9. The mass chromatogram was generated from the source data stored on a disk as the RGC shown in Figure 13.9, but only the abundances for m/z 130 were plotted.

be as great as 100. In practice, at least two or three characteristic ions are monitored for each analyte of interest. Identification of the analyte depends on obtaining a response for each ion monitored. The responses must also maximize at the characterisitc retention time of the analyte. In addition, the peak height or areas of all characteristic ions must be in the same ratios as are observed from the analysis of standards using the same GC–MS conditions. The selected ion monitoring trace is identical in appearance to that of the mass chromatogram of the same ion but at much greater S/N.

5.5. Computerized Library Searching

It is difficult to resolve molecular structures based solely upon the mass fragments observed in mass spectra. An experienced mass spectrometrist can usually determine a molecule's compound class and specific structural features, and often can develop a plausible molecular structure. However, unequivocal identification generally requires matching GC–MS data of the unknown with the GC–MS data obtained from analysis of an authentic standard of the proposed molecule. In addition, it takes many years of training and experience to interpret any but the most simple mass spectra from first principles. Fortunately, the EI mass spectra of many molecules have been cataloged into computerized libraries. These spectra can be statistically compared to those generated from the GC–MS analysis of real samples by matching the m/z values and intensities of individual peaks in the unknown and reference mass spectra. Because EI mass spectra of molecules are usually obtained under similar ion source conditions (10^{-6}-Torr source pressure; 70-eV ionizing electrons), the mass spectra of compounds in the library will match the mass spectra of these same compounds closely if they are detected in the GC–MS analysis of real samples. Almost all commercial systems now have provision for conducting computerized library searches of unknown mass spectra. Large computerized libraries are available that contain more than 70,000 mass spectra (12). Users can also develop specialized libraries such as for drugs or environmental contaminants. Although a close computer match of mass spectra is not definitive proof of a compound's identity, the correct structure will usually resemble that of the library compound.

The major limitation in compound identification by computer library search is the number of reference spectra in available libraries. Although 70,000 may seem large, the number of known organic compounds is still far greater.

6. ENVIRONMENTAL APPLICATIONS

The use of capillary column GC for complex separations has undergone explosive growth in recent years, especially for environmental applications.

The use of mass spectrometric detectors for these applications has paralleled this growth. Because of its low detection limits and ability to identify compounds present in samples at ppb concentrations, GC–MS is often the only technique available for solving complex environmental analytical problems.

6.1. Qualitative Compound Identification

Figure 13.11 is the RGC trace of the dichloromethane extract of sand taken from a dump site. Children were being affected by an unknown chemical contained in cannisters at this site. The sand had been spread around the area to contain the chemical. A single large peak was obtained from the sand extract, as shown in Figure 13.11. The mass spectrum from this peak and the closest matching mass spectra found after a computerized library search are shown in Figure 13.12. The closest matching mass spectrum is almost identical to the unknown mass spectrum. The library compound that produced the best match is [(2-chlorophenyl)methylene]propanedinitrile, and its structure is shown in Figure 13.12. This substance is a strong lachrymator used in the production of tear gas. The effects on the exposed children were entirely consistent with this identification. Further investigation revealed that expired tear gas cannisters had been improperly disposed at this site by the local police force. The use of GC–MS resulted in correct identification of the hazardous material and rapid solution of a potential hazard.

6.2. Quantitative Analysis by Selected Ion Monitoring

The chlorinated dibenzo-p-dioxins (CDDs) are considered to be among the most toxic chemicals synthesized. There are 75 different members of this class, depending on the number and location of chlorine atoms on the basic structure. The most toxic CDD is 2,3,7,8-tetrachlorodibenzo-p-dioxin (2,3,7,8-TCDD). Because of its high toxicity, GC–MS methods are used to achieve the necessary sub-ppb detection limits for 2,3,7,8-TCDD in environmental samples.

For such determinations selectivity is also required, since there are usually many hundreds of other compounds present at higher concentrations than 2,3,7,8-TCDD in sample extracts. Even after the use of chemical cleanup methods designed to remove interferences, GC–MS analysis by selected ion monitoring is required.

Figure 13.13 (top) shows the GC–MS full-scan analysis of a solvent extract from municipal incinerator fly ash. In addition to the huge peak at retention time 14.5 min, many other compounds are present at much lower concentrations. These other sample components cannot be resolved by the GC column

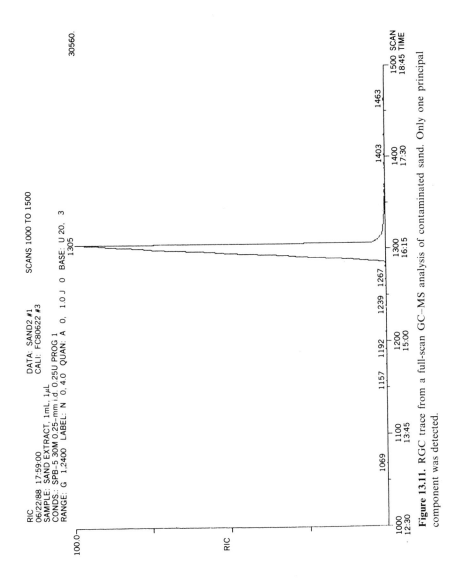

Figure 13.11. RGC trace from a full-scan GC–MS analysis of contaminated sand. Only one principal component was detected.

349

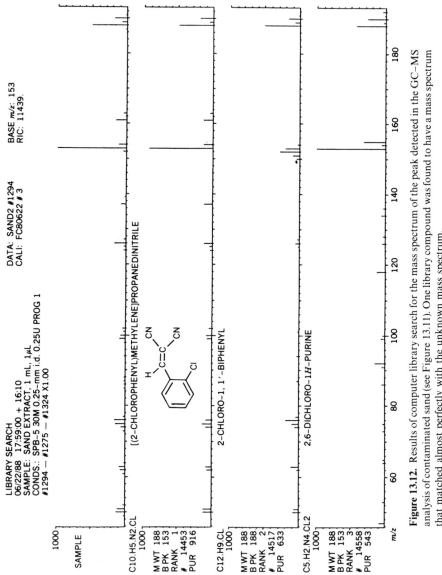

Figure 13.12. Results of computer library search for the mass spectrum of the peak detected in the GC–MS analysis of contaminated sand (see Figure 13.11). One library compound was found to have a mass spectrum that matched almost perfectly with the unknown mass spectrum.

350

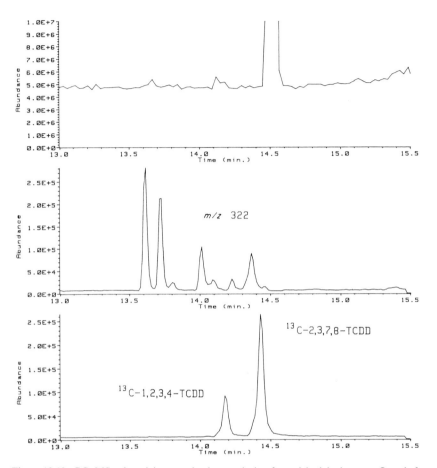

Figure 13.13. GC–MS selected ion monitoring analysis of municipal incinerator fly ash for tetrachlorinated dibenzo-*p*-dioxins (TCDDs). The top trace shows the GC–MS full-scan analysis of the same sample, in which no TCDDs were detected. Several TCDDs were identified by using selected ion monitoring (middle trace). In the same analysis, ^{13}C-labeled TCDDs added to the sample for quality control purposes are detected (bottom trace).

employed and result in the high background observed in the GC–MS full-scan analysis.

By monitoring only a few ions having m/z values characteristic of 2,3,7,8-TCDD, the reanalysis of this sample produced the GC–MS trace shown in Figure 13.13 (middle). Only the trace for m/z 322 is shown. Several peaks are now observed that could not be seen in the full-scan trace, while the principal sample component eluting at 14.5 min in the full-scan analysis is not observed at all in the SIM analysis. The maximum abundance of the largest peak observed in the SIM trace [$\sim 300,000$ counts (arbitrary units)] is more than 10 times less than the average baseline observed in the full-scan trace ($\sim 5 \times 10^6$ counts). However, the TCDD peaks are clearly visible.

Which peak is due to the most toxic 2,3,7,8-TCDD cannot be determined from this analysis. The peaks observed in Figure 13.13 are some of the 22 possible TCDD isomers. By adding a stable isotope analogue of 2,3,7,8-TCDD to the sample (in this case, [$^{13}C_{12}$]2,3,7,8-TCDD), the retention time of native 2,3,7,8-TCDD can be determined. This is shown in Figure 13.13 (bottom). It can be seen that 2,3,7,8-TCDD is not one of the major peaks in this analysis, but the separating power of the 30-m capillary column used is not sufficient to completely isolate 2,3,7,8-TCDD from all other TCDD isomers. Therefore an accurate determination of 2,3,7,8-TCDD cannot be obtained from this analysis. A more selective GC column must be used to accomplish this separation. This application illustrates that although mass spectrometer detectors are capable of producing vast quantities of data useful for trace chemical analysis and compound identification, the importance of the chromatographic separation cannot be ignored.

ACKNOWLEDGMENT

Appreciation is extended to V. Y. Taguchi and S. Jenkins for their assistance in the preparation of this manuscript.

REFERENCES

1. F. W. Karasek, *Res./Dev.* **27**, 42 (1976).

2. F. Karasek and R. E. Clement, *Basic Gas Chromatography-Mass Chromatography–Mass Spectrometry: Principles and Techniques.* Elsevier, Amsterdam, 1988.

3. G. A. Message, *Practical Aspects of Gas Chromatography/Mass Spectrometry.* Wiley (Interscience), Toronto, 1984.

4. Brian S. Middleditch, *Practical Mass Spectrometry: A Contemporary Introduction.* Plenum, New York, 1979.

5. H. M. Rosenstock, M. B. Wallenstein, A. L. Wahlhaftig, and H. Eyring, *Proc. Natl. Acad. Sci. U.S.A.* **38**, 667 (1952).

6. A. G. Harrison and C. W. Tsang, in *Biomedical Applications of Mass Spectrometry* (G. R. Waller ed.), p. 135. Wiley, New York, 1972.

7. K. Levsen, *Fundamental Aspects of Organic Mass Spectrometry*, Prog. Mass Spectrom. Vol. 4. Verlag-Chemie, Weinheim, 1978.

8. I. Howe, D. H. Williams, and R. D. Bowen, *Mass Spectrometry, Principles and Applications*, 1st ed. McGraw-Hill, New York, 1981.

9. A. L. Burlingame, D. Maltby, D. H. Russel, and P. T. Helland, *Anal. Chem.* **60**, 295R (1988).

10. M. E. Rose and R. A. W. Johnstone, *Mass Spectrometry for Chemists and Biochemists.* Cambridge Univ. Press, London, 1982.

11. *NIH/EPA/NBS Mass Spectral Data Base.* National Bureau of Standards, Washington, DC, 1986.

12. E. Stenhagen, S. Abrahamsson, and F. W. McLafferty, *Registry of Mass Spectral Data.* Wiley (Interscience), New York, 1974.

13. D. P. Martinsen and B. H. Song, *Mass Spectrom. Rev.* **4**, 461 (1985).

14. A. G. Harrison, *Chemical Ionization Mass Spectrometry.* CRC Press, Boca Raton, FL, 1983.

15. W. J. Richter and H. Schwarz, *Angew. Chem., Int. Ed. Eng.* **17**, 424 (1978).

16. F. H. Field and M. S. B. Munson, *J. Am. Chem. Soc.* **88**, 2621 (1966).

17. F. H. Field, *MTP Int. Rev. Sci.: Phys. Chem., Ser. One* **5**, (1972).

18. F. W. McLafferty, *Interpretation of Mass Spectra.* University Science Press, Mill Valley, CA, 1980.

19. A. G. Harrison, in *Ionic Processes in the Gas Phase* (M. A. Almaster Ferveria, ed.), p. 23. Reidel, Dordrecht, 1984.

20. P. H. Dawson, *Quadrupole Mass Spectrometry and its Applications.* Elsevier, Amsterdam, 1976.

21. J. E. Campana, *Int. J. Mass Spectrom. Ion Phys.* **33**, 101 (1980).

22. C. Brunee, *Int. J. Mass Spectrom. Ion Proc.* **76**, 125 (1987).

23. C. J. Porter, J. H. Beynon, and T. Ast, *Org. Mass Spectrom.* **16**, 101 (1981).

24. J. R. Chapman, *Practical Organic Mass Spectrometry.* Wiley (Interscience), Toronto, 1985.

25. M. Barber, R. S. Bardoli, G. S. Elliot, R. D. Sedgwickand, and A. N. Tyler, *Anal. Chem.* **54**, 645A (1982).

26. D. A. Landi, C. L. Johlman, R. S. Brown, D. A. Weil, and C. L. Wilkins, *Mass Spectrom. Rev.* **5**, 107 (1986).

27. M. V. Buchman and M. B. Comisarow, in *Fourier Transform Mass Spectrometry, Evolution, Innovation and Applications*, Chapter 1. Am. Chem. Soc., Washington, DC, 1987.

28. R. B. Cody and J. A. Kinsinger, in *Fourier Transform Mass Spectrometry, Evolution, Innovation and Applications*, Chapter 4. Am. Chem. Soc., Washington, DC, 1987.

29. F. H. Llaukien, M. Alleman, P. Bischofberger, P. Grossmann, H. P. Kellerhals, and P. Kofel, *Fourier Transform Mass Spectrometry, Evolution, Innovation and Applications*, Chapter 5. Am. Chem. Soc., Washington, DC, 1987.

30. J. Louris, R. G. Cooks, J. Syka, W. Reynold, and J. Todd, *Anal. Chem.* **59**, 1677 (1987).

CHAPTER

14

DETECTORS FOR CAPILLARY
SUPERCRITICAL FLUID CHROMATOGRAPHY

DARRYL J. BORNHOP

Citation Medical Corp.
Reno, Nevada

BRUCE E. RICHTER

LEE Scientific Division
Dionex Corporation
Salt Lake City, Utah

The attributes of supercritical fluid chromatography (SFC) and supercritical fluid extraction (SFE) are well documented (1–3). Thermally labile species (4) and relatively high-molecular-weight species (5) have been analyzed more easily by SFC than by other techniques. Unlike gas chromatography (GC) or high-performance liquid chromatography (HPLC), where the mobile phase dominates the type of the detection method employed, SFC utilizes mobile phases with physical parameters that allow the implementation of either fluid-phase or gas-phase detectors. This multidetector compatibility makes SFC an attractive separation method for the analysis of complex mixtures.

The first reported use of supercritical fluids as chromatographic mobile phases appeared in 1962 (6). In the last few years, SFC has progressed from a laboratory curiosity to a viable analytical technique that is used to solve many otherwise intractable problems. SFC is used routinely in research labs for problem solving and in industrial facilities for process monitoring. Evidence of the widening acceptance of SFC can be seen in the increased number of publications in the field.

Detectors for Capillary Chromatography, edited by Herbert H. Hill and Dennis G. McMinn.
Chemical Analysis Series, Vol. 121.
ISBN 0-471-50645-1 © 1992 John Wiley & Sons, Inc.

The unique properties of supercritical fluids (high diffusivities, low viscosities, and controllable solvating abilities) that facilitate fast, high-resolution separations at moderate temperatures are contributing factors to the success of SFC. Advancements in instrumentation such as pumping systems and injection techniques have been important, too. Developments in column technology such as nonextractable selective stationary phases for capillary columns have been significant elements as well. Another important reason for the increased acceptance of SFC is its multidetector compatibility.

As analysts collect more information about the sample, or more degrees of informational freedom, they learn more about the identity and the amount of the solutes contained in a particular sample. As the number of informational degrees of freedom is increased, the level of confidence (7) with respect to identification and quantification of a complex matrix is enhanced. Thus there has been a proliferation of hyphenated detection techniques after all chromatographic methods, and in particular capillary column SFC.

In SFC, as in other chromatographic techniques, methods have been developed to effect separation of closely eluting components (4, 8–10). Density or pressure programming are the most commonly used methods and are analogous to temperature programming in GC. Another method that has shown promise (5, 11–13), is the use of organic modifiers as additives to the supercritical fluid mobile phase. These modifiers can significantly change the k' value for a solute and thus effect a previously difficult separation. Modifiers have been employed successfully in packed, packed capillary, and open tubular column SFC. In concentrations of 1–20%, these organic molecules pose a serious background problem to the flame ionization detector (FID). Since organic modifiers are used extensively in SFC, the ultraviolet (UV) detector or another detector that has poor sensitivity for the commonly used organic modifiers is necessary.

This chapter covers the developments in conventional detection systems for SFC including the FID, nitrogen–phosphorus detector (NPD), flame photometric detector (FPD), photoionization detector (PID), electron capture detector (ECD), UV–visible, scanning UV, fluorescence, light scattering, chemiluminescence, Fourier transform infrared spectrometry (FTIR), and mass spectrometry (MS).

1. GC-TYPE DETECTORS

1.1. Flame Ionization Detector

The FID is the most widely used detector in GC, and it is natural that it be used in SFC. The reliability and sensitivity of this detector along with its response to essentially all organic compounds make it a good choice for SFC. Giddings et al. (14) and Bartmann (15, 16) were probably the first to explore the FID in SFC. This early work employed conventional HPLC-size columns (2.0- to 4.6-mm i.d.), and interfacing of the SFC column effluent to the detector presented some difficulties. Giddings et al. (14) were the first to observe that severe detector "spiking" occurs upon decompression at the end of the analytical column. They attributed this phenomenon to ion bursts in the detector resulting from irregular molecular clusters formed by the aggregation of desolvated molecules. There have been some approaches recently aimed at solving this problem.

One method of solving this "clustering" problem is to change the configuration of the detector itself. This approach has been investigated by several researchers, and a patent for an FID specifically for SFC has been issued (17). This patent details the use of a series of diaphragms or orifices to form a short decompression zone to prevent the solutes from forming clusters before being transferred to the flame. Rawdon (18) has also reported on the use of a modified FID. In this case, the flame tip was flattened by using a pair of pliers to act as an orifice or restrictor for the CO_2. The detector was further changed so that the original hydrogen line became an outlet for excess high-pressure CO_2 and the original air line became the hydrogen line. Two lines were installed on the base for air, and a new flame tip was fabricated.

Both of these research groups made the decompression region or restriction zone part of the detector itself. In this way heat can be applied directly and efficiently to the decompression zone to minimize cluster formation. Any unswept volumes are eliminated by having the decompression zone an integral part of the detector because the column can be connected directly to the detector without any connecting tubing. Having the decompression zone close to the flame tip also helps eliminate clustering problems because the solutes are transferred quickly to the flame before having a chance to aggregate.

Although they are of great importance in SFC, packed column (> 1.0 mm) detection approaches will not be discussed here except to note that the high fluid/gas flow rates normally used have forced researchers to employ specific detector designs or even predetector effluent splitting so that system functionality can be achieved.

Recent work by Hirata and co-workers (19) has shown that packed capillary columns (0.25-mm i.d. packed with 5- to 10-μm packing) can be used effectively with an FID. The use of these columns may represent a good compromise in SFC because they have better loadability than open capillaries but lower flow rates and higher permeabilities than packed columns, all of which produces a system with better performance.

As in GC, the column flow rates seen with capillary column SFC present fewer problems when interfacing to an FID. The typical flow rates exiting capillary columns are usually in the range of 0.25–5.00 mL/min. Column head pressure, the column diameter, and the type and amount of restriction used at the end of the column dominate actual flow rates measured.

Fjeldsted and co-workers (20) published the first report in the literature of using open tubular columns in SFC with FID. They observed the same spiking phenomenon with capillary columns that was seen with packed columns when high-molecular-weight compounds were transferred to the detector. Since the spiking consisted mostly of fast transient signals, they used a low-frequency bandpass filter to electronically filter the detector output. The use of the filter smoothed out the chromatographic trace and allowed for data collection. However, electronic filtering cannot be a permanent solution and can result in loss of information.

Two other approaches to this clustering and spiking problem have proven to be better long-term solutions because they address the fundamental cause. The first is to modify restrictor design or control the manner in which the solutes are transferred to the detector from the analytical column.

Four basic restrictor types have been used in capillary SFC (see Figure 14.1). Even though the results on restrictor designs have been obtained using FIDs, the same discussion and design considerations apply to the restrictors for all GC-type detectors and some HPLC detectors. The linear restrictor (Figure 14.1A), short pieces of 5- to 10-μm-i.d. tubing, was used by Fjeldsted et al. (20) and showed significant spiking problems. Performance

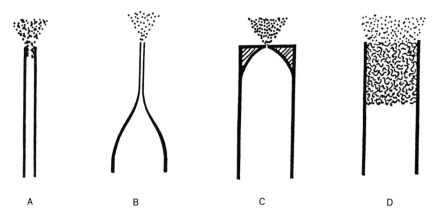

    A   B   C   D

Figure 14.1. The four main types of restrictors commonly used in capillary SFC (the laser-drilled orifice is omitted because of the low level of use and difficulty of manufacture). (A) small-i.d. open capillary tube; (B) robot-drawn capillary tube; (C) integral-type restrictor; (D) ceramic frit restrictor (U.S. Patent 4,793,920). See the text for details.

of the linear restrictor has been found to be less than optimal, especially for high-molecular-weight species and polar materials.

Chester (21) first described the use of a new restrictor design in which a flame was used to soften a piece of fused silica tubing, which was then drawn out to reduce the internal diameter (Figure 14.1B). The process of drawing out the restrictors can be done manually or with dedicated hardware (i.e., a robot or other mechanical means), but the restrictors fabricated by repro-ducible means function better because they are more uniform. This design of restrictor has a short decompression zone, so that the solutes are transferred quickly to the flame before significant clustering can occur. The thin wall of these drawn-out restrictors also allows effective thermal transfer from the detector heating block to the solutes, thus helping to prevent the spiking problem. Chester and co-workers have published several reports on the performance of these restrictors (22–24), and they have shown the detection of compounds with molecular weights in excess of 5000 Da without spiking in the FID. However, in some experimenters' hands, these restrictors can be difficult to use because of their fragile nature. In some cases it is necessary to insert the restrictor from the top of the detector down into the flame jet, which may not always be convenient. If proper care is taken, these restrictors function well and last for several months.

A similar design of restrictor that is more robust was first reported by Guthrie and Schwartz (25) (Figure 14.1C). In this case, the end of the fused silica capillary is closed by heating with a torch to form a conical closure. The end of the column is then filed with fine emery cloth until the proper-sized hole (1- to 2-μm diameter) is obtained. These restrictors perform as well as the other flame-tapered restrictors, and they have the added advantage that they can be fabricated in the end of the analytical column. However, this method does take some practice to obtain good usable restrictors on a routine, reproducible basis.

The fourth type of restrictor (Figure 14.1D) was developed at approxi-mately the same time and first detailed at the 1986 Pittsburgh Conference and Exposition (26). This restrictor is produced by bonding an inorganic porous frit inside a short length of fused silica (25- to 50-μm i.d.). The larger diameter of this frit makes it robust, and the multichannel flow path of the frit is less prone to plugging than a single orifice. Compounds of molecular weights in excess of 5000 Da have been successfully eluted from capillary columns and detected without spiking in an FID using these restrictors (26). However, to date, it has proven impossible to put the frit material directly in the end of the analytical column and thus eliminate the "butt connection," or union of the restrictor to the column. There have been some concerns raised about the frit restrictors because their larger diameter makes it harder to apply heat directly to the internal section of the restrictor.

Raynor and co-workers (27) have reported on ways to make the flame-tapered restrictors more robust and less likely to break by putting sleeves around the tapered end. Direct comparisons of the various restrictor types both from experimental and theoretical considerations have been made (28–32). In all cases, the two types of flame-tapered restrictors and frit restrictors have performed better than the straight linear restrictors. An easier way to fabricate robust restrictors quickly and reproducibly would help SFC progress more quickly.

To further ensure complete transfer of the analytes to the detector, it is common practice to operate all GC-types detectors at temperatures higher than customarily used in GC. For example, it is not uncommon to have an FID at 350–400 °C during operation. Proper restrictor design is important to good performance with an FID as well as with all detectors in which detection takes place in the gas phase rather than in the condensed phase. As will be further covered below in Section 1.7 on SFC–MS, restrictor design, heating, and positioning are crucial to good MS performance.

An alternative approach to improving the transfer of analytes to the detector flame is to modify the design of the detector itself. Richter (33) showed that if more heat were applied to the restrictor by adding an extra heating block and cartridge, then the performance of the linear restrictors could be improved dramatically. Another design alternative is to shorten the entire length of the detector to ensure that the decompression takes place within a zone completely surrounded by the heater block. Most FIDs used in SFC have been designed for use in GC. A decrease in temperature or a cold spot in the detector is not so detrimental to GC performance as it is in SFC. The nonvolatile species analyzed in SFC are much more likely to condense and produce spiking in an FID that has cold spots. Most commercial detectors have regions that remain cooler because of inadequate heating. Either the flame jet extends into the oven cavity, which is cooler, or else the flame jet protrudes out of the heater block, which also makes it cooler. Running the detector hotter than in GC overcomes some of the thermal gradient problems. However, if an FID has severe cold spots, the molecular weight which can be eluted from the column and successfully detected will be limited. The need exists for an FID optimally designed for use in SFC.

Data recently obtained (34) have shown that if the FID is designed with the flame jet entirely surrounded by the heating block, the detection of high-molecular-weight species can be improved significantly. Despite the work that has been reported on the FID in SFC, few descriptions of the performance of the detector have appeared. As part of the investigations and developments in detector technology for SFC being conducted currently (35), the linearity and sensitivity of the FID have been measured under capillary SFC conditions. A detectivity of 1–5 pg C/s for normal alkanes [signal-to-noise ratio

$(S/N) = 3$] has been observed. This corresponds to 25–125 pg of alkane on column as the minimum detectable level. The response is linear over at least 6 orders of magnitude for the alkanes tested.

In practice, if the CO_2 mobile phase is free from organic materials and column bleed is low, there will be little change in the baseline during pressure or density programming. Baseline rises of less than 5 pA can be obtained if the mobile phase and entire pumping system is free from organic contamination. However, getting organic-free CO_2 is still not guaranteed even when "SFC-grade" material is used. If the CO_2 is not free from organic compounds, there will be a baseline rise during pump programming because, as the pressure increases, the column exit flow increases. With an increasing amount of organic compounds in the flame, the baseline signal increases. Cleanup traps loaded with activated charcoal, alumina, or silica gel can be used to remove organic compounds from the CO_2. Even with clean fluid, significant rises in the baseline may occur if the pump and tubing are not totally free from contamination. When the FID or MS are employed, it is absolutely necessary to have a system that is clean of organic contamination, for these compounds will produce background high response in both detectors.

1.2. Nitrogen–Phosphorus Detector

The universal and sensitive response that the FID provides is an essential analytical tool. Yet many applications require a selective detector. In SFC as in capillary GC, the nitrogen–phosphorus detector (NPD), sometimes called the TID (thermionic ionization detector), is the most common sensitive and specific detector. In the simplest terms the NPD is a modified FID. In fact Patterson (36) uses these design similarities to provide the NPD user with great detection flexibility.

Fjeldsted et al. (20) were the first to describe the use of an NPD with SFC. They used both CO_2 and N_2O as mobile phases with the detector; however, the chromatograms shown were obtained in an isoconfertic (or constant-density) mode, indicating that there was a problem with the baseline during pump programming. This was a preliminary report, and no discussion of sensitivity or linearity was provided.

West and Lee (37) published a more complete study of the performance of a new series of thermionic detectors that were originally designed for GC (38, 39). These detectors have several modes of operation depending on the chemical composition of the source, the temperature of the source, and the composition of the gases surrounding the source. In one of the modes, TID-1-N_2 (the source has high cesium content and is surrounded by a nitrogen atmosphere), extremely good sensitivity was obtained. A minimum detectable level of between 1 and 850 pg was reported depending on the nitro-containing compound tested. The linear dynamic range was very narrow (1–2 orders of magnitude); however, the baseline appeared stable during pressure or density programming. The FTID mode was too insensitive to be pursued. The third mode, the TOD-2-H_2/air (cesium and strontium source and H_2/air flow) mode, had sensitivities in the 20 pg–4.5 ng range and had severe problems during pump programming. Apparently the baseline shift and loss of sensitivity during pump programming is caused by increased cooling at the bead (changing reaction conditions) that takes place as the column head pressure and effluent stream velocity increase.

David and Novotny (40) reported on the optimization of a Perkin–Elmer nitrogen thermionic detector for capillary SFC. The beads were specially synthesized to contain 1.6% B_2O_3, 12.4% Na_2O, 74.0% SiO_2, and 12.0% Rb_2O. With this bead composition, the baseline drift and loss of sensitivity seen with other thermionic detectors during pump programming was eliminated. It was also observed that this NPD shows negligible loss of selectivity and sensitivity with modified mobile phase of up to 10% methanol or isopropanol. These results indicate that the bead chemistry is an important factor affecting the use of the NPD in SFC.

It is possible to get sub-picogram sensitivity with a commercially available NPD (41) by compensating for the decline in the baseline signal during pump programming. Campbell (42) has found that the drift can be minimized and sensitivity maximized by careful control of the detector gas flow. Hydrogen flows of 1.0 ± 0.1 mL/min with air at 108 ± 1 mL/min and nitrogen makeup at 13 ± 0.5 mL/min are values obtained by simplex optimization for a commercial NPD (42). Good-quality flow controllers are essential to keep these flows constant with time and hence maintain reproducible sensitivities.

1.3. Flame Photometric Detector

Another selective detector that has been applied to SFC is the FPD. The initial report (43) of the use of FPD with SFC was not extremely promising. A standard dual-flame FPD was used for this study after some modifications were made. These modifications included shortening the detector body and adding a line for makeup gas to the detector base. Relatively poor sensitivity was obtained in sulfur mode: 25 ng for benzo[b]thiophene ($S/N = 2$). The poor sensitivity was attributed to an interfering emission line from CO_2. A rise in baseline was seen during programming owing to the increased amount of CO_2 in the flame. This rise necessitated baseline correction during the analyses. In the phosphorus mode, monitoring emissions at 530 nm, the sensitivity was much better. A detection limit of 0.5 ng was reported for parathion ($S/N = 2$).

Olesik and co-workers (44) published a thorough report on their characterization and optimization of a FPD for SFC. In this work, several variables were investigated to optimize the flame chemistry and sensitivity of the detector. Parameters such as flame gas conditions (using H_2/O_2 instead of H_2/air), hydrogen quenching, restrictor position, and mobile phase flow rate were studied. Greatly improved sensitivity and performance was obtained after optimization. A detection limit of 1.95 ng was reported for benzo[b]thiophene ($S/N = 3$). Detection limits as low as 650 pg and detectivities of 8 pg S/s were demonstrated for dibenzothiophene. As is the case in GC, the FPD response in SFC is nonlinear. As expected of a spectophotometric detector (with a molecular mobile phase) baseline rise still occurred with pressure programming after all optimization. Increasing the hydrogen flow to the detector as the column pressure increases may be a way to correct this problem.

1.4. Chemiluminescence Detector

The need for selective detectors in SFC is particularly evident for the sensitive analysis of sulfur-containing species. These sulfur species are of interest to the environmental community for the analysis of pesticides (45) and atmospheric pollutants (46), as well as to the petroleum industry (47). Various detectors have been employed for the analysis of sulfur. Among these are the Hall electrolytic conductivity detector (ELCD) (48–50), a microwave-induced plasma detector (51), a helium after-flow plasma detector (52), a non-flame-source-induced S_2 fluorescence detector (53), and a direct-current plasma detector (54). The most popular sulfur-selective detector for GC is the FPD (55), which has also been applied to capillary SFC (43, 44).

The sulfur chemiluminescence detector (SCD) has been interfaced successfully with GC, HPLC, and SFC (56–60). Operation of the SCD is based on the detection of the chemiluminescent radiation resulting from the reaction of fluorine (F_2) gas with a sulfur atom contained in the solute. The specific reaction and its dynamics have been reported by Mishalanie and Birks (57), who demonstrated that the high-frequency vibrational overtone emission dominates the chemiluminescence signal produced and that the signal is dependent on the specific reaction conditions. The chemiluminescence produced by the F_2 reaction is well suited to the sensitive and selective detection of chromatographic solute species. Owing to the nature of the chemiluminescence phenomenon, the detection sensitivity can be quite high, whereas the selectivity is dependent on the solute species and its reaction cross section.

Two groups have independently yet simultaneously reported SCD detection with capillary SFC. Both Foreman et al. (59) and Bornhop et al. (60) investigated SCD performance characteristics such as linearity, detection limits, and applicability. The connection of the SFC column to the SCD is elegantly simple. The only two major criteria for an effective transfer line are that (1) the restrictor (frit or integral restrictor) must be positioned in the reaction chamber such that only 1 mm of fused silica protrudes into the chamber and (2) a pressure-tight connection must be made at the entrance point of the transfer inlet.

In the work by Foreman et al. (59), a linear dynamic range for octanethiol of 1 decade was reported. The excellent selectivity was shown through the analysis of the thermally labile sulfur-containing pesticide malathion in the presence of large amounts of xylene and toluene. Detection limits were estimated at 3σ (standard deviations) for three sulfur-containing antioxidants, malathion, methyl parathion, 1-octanethiol, and 1-dodecanethiol. These values ranged from 890 pg S/sec for one of the antioxidants to 6 pg S/s for 1-octanethiol. Pressure programming was performed with little effect on the baseline of the chromatographic trace.

Calibration curves were constructed by Bornhop and co-workers (60) after analysis of standard solutions of isopropylthiol and dodecanethiol. Isopropylthiol was chromatographed at constant pressure and temperature, but it was necessary to employ pressure programming at constant temperature to elute dodecanethiol. Calibration curves for each solute were linear, with correlation coefficients for 0.998 for isopropylthiol and 0.997 for dodecanethiol. Linear calibration extended over 3 orders of magnitude in concentration from the detection limit (5×10^{-3} mg/mL) to >1.5 mg/mL. The solute mass and sulfur mass detection limits, determined at 3σ, were reported to be in the range of 15–18 pg of sulfur and compare well with those obtained in GC–SCD (56).

A further advantage illustrated for the SCD in SFC is that the solute response appears to be independent of pressure. Upon normalization of the response of the SCD for the percentage of elemental sulfur in the solute, and to allow for differences in peak width (solute dilution), the detection limits for isopropylthiol and dodecanethiol were found to be 0.71 and 0.75 pg S/s, respectively. These two numbers are identical within the experimental error of the method. The normalized detection limit of 0.71 pg S/s of sulfur for the isopropyl analysis was performed at a constant pressure of 60 atm, whereas the normalized dodecanethiol detection limit of 0.75 pg S/s was obtained during pressure programming with the solute elution pressure of about 163 atm.

Selectivity is a key issue for selective detection methods. Various definitions can be used to represent the selectivity of a method. We utilize the selectivity ratio (SR) (57) as defined thus:

$$SR = \frac{\text{signal/mole S compound}}{\text{signal/mole interfering compound}} \tag{1}$$

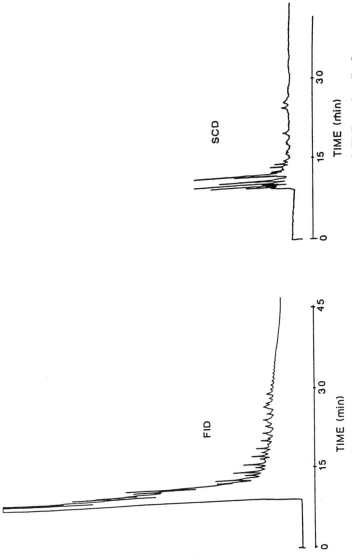

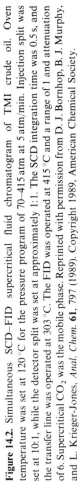

Figure 14.2. Simultaneous SCD–FID supercritical fluid chromatogram of TMI crude oil. Oven temperature was set at 120 °C for the pressure program of 70–415 atm at 5 atm/min. Injection split was set at 10:1, while the detector split was set at approximately 1:1. The SCD integration time was 0.5 s, and the transfer line was operated at 303 °C. The FID was operated at 415 °C and a range of 1 and attenuation of 6. Supercritical CO_2 was the mobile phase. Reprinted with permission from D. J. Bornhop, B. J. Murphy, and L. Krieger-Jones, *Anal. Chem.* **61**, 797 (1989). Copyright 1989, American Chemical Society.

Each of the two solutes isopropylthiol and dodecanethiol were compared to toluene and hexane to determine the selectivity ratio in SCD–SFC. Values of 5.5×10^4 for hexane and 8327 for toluene (each relative to isopropylthiol) were obtained. The selectivity to dodecanethiol vs. hexane and toluene was determined to be 3.8×10^4 and 7950, respectively. SCD with SFC was found to be more selective than SCD with HPLC yet is less selective than SCD detection in GC.

As in the work by Foreman et al., Bornhop and co-workers found that the SCD–SFC provides baseline stability during pressure-programmed runs. Chromatograms were presented with pressure programs from 70 to 415 atm with no evidence of baseline perturbation.

As previously illustrated, Bornhop and co-workers (35) have shown that simultaneous selective and universal detection can be employed in capillary SFC. By using two detectors simultaneously, added solute information can be gained. The simultaneous FID and SCD chromatograms of a crude oil are shown in Figure 14.2. The information gained from the early part of the separation through use of the SCD is dramatic. Various low-bonding sulfur containing compounds are present in the sample that would not be detected if only the FID were employed. The expected oligomeric hydrocarbon distribution is detected by the FID yet is ignored in the sulfur-selective SCD. Two peaks are detected in the SCD and FID at an elution time of about 25 min that are thought to be sulfur-containing aromatic compounds. An advantage of the use of a surfur-selective detector with a universal detector is seen in this example: one chromatographic analysis yields two types of information.

1.5. Photoionization Detector

The PID uses a high-energy UV lamp to ionize the organic species present in the column effluent stream. The PID can be a universal or specific detector depending on the lamp energy used. By careful choice of the lamp energy, compounds that may interfere in a particular determination can be made "transparent" in the PID. In fact, there have been some reports of the use of the PID in LC (61, 62).

Gmür et al. (63) published the first report of the use of a PID with SFC. In this work the authors constructed three different detectors and evaluated their sensitivity with argon and CO_2 as the mobile phases, using benzene in methanol as the test solution. All three detectors were able to be pressurized to ~ 100 bar pressure. In this case, it was reported that the PID with CO_2 mobile phase at 100 bar gives 6–7 orders of magnitude poorer sensitivity than an FID. This poor sensitivity was attributed to UV absorption by the CO_2 and by the large electron affinity of the CO_2, which may cause quenching in the detector. However, either of these factors will be intensified by the high pressure present in the detector during these experiments. With argon mobile phase, the PID was one order of magnitude less sensitive than the FID. Argon is of little interest in SFC because it lacks solvent strength.

Sim et al. (64) evaluated a commercial PID using packed microbore columns. Sensitivities were demonstrated for some polycyclic aromatic hydrocarbons in the upper-picogram levels ($S/N = 3$). These detection levels are not extraordinary, but they did demonstrate that a nonpressurized PID can be made to work well in SFC and have sensitivity comparable to that in GC. This also confirms the idea that the poor sensitivity seen by Gmür and coworkers was due to the high pressure in the detector cell. Linearity is also similar to that obtainable in GC: 4 orders of magnitude.

Recent work has demonstrated that if a PID is used with a capillary system even better mass detection levels can be achieved (65). The minimum detectable level for dichlorobiphenyl sulfone was estimated to be 0.3 pg using timed split injection and a capillary column interfaced to a commercially available PID. It would seem obvious from these results that the PID is a detector that should be investigated more fully.

In terms of baseline changes during programming, if the mobile phase is clean, no rise occurs with the PID because gaseous CO_2 is transparent to the excitation light normally used (CO_2 has an ionization potential of 13.79 eV). The PID may also be of some utility when one is using mixed mobile phases because lamps with energies lower than the ionization energy of the organic component can be employed, thus making these modifier compounds "transparent" to the detector.

1.6. Electron Capture Detector

The ECD is probably one of the most sensitive detectors routinely used in GC. Some preliminary work that demonstrated the potential of coupling SFC with an ECD has recently been published (66). In this work, a capillary column and injection system was connected to a packed column ECD. Makeup gas was used in the detector, but there were some dead-volume problems evident from the shape of the chromatographic peaks. A detection limit of ~ 35 pg of "triazole fungicide metabolite" was reported. This poorer than expected sensitivity may be due to the detector design or it may be due to the background contribution of the CO_2 itself. One of the figures presented shows a significant increase in the baseline when the mobile phase pressure is increased. The large electron affinity of the carbon dioxide may be a contributing factor in the low solute sensitivity and baseline rise. The change in baseline may also be due to trace halogenated impurities in the CO_2 or materials from the fluorocarbon seal components of the system being extracted by the supercritical CO_2.

If 50-μm-i.d. capillary columns are used, the baseline shift can be subtracted. On the other hand, if 1-mm-i.d. packed columns or even 100-μm-i.d. capillaries are used, the flow rates are sufficiently large that the baseline change becomes unmanageable (67). DDT in pork fat has been determined by an ECD coupled with capillary SFC (68) using baseline correction.

Minimum detectable levels in parts per billion (ppb) for a monochlorinated herbicide have been achieved using an ECD coupled with timed split injection and capillary columns (67). Linearity appears to be maintained over at least 4 orders of magnitude in the ECD with capillary columns in SFC. Clearly these data indicate that the ECD has potential as a viable detector for SFC.

1.7. Mass Spectrometric Detector

Indisputedly, MS is the analytical detection technique that provides the most structural information for the least amount of analyte. It is therefore of great interest to interface the MS with SFC as has been done in GC and HPLC. In concept this SFC–MS interface should allow for the analysis of nonvolatile species (as in HPLC) and allow for a simple interface of the chromatograph to the MS (as in GC). As shown below this interfacing task is certainly less difficult in SFC than in HPLC, yet there are limitations to the use of SFC–MS.

Interfacing has been reviewed by various authors (69–71). The first SFC–MS interface was a molecular beam approach reported by Randall and Wahrhaftig (72). This system was complex, impractical, and provided poor sensitivity. The first practical interface was reported by Smith and co-workers (73, 74) and involved directly interfacing a capillary column with the ion source via a short section of 5-μm i.d. restrictor. Both electron ionization (EI) (75) and chemical ionization (CI) (73) spectra were obtained with this system. The direct interfacing of capillary SFC to a Fourier transform mass spectrometer (76) and a bench-top GC–MS system (77) have been reported. In both cases it was the desire of the researchers to obtain EI—and hence library-searchable—spectra.

Various design considerations dominate the particular aspects of the SFC–MS interface. With capillary column SFC, the entire effluent (often amounting to less than 1 mL/min gas) is generally handled by existing GC–MS pumping systems. For example, in most cases the gas-volume load into a mass spectrometer from a 50-μm i.d. capillary used at modest densities is comparable to that of a 300-μm-i.d. GC capillary column. The current drawback to pressure programming with existing restrictor designs is the increased volumetric flow as the mobile phase density is increased. Giorgetti et al. (78) and Anton et al. (79) have built a constant-flow pressure programmable SFC system that could simplify the SFC–MS interface. Currently a vacuum pump with a capacity in the 330 L/s range is required for flexibility in operation.

Interfacing of SFC to a bench-top mass analyzer, a quadrapole MS, and a high-resolution sector instrument all require a specially designed transfer line. Yet in all cases the distance between the chromatograph and MS is greater than interfaces to other detectors; therefore a temperature-regulated interface is needed to keep the transfer line under supercritical fluid conditions. Additionally the fixed restrictor must maintain supercritical conditions inside the column while allowing rapid decompression of the fluid in the low-pressure region inside the MS ion source. Because expansion occurs during the decompression process, rapid cooling occurs producing cold particles rather than heated vapor as in GC–MS. In order to produce EI and CI spectra, it is important that the analyte be in the vapor phase as it enters the ion source.

Several approaches have been employed aimed at production of vapor at the restrictor terminus, yet all involve "heating" the end of the restrictor or the ion source block. It is difficult to transfer thermal energy to the solute molecules under the low-pressure conditions found in the MS ion source. These difficulties are the primary reason for the lack of EI SFC–MS reports (75, 80). Good sensitivity, however, has been reported in the CI mode with SFC–MS (81). This presumably is a result of better heat transfer and more desirable ionization conditions in the higher pressure regime within a CI ion source. Because CI is a "soft ionization" process, CI spectra normally only contain molecular weight information and are lacking the structural information normally provided by the fragmentation produced by EI. Therefore a major goal in the development of a SFC–MS interface is the production of high-sensitivity EI spectra. Recent progress in this area has facilitated operation of a capillary SFC–MS system in the EI-like mode (77). Of course it is possible to produce EI-like spectra when employing a multisector (or triple-quad) instrument via the use of MS/MS techniques, although this is an expensive instrumental approach.

One effective approach taken to produce library-searchable spectra uses a heated capillary restrictor in combination with a CI source to produce charge exchange (CE) (82). CE ionization involves operating the CI ion source pressurized with a reactant gas whose ionization potential is low but is higher than that of each organic analyte of interest. The ionized reagent gas interacts bimolecularly with gaseous organic molecules to cause ionization of the latter by charge exchange. This process will not produce true electron ionization, but rather EI-like spectra comparable to those obtained under low-voltage EI conditions (83). The value of this approach to capillary SFC–MS is that the higher pressure in the ion source (0.5–1.0 Torr) presumably improves heat transfer from the heated ion source to the organic analytes and improves

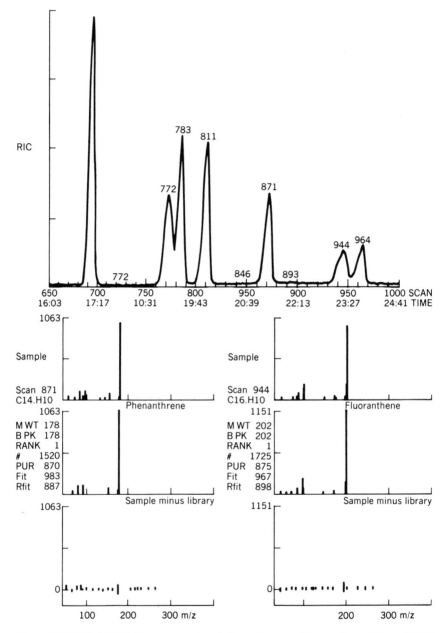

Figure 14.3. SFC–MS. EI-like mass spectra of phenanthrene and fluoranthene. CO_2 at 100 °C, linear density programmed, 5-m × 50-μm-i.d. SB-biphenyl-30 column with frit restrictor heated to 270 °C. Results shown below the total ion chromatogram show that EI-like spectra are obtained because when the library mass spectra (true EI spectra) are subtracted from the sample spectra almost no deviations are seen.

ionization conditions for the bimolecular ion molecule interactions. These conditions, coupled with increased electron energy (230 eV), have produced EI-like mass spectra in capillary SFC–MS with detection limits comparable to those achieved with capillary GC–MS (77).

Owing to the nature and complexity of the instrumentation used in SFC–MS, it would be impossible to cover the topic completely in a summary chapter such as this. Therefore we will give one example of an SFC–MS application and leave the details to a complete treatise on the subject. A relatively simple capillary SFC–MS interface has been developed by Disinger (84) that is compatible with a single-quadrupole instrument. This system employed a modified ion volume for a Finnigan INCOS 50, a special transfer line, and a commercial SFC. As shown in Figure 14.3, EI-like (CE) spectra can be obtained for simple solutes such as phenanthrene and fluoranthene. Library search fits correlate well with standard spectra, and the difference plots further illustrate the correlation. Similar results have been obtained for various pesticides and compounds from the U.S. Environmental Protection Agency (EPA) Appendix VIII and IX lists. As the technology develops, it is certain that MS will become as valuable to SFC as is it to GC.

1.8   Fourier Transform Infrared Detector

FTIR spectrometry has been utilized as a detection system on GC instru-mentation with good success for some time (85). One of the first reports of combining SFC with FTIR came more recently from Shafer and Griffiths (86). A review by Jinno (87) describes many past efforts in the area of FTIR detection in SFC. The advantage of spectrometric detection techniques such as FTIR is that one can obtain molecular structure information from the compounds eluting from the column and not just retention times or area counts, as in the case of nonspectrometric detectors such as the FID. Infrared spectrometry excels in the ability to indicate quickly and unambiguously the presence or absence of specific organic functional groups in a sample molecule. By monitoring the presence of absorbance bands in narrow regions of the spectrum, the FTIR detector functions as a chemically specific detector. For example, monitoring the $1760–1700 \, cm^{-1}$ region allows carbonyl com-pounds to be monitored selectively. This approach is somewhat analogous to single-ion monitoring in MS, with the advantage that an IR absorption band is often more characteristic of a class of compounds than is a single-mass ion. It is therefore feasible to detect and identify specific types of compounds in a chromatographic run while recording complete spectra for subsequent data analysis. These attributes make IR detection especially useful for identifica-tion of unknown analytes.

Two basic approaches have been used for the coupling of FTIR detection to SFC: one involves depositing the column effluent on a solid substrate; the other involves a flow-through cell connected at the end of the column. In the deposition techniques, the column effluent is sprayed onto a substrate, which is either stepped or continuously moved beneath the column effluent spray. This allows the chromatographic peaks to be spatially separated on the substrate for subsequent analysis. The substrate is composed of a material that is IR transparent such as KBr or ZnSe. After deposition, the substrate can be heated to remove the mobile phase if necessary. When CO_2 is used, the mobile phase vaporizes without heating, leaving the solute deposited on the substrate. The deposited sample can then be analyzed by an FTIR beam using a microscope (88–90), beam-condensing optics (91), or diffuse reflectance (92).

The deposition techniques can have advantages over the flow-through technique, depending on the particular application. Increased sensitivity can be obtained with the deposition techniques by signal averaging multiple scans of the trapped solute. The lowest injected quantity from which an identifiable spectrum can be obtained, or the minimum identification limit (MIL), has been reported to be about 1 ng for deposition techniques (93, 94). Deposition techniques also allow the removal of the mobile phase, which may be important when supercritical fluids that have interfering bands in the IR are used as the carrier fluids. The biggest advantage of the deposition techniques is that the spectra obtained are essentially identical to those obtained under standard conditions. This means that libraries of spectra such as the Georgia State Crime Laboratory (GSCL) collection or the Sadtler Standard Spectra Library (Sadtler Research Laboratories, Philadelphia, PA), which contains spectra for more than 100,000 compounds, can be searched to help in the identification for unknown compounds. On the other hand, with flow-through or on-line SFC–FTIR, the IR spectra of many compounds are different from the corresponding vapor-phase or condensed-phase spectra because of shifts of maximum absorbance wavenumber, variations of bandwidth, and modifications of the intensity distribution (95). Vapor-phase spectra are often used successfully for comparison with the spectra obtained in on-line SFC–FTIR; however, some of the applications of SFC that are of most interest involve compounds that are nonvolatile, and the vapor-phase spectra are not applicable. Some researchers are advocating the generation of a library of FTIR spectra collected in supercritical phase CO_2 so that better matches between the spectra of the unknowns and standards can be achieved.

Despite their advantage, current deposition techniques tend to be mechanically complex and have yet to demonstrate the degree of reliability required for broad use in analytical laboratories. Further development and improvement in the deposition technique will undoubtedly improve its viability.

The other method of interfacing SFC to FTIR is the use of a flow cell. This approach is analogous to other on-line detection schemes like GC–FTIR or UV detection. The effluent is monitored in real time as it flows from the column. The flow cell assembly that is attached at the column end consists of a high-pressure IR absorption cell with inlet and outlet lines and IR-transparent windows (usually ZnSe). The FTIR beam analyzes the column effluent as it flows through the detector cell. This approach has been demonstrated for both CO_2 (86, 95–113) and xenon (96, 105) mobile phases.

Flow cell detection analyzes the entire effluent stream so all sample components are detected intact. Provided that the mobile phase does not interfere with the desired sample spectra, this allows for a broad range of applications. The flow-through cell is similar to a cell used in UV detection except that the windows are made of IR-transparent materials. The cell must be able to withstand high pressures as in UV detection schemes in SFC. This approach is, in most cases, mechanically less complicated and more reliable than the apparatus in deposition systems. Other advantages include real-time monitoring of the chromatographic effluent and the ability to use other detection techniques such as MS or flame ionization in series after FTIR detection.

There are two major design considerations for the flow cell in FTIR for SFC. As with any chromatographic instrumentation, dead volume must be minimized to avoid excess broadening and loss of resolution. This usually is done by making the internal diameter of the cell small. However, if the diameter is too small, the amount of energy transmitted by the cell (throughput) is decreased. This decreased throughput causes the noise level in the recorded spectrum to increase. Since detection is based on the S/N of an absorption band in a spectral baseline, sensitivity is reduced. The other consideration is of course the pathlength. Since Beer's Law operates in SFC–FTIR, increasing the cell path length will increase the absorbance (up to a point of no gain, as discussed below in Section 2.1 on UV absorbance) and thus the sensitivity. However, in SFC–FTIR, the mobile-phase absorbance must be considered. Carbon dioxide has Fermi resonance absorbance bands between 1471 and $1225\ cm^{-1}$. In vapor-phase spectra these bands are weak, but they are quite pronounced in supercritical fluid spectra, particularly at high density. Therefore, to be compatible with CO_2 and density or pressure programming, the cell path length must be limited to avoid interference from the mobile-phase absorbance with sample spectra. This absorbance is a function of the mobile phase and not the chromatographic peak volume, so a cell optimized for capillary SFC will also work well for packed column SFC. The cells reported in the literature have been close to 600-μm i.d. with a 5-mm optical path.

Pressure or density programming in SFC presents unique problems for FTIR when flow cells are used. The IR spectrum of CO_2 contains totally absorbing bands from 3800 to 3500 and 2500 to $2200\ cm^{-1}$. The width and intensity of these bands changes with increases in the pressure or density of the CO_2. These spectral features of the mobile phase can easily be removed using conventional FTIR subtraction algorithms. Increasing density also causes a sloping baseline in the Gram–Schmidt reconstruction. That makes it difficult to detect the elution of minor sample components. By including an interferogram vector taken at higher density in the Gram–Schmidt basis set, the density gradient in the reconstructed chromatogram can be removed. This effectively eliminates any problems associated with flow-through cell detection during density-programmed SFC–FTIR.

The main disadvantage of the flow cell approach to SFC–FTIR is the difference in the IR spectra of compounds in supercritical CO_2 compared to those collected in the vapor phase or condensed phase. Stretching vibrations are more sensitive to a change in density than are bending vibrations. Wavenumber shifts in the spectra are relatively moderate for apolar functional groups ($1 < 2 \, cm^{-1}$) but more significant for polar functional groups (up to $10 \, cm^{-1}$) (95). In spite of these differences and shifts, the quality of the spectra collected from on-line techniques is many times sufficient for identification and for distinguishing subtle differences between related compounds.

The on-line coupling of SFC and FTIR does not have the same sensitivity as the deposition techniques. The best reported MIL is 10 ng (107), as compared to 1 ng for the deposition techniques (93, 94).

The advantage of using xenon as a carrier in SFC–FTIR is that xenon is totally transparent in the IR region of the spectrum. This means that absorption bands which are undetectable in CO_2 (for example O—H stretching absorptions) can be detected in this unique mobile phase. The main disadvantage is the cost of the xenon itself. The use of capillary columns rather than packed columns makes the use of xenon more attractive because of reduced solvent consumption.

Applications of SFC–FTIR have included the determination of pesticides and related compounds (108, 110, 112), derivatized monosaccharides (109), polycyclic aromatic hydrocarbons (114), aromatic isocyanate oligomer (114), epoxy acrylate oligomers (114), steroids (113), UV-curable coatings (111), and polymer additives (90).

1.9. Other GC Detectors

Other GC-type detectors have been interfaced to SFC. They include the ion mobility detector (IMD), hydrogen atmosphere flame ionization detector (HAFID) (116), thermal energy analysis detector (TEA) (117), and microwave-induced plasma detector (MIP) (118, 119).

As a summary of the data presented, Table 14.1 contrasts the minimum detection limits, linearity, and pump programming effect on the baseline for the five GC detectors detailed herein. Just as in GC, the selective detectors

Table 14.1. GC-Type Detector Comparison

Detector	Reported MDL[a]	Linearity	Baseline Effect[b]
FID	25–125 pg	10^6	None
TID-1	1–850 pg	10^2	None
TID-2	0.02–4.50 ng	10^1	Yes—down
TID-3	20–100 pg	10^4	None
FPD	0.65–1.90 ng	—[e]	Yes—up
PID	1–100 pg	10^4	None
ECD	1–10 pg	10^4	Yes—up

[a]Approximate minimum detectable level taken from referenced papers ($S/N = 3$). This is the amount of material on column.
[b]Direction of baseline change during pump programming.
[e]Response is nonlinear.

can be used to achieve low detection limits, but they are often dependent on the molecular structure of the analyte. Much work needs to be done to better understand and optimize the reaction chemistries in these detectors for SFC.

Most of the GC-type detectors currently used on commercial instruments were originally designed for use in GC and have not been optimized for use in SFC. For many applications, they function well. However, to take full advantage of the benefits of SFC, these detectors must be optimized for use with supercritical fluids as mobile phases and have the ability to detect higher-molecular-weight materials. Since the number of fluids that do not respond in GC-type detectors is limited, not all analyses can be accomplished. This is especially true when polar compounds are involved and organic modifiers are used in the mobile phase.

2. LC-TYPE DETECTORS

2.1. Multiple- or Fixed-Wavelength UV–Visible Detectors

Theoretical and Practical Considerations

Conventional optical detection methods such as UV absorbance have been used in SFC since the first report appeared (8) and are still among the most popular detection techniques. The FID responds to practically any molecule containing a carbon–hydrogen bond, whereas the UV detector responds only to solutes that have UV absorbing functional groups. Unmodified inorganic mobile phases such as CO_2 exhibit little background absorbance in the 200–

800 nm region and therefore produce a low background for high-sensitivity absorbance detection. Unlike the FID, which is a nonspecific and destructive detector, the UV detector can be tuned to the absorption wavelength of the analyte of interest in order to provide compound-specific information. Additionally, a scanning UV–visible monitor may be employed to gain solute spectral information. The UV detector is of particular value when an organic modifier has been added to the supercritical fluid.

Capillary Columns. The use of capillary columns in SFC imposes the same flow cell design requirements as packed columns, with the added dimension of a stringent detector dead-volume limitation. Peaden and Lee (2) have developed an expression to estimate allowable detector volume in capillary SFC:

$$V_i = 0.86\pi d_c^2 (LH)^{1/2} [1/(1-R_s)^2 - 1]^{1/2} (1+k') \tag{2}$$

where V_i is the volume of the detector; R, the fractional resolution loss; L, the column length; H, the plate height; k' the capacity factor; and d_c the internal column diameter. For example, the allowable detector volume producing a 1% resolution loss for a 20-m-length, 100-μm-i.d. column, with a plate height of $0.6\,d_c$ and a k' value of 1, is 270 nL. Employing the same parameters for a 50-μm-i.d. column, the size most commonly used in capillary SFC, the required cell volume is now only 50 nL. These small volumes can be attained, but at the expense of detector path length. A more detailed discussion on sensitivity as it relates to path length is found later in this section.

One approach taken for UV detection in capillary SFC (120) is similar to that employed for capillary HPLC (121). The use of on-column or pseudo-on-column detection has become a common procedure in many other capillary separation techniques (122). In this technique a small section of the polymer coating is removed from the analytical capillary column or from a short piece of capillary tubing that serves as the detector cuvette. As shown in Figure 14.4, two butt connectors connect the detector capillary cuvette directly to the capillary column and to the restrictor. In this configuration a 250-μm-i.d. capillary is the detector cell. Certainly a portion of the 50- or 100-μm-i.d. separation column could be employed directly, but only at the expense of detector sensitivity. Use of a 250-μm-i.d. capillary performs well by minimizing the dead volume and providing acceptable sensitivity. A 250-μm-i.d. capillary produces a detector volume of about 60 nL. Resolution losses due to detector band broadening when the on-column detection approach is used have been calculated for HPLC microcolumn chromatography (123) and for capillary SFC (2). In a study performed by Later and co-workers (124) it was shown that the practical upper limit for the inner

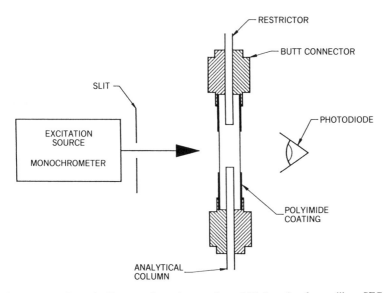

Figure 14.4. Schematic diagram of pseudo-on-column UV detection for capillary SFC.

diameter of the capillary tubing used as a flow cell in conjunction with 50-μm-i.d. columns is about 320 μm, yet the best chromatographic performance, while retaining reasonable sensitivity, was achieved with a 250-μm-i.d. flow cell.

Although a 0.25-mm path length is quite small, adequate sensitivities can be obtained. The measured noise for a capillary UV–SFC system at CO_2 density of 0.24 g/mL is $\sim 2 \times 10^{-5}$ AU, whereas the static, short-term noise is measured to be $\sim 1.2 \times 10^{-5}$ AU.

The dependence of absorbance detector sensitivity on path length has been investigated by various workers (125, 126). Poppe derived an equation that has a somewhat surprising implication in that, at a specified cell volume, longer cells result in worse detection limits. In his expression, movement to smaller path lengths (wide, short cells), which would favor detectability, is limited by the instability of the source and the receptor. The work of Baumann and Fresenius (127) on a number of conventional excitation sources reaffirms that a volume of 10 μL and a path length of 1.0 cm is the optimum for the existing technology. In particular, conventional deuterium and low-pressure mercury lamps limit the number of photons per unit volume, and thus cells with a volume of less than 1.0 μL can be used only at the expense of detection limit. Interestingly, in the shot-noise-limited case, the detection limit in amount (minimum detectable quantity) depends on the source exclusively and not on the volume of the cell. Ultimately the analyst strives for the maximum S/N that can be obtained by the instrument. Because of these limitations, researchers have sought to develop absorbance detection technology utilizing the high photon flux that a laser can produce in order to push the detection limits lower (128–131). These nonconventional absorbance methods will not be considered in this chapter, yet it is important to note that the thermal lens, one of the nonconventional absorbance measurements, has been used in HPLC (128–131) and in supercritical fluids (132). Leach and Harris (132, 133) found that near the critical point a significant enhancement of the thermal lens signal is realized. Thus, it appears possible to design an absorbance detector with no path length dependence and superior sensitivity.

In view of the basic physical limitations of a conventional UV system, the on-column approach has proven quite successful. Because of the mass-sensitivity enhancement as a result of the small column diameter and the low flow rates used in the capillary techniques, reasonable detection limits are obtained, even with the short path length provided by a capillary cuvette. Solutes such as anthracene and pyrene can be detected in the mid-picogram range (134).

Recently Chervet and co-workers (135) reported a novel approach to microcolumn detection for HPLC and SFC. In this work a 2-cm (20,000-μm) path length is obtained while the detector volume is kept to less than 90 nL. Figure 14.5 illustrates how this long path length measurement is made in a small volume. A dynamic noise level of 1×10^{-4} AU with a 2.0-s detector response time is reported. This value is about an order of magnitude poorer than that reported by Bornhop and co-workers for the 0.25-mm on-column approach. Although not postulated in the original report, it assumed that this high noise level is a result of light scattering at the exit and entrance points of the flow cell. The bent capillary produces various curved surfaces that cause

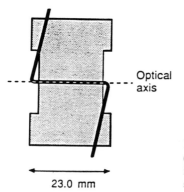

Optical axis

23.0 mm

Figure 14.5. Cross-sectional view of the flow cell holder (template) with longitudinal capillary flow cell. The cell is designed for long path length while maintaining a small volume.

the excitation beam to be scattered, reflected, and vignetted. Band-broadening effects appear to be well controlled in this flow cell arrangement; thus chromatographic performance can be maintained. The maximum pressure rating for this cell was postulated to be about 500 bar.

Chevret and co-workers report a mass detection limit, across a 20-mm path, of 7 pg for phenanthrene (135). This value should be compared to a mass detection limit of 5.8 pg for anthracene, as determined by Weinberger and Bornhop (136), while employing a 0.25-mm path length on-column flow cell with a low noise (2×10^{-5} AU) scanning UV monitor. In this example an instrument with an 80-fold increase in path length produced similar results to an instrument with five times less noise.

Density Programming. Density programming poses another problem for the UV detector, which is manifested in the presence of a baseline shift.

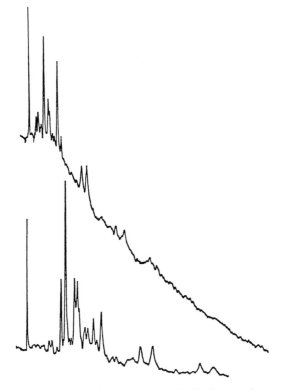

Figure 14.6. UV–SFC without (top chromatogram) and with (bottom chromatogram) baseline correction. CO_2 at 100 °C, UV at 254 nm, 3-m × 50-μm-i.d. SB-methyl-100 column. Sample: PAH species extracted from a crude oil sample.

When CO_2 is the mobile phase this baseline offset is seen as a negative deflection in the chromatographic trace, whereas Novotny reported a positive shift in n-butane (137). It is postulated that density-related refractive index (RI) changes are the main contributor to this annoyance (120). The magnitude of this effect has been quantified for supercritical CO_2, with values of $\sim 1.3 \times 10^{-2}\,AU \cdot g^{-1} \cdot mL^{-1}$ typically encountered when a 250-μm-i.d. capillary is the detector cuvette. By temperature thermostating the cell with cool H_2O, as much as a 55% improvement in RI sensitivity can be realized. Another way to correct for the baseline shift is to program the detector to perform a baseline compensation. Figure 14.6 shows a typical separation that employs density programming before and after baseline correction.

The sensitivity of the bent capillary flow cell (135) to density or pressure programming is unknown because only constant pressure runs were presented. It can be postulated that this perturbation will be significant since over long path length the incoherent excitation beam would interact with the walls of the capillary, causing RI sensitivity.

Modifiers. Another method available in SFC is the addition of organic modifiers to the mobile phase. Formic acid can be added at low concentrations to effect changes in solute retention while only moderately affecting the FID response, yet most of the other modifiers that show potential normally produce an unacceptable background in the FID. These modifiers (e.g. isopropyl alcohol) overwhelm the FID detector. Modifiers such as isopropyl alcohol, acetonitrile, ethanol, propylene carbonate, and methanol can be employed with SFC–UV because of their relative transparency to the UV excitation light.

Figure 14.7 presents a chromatogram of a 50% phenylmethylpolysiloxane sample (138) that has been eluted from a packed capillary column by the

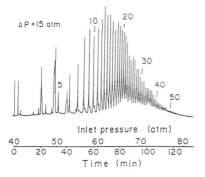

Figure 14.7. Supercritical fluid chromatogram of OV-17 (50% phenylmethylpolysiloxane) on a packed capillary column. Column: 50cm × 200 μm Finesil C_{18}-10. Mobile phase: supercritical 10% ethanol in hexane. Pressure programs: as shown. Temperature: isothermal, 260 °C detection UV at 210 nm. Reprinted with permission from Y. Hirata, F. Nakata, and M. Kawasaki, *HRC & CC, J. High Resolut. Chromatogr. Chromatogr. Commun.* **9**, 633 (1986). Copyright 1986, Dr. Alfred Huethig Verlag.

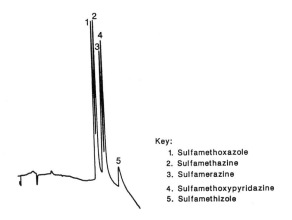

Key:
1. Sulfamethoxazole
2. Sulfamethazine
3. Sulfamerazine
4. Sulfamethoxypyridazine
5. Sulfamethizole

Figure 14.8. Organic modified CO_2 capillary column SFC of sulfa antibiotics. Column: SB-cyanopropyl-25, 7 m × 50-μm i.d. Mobile phase: 8% isopropanol in supercritical CO_2 UV detection at 254 nm. Temperature: isothermal 100 °C. Density program 0.34–0.77 g/mL at 0.015 g·mL^{-1}·min^{-1}. Reprinted with permission from S. Schmidt, L. G. Blomberg, and E. R. Campbell, *Chromatographia* **25**, 775 (1988). Copyright 1988, Frieder. Vieweg & Son.

use of a 10% ethanol in *n*-hexane mobile phase. This analysis would have been quite difficult without the use of an organic modifier. Additionally, the analysis would have been impossible to perform had the FID been used as the detector.

Since several researchers have used modifiers in capillary SFC with reasonable success (5, 139), it is certain that the modifier effect is more involved than a simple surface effect on packed column particles. However, it does appear that higher levels of modifier are necessary to effect a result in capillary SFC. Regardless of the mode of activity, organic modifiers can be quite valuable in capillary SFC, as shown in the separation presented in Figure 14.8. Separation of the sulfur-containing antibiotics was attempted on a packed column and on a capillary column with 100% CO_2, but the best separation was obtained by use of a cyanopropyl capillary column and supercritical CO_2 containing 8% isopropanol. This example illustrates the utility of modified mobile phase SFC with UV detection.

2.2.   Scanning UV–Visible Detectors

Background and Practical Considerations. Multiwavelength (scanning) UV–Visible detection provides the advantages that tunable UV–Visible detection does in SFC, with the added benefit of spectral information. As discussed previously, this added gain in solute information often facilitates the identification of specific species through spectral matching. Collection of on-the-fly spectra to assist in compound identification has been realized in SFC with the use of a photodiode array–UV (PAD–UV) detector and with a galvanometer-driven monochromator (GDM) UV–visible detector. Both approaches to scanning detection offer advantages and trade-offs. Scanning UV and UV–visible monitors offer the same advantages as the fixed-wavelength detectors, with the gain of additional solute information.

Miniaturization of a scanning photodiode array instrument for analysis in capillary HPLC (140) has been reported. While it is a nontrivial task to apply these detectors to capillary SFC, France and Voorhees (141) have used the PDA in capillary SFC for the analysis of some pesticides and herbicides. The use of a 320-μm-i.d. capillary as a flow cell cuvette allowed acceptable chromatographic performance. A detection limit of about 3.8 ng of bendiocarb ($S/N = 5$) was reported, with an on-column mass linear dynamic range extending from 3.8 to 150 ng. The reported noise level was found to be about 3×10^{-3} AU during broad band integration. Although this approach was successful in obtaining spectral information, the modest sensitivity values, short linear dynamic range, and relatively high noise levels observed are due to optical constraints associated with direct on-column pseudo-on-column detection in PDAs.

An alternate approach to multiple-wavelength detection on a capillary scale is obtained through the use of a forward optical bench in conjunction with a rapid-scanning GDM. Such a design minimizes propagational distances between the flow and phototransducer, thus limiting the overall variation in light distribution secondary to flow cell RI change. Additionally, forward optical designs exhibit lower stray light and can utilize a more efficient double-beam scheme, resulting in superior signal-to-noise performance when compared to reverse optical systems.

Weinberger and Bornhop (136) used this recently developed monitor to successfully interface capillary SFC to a "scanning" UV–visible detector. A pseudo-on-column detection approach was used with a 250-μm-i.d. capillary as the detector flow cell cuvette. The noise level in the rapid scanning mode was $\sim 1 \times 10^{-4}$ AU under capillary SFC conditions when CO_2 was used as the mobile phase. On-column mass detection limits of 32 pg for naphthalene while operating in the high-speed scanning mode were reported, with the linear dynamic range extending over more than 3 orders in magnitude. Relatively high-quality spectra were obtained during capillary SFC operation.

In a report by Jinno and co-workers (142), a multichannel UV detector was employed with computer enhancement in order to monitor the separation of coeluted solutes. Packed columns with inner diameters of 4.6 and 1.0 mm were used to analyze perdeuteriotoluene and perprotiotoluene. Completely overlapping peaks are separated spectroscopically by the use of deconvolution techniques functioned in the UV multichannel detector.

Other advantages of using UV–visible detection in SFC are (1) the ability to tune the detector to a particular wavelength for compound specificity; (2) the ability to determine peak purity by using a sample duochrome method with tunable UV–visible detectors; and (3) the ability to make direct comparisons between the SFC–UV chromatogram and the analysis performed by a HPLC–UV method. Additionally, in capillary SFC it has been shown that the FID and UV–visible detector can be used simultaneously (134) to gain two forms of information (universal and selective) in order to characterize complex matrices.

2.3. Fluorescence Detectors

Fluorescence detection is a sensitive method and can be selective for the analysis of species that either show native fluorescence or can be derivatized to form a fluorescent species. The technique has been applied to HPLC, SFC, and capillary zone electrophoresis. Its sensitivity is a result of the fact that the signal is produced in an information region containing little noise and often no background. Because the sample must be fluorescent, the number of compounds that can be detected is somewhat limited. Fluorescence does lend itself fairly well to detection flow cell miniaturization, and small volumes can be probed when a laser excitation source is used.

Fjeldsted and co-workers (143) successfully modified a commercial scanning fluorescence detector for use with capillary SFC. A novel fiber-optic-aided flow cell minimized noise due to scattering at the surface of the capillary during pseudo-on-column detection. In order to obtain a full spectra, it was necessary to introduce at least 1 ng of pyrene. Picogram detection limits seem achievable with this device. The authors showed that a carbon black extract contained various polycyclics and that they could be detected and tentatively identified by capillary SFC with fluorescence detection.

Novotny and co-workers (144) constructed a high-sensitivity multichannel fluorescence detector that can be employed with HPLC or SFC. By use of an intensified linear photodiode array to monitor fluorescence emission, on-line spectral information was obtained over a linear range of 3 decades. Minimum detectable quantities were reported to be less than 2×10^{-11} g. Standard polycyclic aromatic hydrocarbons were separated with high efficiency while retaining the spectral scanning capabilities.

Bruno et al. (145) have reported on the design and optimization of a fiber-optic-aided flow cell for the use of either UV–visible absorbance or fluorescence detection. In their design they employed an in-house optical ray tracing program to optimize light throughput and minimize stray light and scattering. The flow cell assembly used fiber optics to deliver and collect light. After optimization of all parameters, the density/pressure-program-related absorbance shift was found to be about 2.5×10^{-3} AU over a 265-bar pressure program. This is approximately 1 order of magnitude less than for UV–visible detection.

Since these reports, little work has been done on fluorescence in SFC. Considering the advances that have been made in laser fluorescence (146) and in commercial instrumentation (147), the potential for improvements in fluorescence as a detection method for SFC is great. In a recent paper, Kuhr and Yeung (148) have optimized the indirect fluorescence detection method for maximized sensitivity in capillary zone electrophoresis. The technique allows for on-column fluorescence detection of species that are not fluorophores. Although linear dynamic range is limited, this method does offer another technique for the sensitive detection of small quantities of analyte probed in small-volume flow cells and should have utility in SFC. Fluorescence, scanning, or dedicated solute-specific detection still shows great potential in SFC, in particular as the chromatographic methods become available for the analysis of biological samples.

2.4. Light Scatter Detectors

In the search for a universal detector for HPLC that does not have the limitations of the RI detector or the UV–visible detector, the light scatter detector (LSD) has been investigated by various workers (149–151). The LSD has also been demonstrated to be compatible with packed capillary HPLC (152). Atomization of the column effluent into a gas stream is accomplished via a Venturi nebulizer. A heated tube applies the necessary energy to evaporate the solvent, yielding an aerosol of nonvolatile solute, and, finally, the particle scatters light, which is measured by a photomultiplier tube (PMT). The sensitivity of the detector is varied by changing the integration time of the PMT. Ultimately, the performance of the light scatter measurement is dependent upon the size of the scattering particles relative to the wavelength of the incident light. With a liquid phase (HPLC), particle size may be changed by altering the physical parameters of the system such as density, viscosity, surface tension, velocity of the mobile phase, and the initial solute concentration. Although the physical parameters that determine particle size for LSD in packed column SFC have not been investigated in detail, LSD can be used in SFC because the nebulization is realized by use of a restrictor only (153–155), thus making the LSD a detector well suited to SFC.

Because of the necessity to add organic modifiers to CO_2 to enhance solubility or to tie up packed column active sites, there continues to be significant interest in interfacing universal detectors to SFC that are compatible with modifiers. Carraud and co-workers (153) have shown that the LSD will work as a universal detection method when methanol is an organic modifier added to CO_2 in concentrations as high as 10% w/w. It was found that with 4.6-mm-i.d. packed columns the LSD response was not hindered by the modifier. A simple interface was employed in order to connect the commercial LSD to a supercritical fluid chromatograph. As previously determined experimentally for liquid phases (156), Carraud and co-workers found that the LSD response in SFC is nonlinear. Upon utilizing a logarithmic coordinate for the concentration (mass) axis, a straight line was achieved. A slope of 1.66 was reported for the response of pyrene and is a similar value to that reported for the LSD with HPLC. Detection limits at $S/N = 3$ for docosane ($C_{22}H_{46}$) and trimyristine of 75 and 40 ng, respectively, were reported. It has been speculated that a slope greater than 1 is particularly interesting for trace component analysis (156), because both detection limit and resolution are increased. Although a nonlinear calibration curve is less than ideal, this type of response can be dealt with for analytical analysis. The major drawback of the LSD in SFC is response factor variation vs. CO_2 velocity. In other words, the use of uncontrolled-flow pressure programming produces a change in the peak area for a constant concentration of solute introduced. Because the addition of volatile polar modifier allows for significant increase of the overall polarity of the mobile phase, the authors note that instability to pressure programming only limits the use of LSD with SFC, but will not prohibit the application of the method.

In fact, in a recent paper, Hoffmann and Greibrokk have applied pressure programming with packed capillary columns in LSD–SFC analysis (155). Pressure programs from 9 MPa to as high as 30 MPa were used with n-propanol–modified CO_2 in order to elute trimyristin and Irgafos-168 [tris-(2,4-di-$tert$-butylphenyl)-phosphite]. N-Propanol was used in mole percent values of 2.9% and 5.6%. At constant modifier concentration, with pressure programming, Hoffmann and Greibrokk report peak area reproducibility values of less than 4% RSD. Retention time reproducibility was found to be better than 0.5%. They note that the use of packed capillary columns offer an excellent compromise between large-i.d. packed and open tubular capillary columns. Packed capillaries allow for high efficiency per unit time and high solute loading, while producing little back pressure and modest solvent flow requirements. The authors note that the nebulization producer—and thus the design of the interface and restrictor—is critical for optimum performance. To generate large drops, the restrictor temperature should be kept at a minimum, yet ice crystal formation must be avoided. A

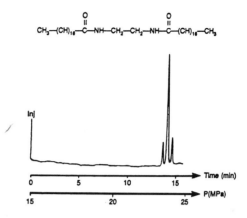

Figure 14.9. 120 ng EBS. Mobile phase: 5.6% *n*-propanol in CO_2 detected by light scatter detector. Oven temperature: 145 °C. Detector temperature. 90 °C. Reprinted with permission from S. Hoffman and T. Greibrokk, *J. Microcolumn Sep.* **1**, 35 (1989). Copyright 1989, Aster Publishing Corporation.

detection limit for Irgafos-168 and trimyristin was determined to be about 5 ng. Nonlinear calibration curves were observed, and slopes of > 1 were reported.

In order to illustrate the utility of the SFC–LSD system for the analysis of a difficult problem, Figure 14.9 presents the determination of *N,N'*-ethylene bisstearamide (EBS). It is postulated that previous attempts to elute EBS with pure CO_2 on an open tubular column failed owing to the low solubility of the species in CO_2. Note that the 5.6 mol% *n*-propanol–modified CO_2 provided elution and separation of the sample. It is certain that the LSD can be used in SFC analysis, particularly when the solute does not contain a UV-absorbing chromophore and when it is necessary to employ significant concentrations of organic modifiers. Further studies undoubtedly will provide better detection limits, while nonlinearity will continue to be a problem.

3. CONCLUSIONS

Table 14.2 presents a comparison of the commonly used "optical" detection methods employed in SFC. Note that the values presented in the table are accurate as of early 1990; improvements in instrumentation are made very rapidly.

As mentioned throughout this chapter and as shown by the large number of detection techniques compatible with capillary SFC, one of the true advantages of SFC is "multidetector compatibility." Because decompression

Table 14.2. Comparison of Commonly Used Optical Detection Methods for Capillary SFC

Detection Method	Sample Detected	Detection Limit	Flow Cell Path Length	Modifier Compatible
UV–visible	Naphthalene	1.0 ppm/30 pg	0.25 mm	Yes
Scanning UV	Naphthalene	30 pg (high-sensitivity mode)	0.25 mm	Yes
Fluorescence	Pyrene	8 pg		Yes
Light scattering	Trimyristin	5 ng (injected mass)		Yes
Chemiluminescence	Benzenethiol	18 pg (sulfur)		Yes/No
FTIR		50/10 ng^a		Yes/No

^{a}Amount necessary to give a quality spectrum with a flow cell or with solvent elimination.

can be performed either before or after the detector, various gas-phase or fluid-phase methods are possible. A great future lies in store for capillary SFC as an analytical method as a result of this detector flexibility. Multiple-detector schemes provide added information over single-detector systems. This solute information allows the scientist to analyze more complex matrices with confidence, thereby easing the task of solute analysis. As spectroscopic detectors are modified to be more compatible with capillary SFC, additional information will be available and capillary SFC will develop further.

REFERENCES

1. J. C. Fjeldsted and M. L. Lee, *Anal. Chem.* **56**, 619A (1984).
2. P. A. Peaden and M. L. Lee, *J. Chromatogr.* **259**, 1 (1983).
3. M. Novotny, S. R. Springston, P. A. Peaden, J. C. Fjeldsted, and M. L. Lee, *Anal. Chem.* **53**, 407A (1981).
4. S. M. Fields and M. L. Lee, *J. Chromatogr.* **349**, 305 (1985).
5. C. E. Yonker and R. D. Smith, *Anal. Chem.* **59**, 727 (1987).
6. E. Klesper, A. H. Corwin, and D. A. Turner, *J. Org. Chem.* **27**, 700 (1962).
7. T. Hershfield, *Anal. Chem.* **48**, 16A (1976).
8. E. Klesper, A. H. Corwin, and D. A. Turner, *J. Org. Chem.* **27**, 700 (1962).
9. K. H. Linnemann, A. Wilsch, and G. M. Schneider, *J. Chromatogr.* **369**, 39 (1986).
10. J. C. Fjeldsted, P. A. Peaden, and M. L. Lee, *J. Chromatogr. Sci.* **21**, 222 (1983).

11. A. L. Blilie and T. Greibrokk, *Anal. Chem.* **57**, 2239 (1985).

12. C. R. Yonker and R. D. Smith, *J. Chromatogr.* **361**, 25 (1986).

13. J. M. Levy and W. M. Ritchey, *J. Chromatogr. Sci.* **24**, 242 (1986).

14. J. C. Giddings, M. N. Myers, L. McLaren, and R. A. Keller, *Science* **162**, 67 (1968).

15. D. Bartmann, *Ber. Bunsenges. Phys. Chem.* **76**, 336 (1972).

16. D. Bartmann, Ph. D. Dissertation, University of Bochum, Germany 1972.

17. O. Vitzthum, P. Hubert, and M. Barthels, U. S. Patent 3,827,859 (1974).

18. M. G. Rawdon, *Anal. Chem.* **56**, 831 (1984).

19. Y. Hirata, F. Nakata, and M. Horihata, *HRC & CC, J. High Resolut. Chromatogr. Chromatogr. Commun.* **11**, 81 (1988).

20. J. C. Fjeldsted, R. C. Kong, and M. L. Lee, *J. Chromatogr.* **279**, 449 (1983).

21. T. L. Chester, *J. Chromatogr.* **299**, 424 (1984).

22. T. L. Chester, D. P. Innis, and G. D. Owens, *Anal. Chem.* **57**, 2243 (1985).

23. T. L. Chester and D. P. Innis, *HRC & CC, J. High Resolut. Chromatogr. Chromatogr. Commun.* **9**, 178 (1986).

24. T. L. Chester and D. P. Innis, *HRC & CC, J. High Resolut. Chromatogr. Chromatogr. Commun.* **9**, 209 (1986).

25. E. J. Guthrie and H. E. Schwartz, *J. Chromatogr. Sci.* **24**, 236 (1986).

26. B. E. Richter, presented at the *Pittsburgh Conf. Expos. Anal. Chem. Spectros, Atlantic City, N.J. 1986* Paper 514 (1986).

27. M. W. Raynor, K. D. Bartle, I. L. Davies, A. A. Clifford, and A. Williams, *HRC & CC, J. High Resolut. Chromatogr. Chromatogr. Commun.* **11**, 289 (1988).

28. J. Köhler, A. Rose, and G. Schomburg, *HRC & CC, J. High Resolut. Chromatogr. Chromatogr. Commun.* **11**, 191 (1988).

29. R. W. Bally and C. A. Cramers, *HRC & CC, J. High Resolut. Chromatogr. Chromatogr. Commun.* **9**, 626 (1986).

30. R. D. Smith, J. L. Fulton, R. C. Petersen, A. J. Kopriva, and B. W. Wright, *Anal. Chem.* **58**, 2057 (1986).

31. E. R. Campbell, D. W. Later, and B. E. Richter, presented at the *Pittsburgh Conf. Expos. Anal. Chem. Spectros. New Orleans, LA, 1988* Paper 423 (1988).

32. T. A. Berger, *Anal. Chem.* **61**, 356 (1989).

33. B. E. Richter, *HRC & CC, J. High Resolut. Chromatogr. Chromatogr. Commun.* **8**, 297 (1985).

34. B. E. Richter and E. R. Campbell, Lee Scientific, personal communication, 1989.

35. J. W. Wangsgaard, B. E. Richter, and D. J. Bornhop, presented at the *1989 Pittsburgh Conf.*

36. P. L. Patterson, Detector Technology, personal communication, 1989.

37. W. R. West and M. L. Lee, *HRC & CC, J. High Resolut. Chromatogr. Chromatogr. Commun.* **9**, 161 (1986).

38. P. L. Patterson, R. A. Gatten, and C. Ontiveros, *J. Chromatogr.* **20**, 97 (1982).

39. P. L. Patterson, *Chromatographia* **16**, 107 (1982).

40. P. A. David and M. Novotny, *J. Chromatogr.* **452**, 623 (1988).

41. D. W. Later, B. E. Richter, W. D. Felix, M. R. Andersen, and D. E. Knowles, *Am. Lab.* **18**, 108 (1986).

42. E. R. Campbell, Lee Scientific, unpublished results (1990).

43. K. E. Markides, E. D. Lee, R. Bolick, and M. L. Lee, *Anal. Chem.* **54**, 740 (1986).

44. S. V. Olesik, L. A. Pekay, and E. A. Paliwoda, *Anal. Chem.* **61**, 58 (1989).

45. J. K. Nelson, Ph. D. dissertation, University of Colorado, Boulder, 1984.

46. J. O. Nriagu, ed., *Sulfur in the Environment, part 1*, Wiley, New York, 1978.

47. B. E. Wenzel, *J. Chromatogr. Sci.* **20**, 409 (1974).

48. R. C. Hall, *J. Chromatogr. Sci.* **12**, 152 (1982).

49. B. J. Ehrlich, R. C. Hall, and H. G. Cox, *J. Chromatogr. Sci.* **19**, 245 (1981).

50. S. Gluck, *J. Chromatogr. Sci.* **20**, 103 (1982).

51. J. L. Genna, W. D. McAnnich, and R. A. Reich, *J. Chromatogr.* **238**, 103 (1982).

52. R. J. Skelton, Jr., K. E. Markides, P. B. Farnsworth, and M. L. Lee, *HRC & CC, J. High Resolut. Chromatogr. Chromatogr. Commun.* **11**, 75 (1988).

53. D. R. Gage and S. O. Farwell, *Anal. Chem.* **52**, 2422 (1980).

54. D. R. Treybig and S. R. Ellebracht, *Anal. Chem.* **52**, 1633 (1980).

55. C. H. Burnett, D. F. Adams, and S. O. Farwell, *J. Chromatogr. Sci.* **16**, 68 (1978).

56. J. K. Nelson, R. H. Getty, and J. W. Birks, *Anal. Chem.* **55**, 1767 (1983).

57. E. A. Mishalanie and J. W. Birks, *Anal. Chem.* **54**, 918 (1986).

58. R. J. Glinski, E. A. Mishalanie, and J. W. Birks, *J. Photochem.* **37**, 217 (1987).

59. W. T. Foreman, C. L. Shellman, J. W. Birks, and R. E. Sievers, *J. Chromatogr.* **465**, 23 (1989).

60. D. J. Bornhop, B. J. Murphy, and Lisa Krieger-Jones, *Anal. Chem.* **61**, 797 (1989).

61. J. T. Schmermund and D. C. Locke, *Anal. Lett.* **8**, 611 (1975).

62. D. C. Locke, B. S. Dhingra, and A. D. Baker, *Anal. Chem.* **54**, 447 (1982).

63. W. Gmür, J. O. Bosset, and E. Plattner, *Chromatographia* **23**, 199 (1987).

64. P. Sim, C. Elson, and M. Quillaim, *J. Chromatogr.* **445**, 239 (1988).

65. R. J. Wall, Dionex Corporation, personal communication.

66. S. Kennedy and R. J. Wall, *LC-GC* **6**, 930 (1988).

67. M. R. Andersen, N. L. Porter, and J. T. Swanson, Lee Scientific, Salt Lake City, UT, unpublished results (1989).

68. B. E. Richter, D. J. Bornhop, J. T. Swanson, J. G. Wangsgaard, and M. R. Andersen, *J. Chromatogr. Sci.* **27**, 303 (1989).

69. B. W. Wright, H. T. Kalinoski, H. R. Udseth, and R. D. Smith, *HRC & CC, J. High Resolut. Chromatogr. Chromatogr. Commun.* **9**, 145 (1986).

70. D. E. Games, A. J. Berry, I. C. Mylchreest, J. R. Perkins, and S. Pleasance, *Eur. Chromatogr. News* **1**, 10 (1987).

71. D. E. Games, A. J. Berry, I. C. Mylchreest, J. R. Perkins, and S. Pleasance, in

Supercritical Fluid Chromatography (R. M. Smith, ed.) Royal Society of Chemistry, London, 1988.

72. L. G. Randall and A. L. Wahrhaftig, *Rev. Sci. Instrum.* **52**, 1285 (1981).

73. R. D. Smith, J. Fjeldsted, and M. L. Lee, *Int. J. Mass Spectrom. Ion Phy.* **46**, 217 (1983).

74. R. D. Smith, W. D. Felix, J. Fjeldsted, and M. L. Lee, *Anal. Chem.* **247**, 231 (1982).

75. R. D. Smith, H. R. Udseth, and H. T. Kalinoski, *Anal. Chem.* **59**, 2971 (1984).

76. A. J. Berry, D. E. Games, and J. R. Perkins, *J. Chromatogr.* **363**, 147 (1986).

77. E. D. Lee and J. D. Henion, *Pittsburgh Conf. Expos. Anal. Chem. 1987* Abstr. No. 814 (1987).

78. A. Giorgetti, N. Pericles, H. M. Widmer, K. Anton, and P. Datwyler, *J. Chromatogr. Sci.* **27**, 318 (1989).

79. K. Anton, N. Pericles and H. M. Widmer, *HRC & CC, J. High Resolut. Chromatogr. Chromatogr. Commun.* **12**, 394 (1989).

80. G. Holzer, S. Deluca, and K. J. Voorhees. *HRC & CC, J. High Resolut. Chromatogr. Chromatogr. Commun.* **8**, 528 (1985).

81. R. D. Smith and H. R. Udseth, *Anal. Chem.* **59**, 13 (1987).

82. D. F. Hunt and P. J. Gale, *Anal. Chem.* **54**, 1111 (1984).

83. A. G. Harrison, *Chemical Ionization Mass Spectrometry*, CRC Press, Boca Raton, FL, 1983, p. 71.

84. J. Disinger, unpublished data, SFC-MS Interface for Lee Scientific, Salt Lake City, UT, 1988–1989.

85. W. Herres, *HRGC–FT–IR: Capillary Gas Chromatography–Fourier Transform Infrared Spectroscopy*; Huethig, Heidelberg, 1987.

86. K. H. Shafer and P. R. Griffiths, *Anal. Chem.* **55**, 1939 (1983).

87. K. Jinno, *Chromatographia* **23**, 55 (1987).

88. S. L. Pentoney, K. H. Shafer, and P. R. Griffiths, *J. Chromatogr. Sci.* **24**, 230 (1986).

89. K. H. Shafer, S. L. Pentoney, P. R. Griffiths, and R. Fuoco, *HRC & CC, J. High Resolut. Chromatogr. Chromatogr. Commun.* **9**, 168 (1985).

90. M. W. Raynor, K. D. Bartle, I. L. Davies, A. Williams, A. A. Clifford, J. M. Chalmers, and B. W. Cook, *Anal. Chem.* **60**, 427 (1988).

91. C. Fujimoto, Y. H. Hirata, and K. Jinno, *J. Chromatogr.* **332**, 47 (1985).

92. K. H. Shafer, S. L. Pentoney, and P. R. Griffiths, *HRC & CC, J. High Resolut. Chromatogr. Chromatogr. Commun.* **7**, 707 (1984).

93. P. R. Griffiths, presented at *Supercritical Fluid Chromatogr. Symp./Workshop, Snowbird, UT, 1989.*

94. *P. R. Griffiths, Development of Supercritical Fluid Chromatography for the Separation and Identification of Nonvolatile Organics,* Final Report on Cooperative Agreement Number CR-812258, Environ. Monit. Syst. Lab., Off. Res. Dev., USEPA, Las Vegas, NV.

95. P. Morin, B. Beccard, M. Caude, and R. Rosset, *HRC & CC, J. High Resolut. Chromatogr. Chromatogr. Commun.* **11**, 697 (1988).

96. S. V. Olesik, S. B. French, and M. Novotny, *Chromatographia* **18**, 489 (1984).

97. C. C. Johnson, J. W. Jordon, L. T. Taylor, and D. W. Vidrine, *Chromatographia* **20**, 717 (1985).

98. M. E. Hughes and J. L. Faschig, *J. Chromatogr. Sci.* **24**, 535 (1985).

99. J. W. Jordon and L. T. Taylor, *J. Chromatogr. Sci.* **24**, 82 (1986).

100. R. J. Skelton, C. C. Johnson, and L. T. Taylor, *Chromatographia* **21**, 3 (1986).

101. Ph. Morin, M. Claude, M. Richard, and R. Rosset, *Chromatographia* **21**, 523 (1986).

102. P. Morin, M. Caude, M. Richard, and R. Rosset, *Analusis* **15**, 117 (1987).

103. J. W. Hellgeth, J. W. Jordon, L. T. Taylor, and M. A. Khorassani, *J. Chromatogr. Sci.* **24**, 183 (1986).

104. P. Morin, M. Caude, and R. Rosset, *J. Chromatogr.* **407**, 87 (1987).

105. S. B. French and M. Novotny, *Anal. Chem.* **58**, 164 (1986).

106. R. Wieboldt and D. A. Hanna, *Anal. Chem.* **59**, 1255 (1987).

107. R. C. Wieboldt, G. E. Adams, and D. W. Later, *Anal. Chem.* **60**, 2422 (1988).

108. R. C. Wieboldt, K. D. Kempfert, D. W. Later, and E. R. Campbell, *J. High Resolut. Chromatogr.* **12**, 106 (1989).

109. E. M. Calvey, L. T. Taylor, and J. K. Palmer, *HRC & CC, J. High Resolut. Chromatogr. Chromatogr. Commun.* **11**, 739 (1988).

110. S. Shah and L. T. Taylor, *J. High Resolut Chromatogr.* **12**, 599 (1989).

111. M. W. Raynor, A. A. Clifford, K. D. Bartle, C. Reyner, and A. Williams, *J. Microcolumn Sep.* **1**, 101 (1989).

112. R. C. Wieboldt and J. A. Smith, *ACS Symp. Ser.* **366**; 229 (1988).

113. S. Shah, M. A. Khorassani, and L. T. Taylor, *Chromatographia* **25**, 631 (1988).

114. M. W. Raynor, G. F. Shilstone, K. D. Bartle, A. A. Clifford, M. Cleary, and B. W. Cook, *J. High Resolut. Chromatogr.* **12**, 300 (1989).

115. R. L. Eatherton, M. A. Morrissey, W. F. Siems, and H. H. Hill, Jr., *HRC & CC, J. High Resolut. Chromatogr. Chromatogr. Commun.* **9**, 154 (1986).

116. M. A. Morrissey and H. H. Hill, Jr., *HRC & CC, J. High Resolut. Chromatogr. Chromatogr. Commun.* **11**, 375 (1988).

117. J. M. F. Douse and R. J. Wall, Dionex Corporation, UK Limited, Survey, England, unpublished data (1978).

118. D. R. Luffer, L. J. Galante, P. A. David, M. Novotny, G. M. Hieftje, *Anal. Chem.* **60**, 1365 (1988).

119. L. J. Galante, M. Shelby, D. R. Luffer, G. M. Hieftje, and M. Novotny, *Anal. Chem.* **60**, 1370 (1988).

120. S. M. Fields, K. E. Markides, and M. L. Lee, *Anal. Chem.* **60**, 802 (1988).

121. F. J. Yang, *HRC & CC, J. High Resolut. Chromatogr. Chromatogr. Commun.* **4**, 83 (1981).

122. M. V. Novotny and D. Ishii, eds., *Microcolumn Separation Methods*. Elsevier, Amsterdam, 1985.

123. F. J. Yang, *J. Chromatogr.* **236**, 265 (1982).

124. D. W. Later, D. J. Bornhop, E. D. Lee, J. D. Henion, and R. C. Weibolt, *LC/GC* **5**, 804 (1987).

125. H. Pope, *Anal. Chim. Acta.* **145**, 17 (1983).

126. M. V. Novotny and P. Kucera, ed., *Microcolumn High-Performance Liquid Chromatography*, p. 194. Elsevier, Amsterdam, 1984.

127. W. Baumann and Z. Fresenius, *Anal. Chem.* **31**, 284 (1977).

128. J. M. Harris and N. J. Dovichi, *Anal. Chem.* **52**, 695A (1980).

129. M. D. Morris and K. Peck, *Anal. Chem.* **58**, 811A (1986).

130. T. G. Nolan, B. K. Hart, and N. J. Dovichi, *Anal. Chem.* **57**, 2703 (1985).

131. D. J. Bornhop and N. J. Dovichi, *Anal. Chem.* **58**, 1632 (1987).

132. R. A. Leach and J. M. Harris, *J. Chromatogr.* **218**, 15 (1981).

133. R. A. Leach and J. M. Harris, *Anal. Chem.* **56**, 1481 (1984).

134. D. J. Bornhop, S. Schmidt, and N. L. Porter, *J. Chromatogr.* **459**, 193 (1988).

135. J. P. Chervet, M. Ursem, J. P. Salzmann, and R. W. Vannoort, *J. High Resolut. Chromatogr.* **12**, 278 (1989).

136. S. R. Weinberger and D. J. Bornhop, *J. Microcolumn Sep.* **1**, 90 (1989).

137. M. Novotny, *HRC & CC, J. High Resolut. Chromatogr. Chromatogr. Commun.* **9**, 137 (1986).

138. Y. Hirata, F. Nakata, and M. Kawasaki, *HRC & CC, J. High Resolut. Chromatogr. Chromatogr. Commun.* **9**, 633 (1986).

139. S. Schmidt, L. G. Blomberg, and E. R. Campbell, *Chromatographia* **25**, 775 (1988).

140. T. Takeuchi and D. Ishii, *HRC & CC, J. High Resolut. Chromatogr. Chromatogr. Commun.* **7**, 151 (1984).

141. J. E. France and K. J. Voorhees, *HRC & CC, J. High Resolut. Chromatogr. Chromatogr. Commun.* **11**, 692 (1988).

142. K. Jinno, T. Hoshino, and T. Hondo, *Anal. Lett.* **19**, 1001 (1986).

143. J. C. Fjeldsted, B. E. Richter, W. P. Jackson, and M. L. Lee, *J. Chromatogr.* **279**, 423 (1983).

144. J. C. Gluckman, D. C. Shelly, and M. V. Novotny, *Anal. Chem.* **57**, 1546 (1985).

145. A. E. Bruno, E. Gassmann, N. Pericles, and K. Anton, *Anal. Chem.* **61**, 876 (1989).

146. Y. F. Cheng and N. J. Dovichi, *Science* **242**, 562 (1988).

147. S. R. Weinberger and L. Hlousek, *Int. Labmate* **9**, 9 (1988).

148. W. G. Kuhr and E. S. Yeung, *Anal. Chem.* **60**, 1832 (1988).

149. A. Stolyhwo, H. Colin, and G. Guiochon, *J. Chromatogr.* **265**, 1 (1983).

150. A. Stolyhwo, H. Colin, M. Martin, and G. Guiochon, *J. Chromatogr.* **288**, 253 (1984).

151. P. H. Morurey and L. E. Oppenheimer, *Anal. Chem.* **50**, 1414 (1984).

152. S. Hoffman, H. R. Norli, and T. Greibrokk, submitted for publication.

153. P. Carraud, D. Thiebaut, M. Caude, R. Rosset, M. Lafosse, and M. Dreux, *J. Chromatogr. Sci.* **25**, 395 (1987).

154. M. Lafosse, M. Dreux, and L. Morin-Alloy, *J. Chromatogr.* **95**, 404 (1987).

155. S. Hoffman and T. Greibrokk, *J. Microcolumn Sep.* **1**, 35 (1989).

156. L. T. E. Oppenheimer and T. H. Mourey, *J. Chromatogr.* **323**, 297 (1985).

CHAPTER

15

DETECTION IN MICROCOLUMN LIQUID CHROMATOGRAPHY

JAMES W. JORGENSON

Department of Chemistry
University of North Carolina
Chapel Hill, North Carolina

JOS DE WIT

Tennessee Eastman Co.
Kingsport, Tennessee

The type of column dimensions that have become established over the past 20 years as traditional in high-performance liquid chromatography (HPLC) are 4.6-mm i.d. by 25-cm length, with some variation on the length to shorter columns. "Microcolumn LC" might therefore refer to all columns of inner diameter smaller than 4.6 mm. Columns of approximately 1- to 2-mm i.d. have become known as "microbore" columns. These microbore columns operate with mobile phase flow rates and detection volumes some 5- to 20-fold smaller than traditional 4.6-mm-i.d. columns. Although this represents a significant reduction in solvent consumption, columns of these dimensions do not require a radical departure in injection and detection technology from that used with conventional columns. Microbore columns simply require appropriate miniaturization of the usual injectors and detectors in order to function.

For the purposes of this chapter, microcolumns will refer to those columns with inner diameters less than 1 mm. Such columns are of interest for several reasons. Microcolumns require still smaller amounts of mobile phase than microbore columns, but the amounts used in microbore columns are already so low that this advantage rarely has any significance. The real advantages of microcolumns are that they can be prepared in lengths that yield much higher separation efficiencies, they require far less sample for a full-volume injection,

Detectors for Capillary Chromatography, edited by Herbert H. Hill and Dennis G. McMinn.
Chemical Analysis Series, Vol. 121.
ISBN 0-471-50645-1 © 1992 John Wiley & Sons, Inc.

and they preserve the concentration of this small sample volume better. Furthermore, the low flow rates can be advantageous when the columns are coupled to certain detection modes such as mass spectrometers and gas-phase ionization detectors where the large volumes of mobile phase effluent produced by conventional columns are problematic.

1.   SENSITIVITY AND DETECTION LIMITS

It is important to be able to compare sensitivity and detection limits of micro-column systems to those of conventional column systems. Detection limits are usually reported in one of two ways: minimum detectable quantity (MDQ) or minimum detectable concentration (MDC) of an analyte. Since microcolumns require far less sample volume, they tend to offer lower detection limits in terms of quantity of analyte injected onto the column. This permits realization of fairly routine detection limits of femtomoles and attomoles with micro-columns. This does not, however, mean that concentration detection limits are also lower. In fact the concentration detection limits are often likely to be poorer than with conventional columns. For example, conventional columns permit use of longer optical path lengths in optical detectors and thus can usually attain lower concentration detection limits. Nonetheless, microcolumns occasionally permit better sensitivity to be obtained even in terms of concen-tration detection limits, owing to design possiblities peculiar to microcolumns. This is true, for example, in the case of on-column electrochemical detection. The basic sensitivity advantage of microcolumns still concerns mass detection limits. For this reason, microcolumns are advantageous when an analytical problem involves a sample which is limited in volume to amounts signifi-cantly less than a microliter.

Injecting nanoliter volumes onto a conventional column, while perhaps feasible, results in extreme dilution of the sample. Doing this converts the problem from one of ultramicroanalysis (analysis of extremely small volumes) into one of ultra-trace analysis (analysis of extremely low concentrations). In general it is easier to do ultramicroanalysis than ultra-trace analysis, as ultra-trace analysis is limited ultimately by the purity of solvents and reagents.

2.   COLUMN TYPES

Microcolumns are prepared in a variety of ways and sizes, and different names have evolved to describe these column types (1). Some of their characteristics are summarized in Table 15.1. The first are columns known as "micropacked" columns, which involve the packing of conventional HPLC particles into

Table 15.1. Column Types in Microcolumn Liquid Chromatography

Column Type	Inner Diameter (mm)	Length (cm)	N (plates)	V_{peak} (μL)
Conventional	4.60	25	15,000	65.
Microbore	1.00	25	15,000	3.0
Micropacked	0.25	100	100,000	0.3
Packed capillary	0.05	100	100,000	0.012
Open tubular	0.01	300	300,000	0.002

columns (usually fused silica tubes) with internal diameters ranging from 20 to 500-μm. Another, still smaller column has been called the "packed capillary," referring to columns made by packing silica or alumina into glass tubes, which are then drawn down to 50- to 100-μm i.d. on a glass capillary drawing machine. The packing in these columns can subsequently be modified with silanes to produce reversed phase and other common stationary phases. Finally, there are open tubular liquid chromatography (OTLC) columns, in which a capillary (usually less than 30-μm i.d.) has a stationary phase coated on its wall. In the case of these three kinds of microcolumns, the diameters are so small that conventional injectors and detectors are not suitable. Table 15.1 outlines some of the characteristics of these column types in terms of typical column dimensions, flow rates, column efficiencies, and peak volumes [4σ (standard deviations) for an unretained peak]. It is clear from this table that miniaturization of conventional detectors would be difficult for such small columns and peak volumes.

3. DETECTOR TYPES

The basic requirements for detectors for microcolumns are low dispersion and high sensitivity. The larger micropacked columns are more forgiving in this regard than the smallest of the open tubular columns, but a general lack of suitable commercial equipment for working with any of these column types is a factor they have in common.

The most direct way to try to minimize band broadening from detection is to do the detection directly on the capillary column itself (i.e., focus light into the column or insert an electrode into the end of the column). This detection can be done in a region of the column that actually contains stationary phase ("in-column," "true on-column,"), or it can be done in the capillary in the region just past the packing material ("on-column"). In either case, elimination of a formal postcolumn detection cell simplifies the system as well as minimizing postcolumn band broadening. Indeed, while one might consider designing a miniature postcolumn detector for some micropacked columns, it is difficult to imagine doing this in the case of an open tubular column with an inner diameter of less than 10 μm. Attempting to align a 10-μm i.d. open tubular column with a connecting tube of similar dimension would be a heroic act. An alternative to on-column detection is to use makeup flows to increase the peak volume and then utilize a detector of more conventional volume. The obvious draw back of this approach is that the analytes are seriously diluted by the additional makeup volume, and so this approach has not been used to any great extent. A second alternative is demonstrated in the case of detection with mass spectrometers and ionization detectors. Here the effluent can be vaporized and the analytes dealt with in a gas-phase flow or in vacuum. Alternatively, the effluent can be deposited on a belt or wire, the mobile phase evaporated, and the analyte transported to the detection region. Despite these alternatives, the greatest amount of work has been done with various versions of on-column detection.

The modes of detection that will be reviewed in this chapter are ultraviolet (UV) absorption, fluorescence, chemiluminescence, refractive index, photothermal, optical activity, electrochemical, photoionization, flame ionization, mass spectrometry (MS), infrared (IR) absorption/Raman, and inductively coupled plasma emission.

3.1. UV Absorption Detection

UV absorption detectors are the most commonly used detectors in conventional HPLC, so it is no surprise that this mode of detection was the first to be used in microcolumn LC. The development of UV absorption detection in micro-LC and OTLC has concentrated on the miniaturization of the flow cell volumes. Attention has also been paid to eliminating or minimizing dead volumes in connecting tubing between the column and the detector. Low flow cell volumes are needed to avoid dispersion of the small volume peaks eluting in micro-LC and OTLC. This reduction of flow cell volumes can cause a decrease in sensitivity due to a shorter optical pathlength or an increase in noise due to a smaller light throughput (the compromise between chromatographic fidelity and sensitivity). The detection limit is also a function of the column efficiency and overall chromatographic performance: sharper peaks will provide a higher analyte concentration and greater absorbance (2). Some of the optical approaches encountered in conventional LC–UV absorption detection, such as the use of photodiode arrays and variable wavelength detection, have also been applied to micro-LC and OTLC.

Ishii et al. (3) constructed a micro flow cell for micro-LC by directly inserting a quartz tube (as a flow cell) into the end of the microcolumn. By using quartz tubes of different inner diameters, these authors could vary the optical path length and the volume of the flow cell from 0.1 to 0.3 μL. The direct connection between flow cell and column also helped reduce extra column peak broadening. This flow cell was employed with several different commercially available spectrophotometers, but owing to the short path length of their miniaturized cells and the associated low sensitivity, the best results were obtained using a high-sensitivity UV detector designed for HPLC. The capabilities of variable wavelength detection were demonstrated with a separation of 1 μg of benzene, 10 ng of napthalene, 10 ng of biphenyl, and 2 ng of anthracene. Spectra from 200 to 300 nm with good signal-to-noise ratio (S/N) were obtained by stopping the flow during the elution of the peaks and scanning the monochromator.

Baram et al. (4) reported on a microcolumn liquid chromatograph with double-beam multiwavelength detection. A cyclic program of up to 18 wavelengths from 190 to 360 nm can be selected with this instrument. The wavelengths are changed with a fast stepper motor driving the monochromator. The commercial nonmodified flow cell has a Z-type geometry, with an optical pathlength of 1.6 mm. They demonstrated the detector's capabilities with a separation of four amines, detected at 210, 230, 240, 250, 260, 300, and 360 nm. Since 1.2 s was spent monitoring at each of these seven wavelengths, this approach is not possible for temporally narrow peaks.

Rokushika et al. (5) used variable wavelength UV absorption detection with microcolumn anion chromatography. The column was constructed by packing 10-cm-long by 50-μm-i.d. fused silica tubing with an anion exchange resin. A flow cell was made by removing part of the polyimide coating of a second piece of fused silica and connecting this with a third (sleeve) section of fused silica tubing and epoxy resin to the end of the chromatographic column. A porous poly(tetrafluoroethylene) (PTFE) frit was used to keep the stationary phase in place. This approach greatly reduced extracolumn band broadening. The fragile cell window region was protected by stainless steel tubing, which had an aperture for the light path. This assembly was then inserted into a commercial UV detector. Calibration of the detector was performed by injecting a series of standard anion mixtures (NO_2^-, NO_3^-, Br^-), and linearity was obtained over more than 2 orders of magnitude (5–2000 ppm). The separation of 50 pmol each of four nucleotides, with detection at 260 nm, was also demonstrated.

The use of UV detection for OTLC emphasizes the compromise between sensitivity and chromatographic fidelity even further. With columns varying from 70-μm i.d. (common in the earlier stages of the development of OTLC) down to 2-μm i.d. (state of the art), even 0.5-μL flow cell volumes, acceptable for the larger micropacked LC columns, are no longer usable without causing significant peak broadening. Two effects in going to smaller columns are taking place: higher efficiencies, causing sharper peaks (and thus smaller peak volumes); and smaller column volumes. Standard detectors outfitted with small silica 0.1-μL flow cells were sufficient with 70-μm-i.d. columns and separation times as long as 23 h (6). Ishii and Takeuchi (7) used 30-μm-i.d. columns with a miniaturized flow cell with an internal volume of 0.03–0.10 μL. This flow cell was constructed from a 1-cm-long, 170-μm-i.d. piece of silica tubing. No detection limits were reported with this system.

A second approach sometimes used with UV detection in OTLC is the use of a "makeup" flow, as shown by Krejci et al. (8). This will decrease sensitivity, but peaks eluting with volumes smaller than the detector cell volume can still be detected without noticeable distortion of the peak.

Kucera and Guiochon (9) evaluated five commercial UV detector flow cells, of which two were regular cells, two were micro cells, and one was a modified micro cell. The micro cell was modified by inserting a PTFE disk, which reduced the cell volume from 0.5 to 0.05 μL but left the optical path length the same at 1 mm. Owing to an increased noise level, this modified cell had the poorest detection limit of the five, but it caused the least band broadening.

Knox and Gilbert (10) showed from a viewpoint of kinetic optimization that columns with inner diameters smaller than 30 μm need detector flow cell volumes on the order of 1–10 nL. This cannot be achieved with the conventional postcolumn flow cell approach.

Yang (11, 12) used the approach of on-column detection. In this work, about 1 cm of the polyimide coating was removed from the chromatographic fused silica column, and this piece was positioned between the exit slit of the monochromator and the optical detector. For a 10-μm-i.d. column, the internal volume is only 0.24 nL. With this approach, the loss in peak resolution for columns with inner diameters smaller than 150 μm will be less than 0.02%. A comparison between the MDQ obtained using OTPC and a 50-μm-on-column flow cell, on the one hand, and conventional LC with a 1-cm post-column flow cell, on the other, favors the on-column detector by a factor of 20. With a detector noise level of 1.5×10^{-5} absorbance units, the MDQ for benzene in a 6-nL on-column cell was 3 pmol, corresponding to a MDC of 2×10^{5} M.

3.2. Fluorescence Detection

Fluorescence detection has probably been used more successfully than any other mode of detection in microcolumn LC. This is because fluorescence comes close to being a true "zero background" measurement and thus offers inherently low detection limits. The widespread availability and use of fused silica (instead of glass) as the capillary material further diminishes the luminescence background in direct on-column detection and enhances the detection limit. The advent of reliable, relatively simple and inexpensive lasers operating in the blue and near-UV spectral regions (specifically the helium/cadmium laser and the argon ion laser) has provided usable light sources in fluorescence detection. The ability to focus laser beams to micrometer-sized spots and image this light directly into the fused silica capillaries makes for a nearly ideal marriage of microcolumn LC with laser-induced fluorescence.

Direct on-column fluorescence detection using an arc lamp UV source with 63-μm-i.d. borosilicate glass OTLC columns was demonstrated by Guthrie and Jorgenson (13). This system provided a reasonable MDQ of 50 fmol for perylene, corresponding to a MDC of 1×10^{-7} M. These authors also pointed out an advantage of true on-column detection (detection in a region of column containing stationary phase). In this case the detector signal is enhanced by a factor of $k' + 1$ over detection in a region of capillary without stationary phase, owing to the higher concentration of solute in a partitioning region. This observation does not take into account any additional beneficial spectral effects (i.e., changes in quantum efficiency or spectral shifts) of the solute being detected in the stationary phase as opposed to in the mobile phase. This k'-dependent signal enhancement was verified by Takeuchi and Yeung (14).

Gluckman and Novotny (15, 16) developed a detection system, based on arc lamp excitation and a photodiode array detection, which permitted acquisition of emission spectra from analytes. This system used a novel flow cell and fiber optic to couple the emission light to the emission spectrograph (polychromator). This design permitted low dispersion of chromatographic peaks while allowing for collection of a large solid angle of fluorescence emission.

Folestad et al. (17) demonstrated the use of laser excitation for fluorescence detection in capillaries. They made the useful observation that if the laser beam is incident on a capillary at a right angle, then nearly all of the scattered laser light is in a plane perpendicular to the capillary. As the fluorescence emission radiates relatively uniformly in all directions, the fluorescence collection optics can be placed at an angle out of this plane of scattering, thereby significantly reducing at least one source of background.

Using this simple but effective geometry, Guthrie and co-workers (18) developed a laser-induced fluorescence detection system and used it with OTLC columns. This system utilized a helium–cadmium laser operating on either the blue 442-nm line or the UV 326-nm line. An MDC of riboflavin in a 25-μm-i.d. fused silica capillary with 442 nm excitation was $\sim 1 \times 10^{-9}$ M. This detection system was later used in combination with a 1.7-μm-i.d. OTLC column, providing a detection limit of $\sim 200,000$ molecules of riboflavin (19). The dominant source of noise in this detection system was the shot noise caused by the low photon background.

Van Vliet and Poppe (20) carried out an investigation of detection cell designs for OTLC with laser-induced fluorescence detection. Laser excitation was perpendicular to the capillary, with the emission collection optics placed perpendicular to both the capillary and the excitation beam. These authors investigated on-column detection and coupled capillary cells as well as a sheath flow design. While all these designs were workable, the direct on-column approach was found to be simplest and yielded detection limits comparable to the other two designs. The investigators further determined that the main source of noise was fluctuation in the background caused by laser amplitude noise. Van Vliet et al. (21) developed a postcolumn reaction system for use with OTLC with laser-induced fluorescence. They used o-phthalaldehyde and mercaptoethanol to react with analytes with primary amine functionality to generate the fluorescent isoindole derivatives. A MDQ of ~ 10 fmol and a corresponding MDC of $\sim 2 \times 10^{-7}$ M were found.

3.3. Chemiluminescence Detection

Chemiluminescence (CL) is related to fluorescence, but chemical energy instead of the energy of a photon is used to excite a luminescent molecule, which emits light upon returning to the ground state. Like fluorescence, chemiluminescence offers the promise of a low background and thus high sensitivity. Furthermore, the expense of a light source is eliminated, or–more accurately–replaced by postcolumn chemistry and reagent pumping systems. (The cost trade-off of a light source for some form of pump makes cost a relatively unimportant consideration here.) The key challenge is to provide effective and thorough mixing of the column effluent with the chemiluminescence reagent, while not introducing band broadening or excessive postcolumn dilution of the analytes.

Peroxyoxalate-generated 1,2-dioxetanedione (C_2O_4) has been the chemiluminescent reagent of choice of most workers in this field. This reagent can excite a variety of luminophores, especially those that are reasonably good electron donors, as the mechanism of excitation is believed to involve reversible electron transfer from the luminophore to peroxyoxalate. Imai and co-workers [see Miyaguchi et al. (22, 23) and Honda et al. (24)] have done the most work on chemiluminescence detection, although their work is mainly with packed columns of millimeter bore. They used a cyclonic mixing chamber followed by a delay line to permit chemiluminescence to develop to a useful level. The detection cell was in a commercial chemiluminescence detector. Detection limits for dansyl amino acids were reported to be ~ 200 amol injected into the column (22).

De Jong et al. (25) used a novel mixing system with micro-packed columns of 320-μm i.d. In their system the 1,2-dioxetanedione was formed by mixing bis(2,4-dinitrophenyl)oxalate (DNPO), base, and peroxide immediately prior to combining with the column effluent in a coaxial mixing arrangement. The combined effluent then flowed through coiled tubing in front of a photo-multiplier tube in a commercial spectrophotomer. Again, detection limits were reported as being in the hundreds of attomoles for dansylated compounds. This system exhibited negligible band broadening with 300-μm-i.d. packed columns.

Weber and Grayeski (26) used micropacked columns of 200-μm i.d. by 20- to 30-cm length. Peroxyoxalate-induced CL was employed, and mixing was again carried out in a coaxial system with reagent being introduced around the column effluent. The detector itself was a commerical fluorescence detector with the source turned off. Detection limits for 2- aminoanthracene and perylene were in the hundreds of femtomoles.

Dluzneski (27) also reported work with a coaxial postcolumn mixing system with OTLC columns. He used constant gas pressure and calibrated flow restrictor capillaries to provide for steady delivery of the CL reagents. The MDQ was 1 fmol, and the MDC was 6×10^{-8} M for 3-aminofluoranthene. In the best cases, CL appears to offer better detection limits than does fluorescence excited by conventional light sources (i.e., arc lamps). But CL does not appear to be competitive with the state of the art in laser-induced fluorescence detection, which provides detection limits several orders of magnitude lower than has been achieved with CL.

3.4. Refractive Index Detection

Refractive index detection offers the same possibilities for microcolumns as it does for conventional HPLC columns, namely, a detector that will respond to essentially all analytes but is relatively insensitive (a mediocre MDC) and drifts severely when used with gradient elution. Fujimoto and co-workers (28) modified a commercial refractive index detector by simply replacing the original cell gasket with a thinner (20-μm) gasket, yielding a cell volume of 0.8 μL. This cell volume was found to be useful with packed capillaries of 0.5-mm i.d. The MDQ for di-n-pentylphthalate was ~ 10 pmol, corresponding to an MDC of $\sim 3 \times 10^{-5}$ M. As anticipated, this is not a particularly low value, but the detector was demonstrated to be useful with difficult-to-detect analytes that lack a good chromophore, such as sugars.

Bornhop et al. (29) developed a novel laser-based refractive index detector, similar to a design described by Lukacs (30), and used it for on-column detection in capillary electrophoresis. These detectors use the refractive index dependence of the interaction (refraction and scattering) of a laser beam with a fluid-filled capillary. As the refractive index of the fluid in the capillary changes, the scattering pattern of a laser beam intersecting the capillary is altered, and this change may be detected by use of a photodiode placed in a position between a scattering maximum and minimum. This design is extremely simple and inexpensive, as a helium-neon laser may be used and the required optics are minimal. Bornhop et al. (29) used their detector with micropacked columns of 250-μm i.d. found an MDQ for sugars of approximately 200 pmol. This corresponds to an MDC of 7×10^{-5} M, similar to the value reported by Fujimoto et al. (28) with the modified commercial refractive index detector.

3.5.   Photothermal Detection

Photothermal detection is a variation on absorption detection, where the absorption of photons is monitored by the increase in temperature of the absorbing medium instead of by the decrease in optical transmittance of the medium. The temperature increase is conveniently monitored through the decrease in refractive index by optical means, using laser-based refractive index monitoring not unlike that described in the previous subsection. Laser beams are used both as the "pump" beam (a high-power beam whose absorption leads to the photothermal effect) and the probe beam (a weaker beam used to measure the heating by its effect on refractive index). The real advantages of lasers and the photothermal effect are that lasers can be focused into small-diameter capillaries for on-column detection and that the sensitivity of the photothermal effect is much less dependent on optical path length.

Sepaniak et al. (31) demonstrated the use of photothermal detection with 20-μm-i.d. OTLC columns and with an argon ion laser, operated at 458 nm with a power of 500–800 mW, focused onto a square-cross-section capillary detection cell with 100- or 200-μm sides. In this work, a single laser beam was used to both generate the photothermal effect and detect it. This was based on the defocussing ("blooming") of the transmitted laser light as the liquid medium was heated by photon absorption (thermal lens effect). The laser beam was chopped at 500 Hz in order to permit gated detection and observation of the rate of production of the thermal lens. The baseline noise level for their system was 3×10^{-5} absorbance units. This yielded an MDQ of ~ 300 fmol for o-nitroaniline and a corresponding MDC of 6×10^{-5} M injected.

In later work, Kettler and Sepaniak (32) used a two-laser system (crossed-beam photothermal refraction), employing an eximer pumped dye laser (operated at 400 nm) for the pump beam and a helium–neon laser for the probe beam. This detection system was used in conjunction with packed capillary LC columns with 200-μm-i.d. The MDQ was 80 fmol of 1-nitropyrene, corresponding to an injected concentration of 4×10^{-6} M analyte.

Nolan et al. (33) used a two-laser system with a 3-mW helium–cadmium laser operated at 442 nm as the pump beam and a helium–neon laser for the probe beam in a "crossed-beam thermal lens" detection system. The beams intersected in a detection cell of square cross section and 80-μm i.d. The pump beam was choped at 128 Hz, and the detection electronics included a lock in amplifier. This system was used with a packed capillary column of 250-μm i.d. The MDQ of a dinitrophenyl hydrazine derivative of acetone was 120 fmol injected onto the column, corresponding to an injected concentration of 2×10^{-5} M.

Photothermal detection has been demonstrated as a viable means of detection but has not demonstrated any great superiority over conventional absorption detection and is subject to the following limitation. The photothermal effect works best with lasers operating in the visible region. Operation in the UV region would seem to offer even better prospects for sensitive detection, owing to the greater proportion of molecules with significant molar absorptivities in the UV region. But most solvents have trace absorptions in the UV, which serve to increase the photothermal background signal and noise. Thus performance in the UV will not offer as low a noise level as in the visible region. Furthermore, the beam qualities of lasers available for operation in the visible region are far superior to those in the UV. Unfortunately, few compounds of interest have significant absorptions in the visible region. In order to detect a wider range of compounds, colored (visible-wavelength absorbing) derivatives of these compounds must be made. But if one goes to the trouble of making derivatives, one might ask why not make fluorescent derivatives and use a truly sensitive mode of detection—laser-induced fluorescence?

3.6. Optical Activity Detectors

Bobbitt and Yeung (34) and Synovec and Yeung (35) have developed sophisticated laser-based detectors for measuring optical activity and circular dichroism in the effluent from microbe columns of 1-mm bore. Although these columns are slightly large to be included in this discussion, the detectors are interesting and may eventually be used with micropacked capillary columns. In the optical rotation detector (34), 20 mW of light of 488 nm from an argon ion laser was modulated at 1 kHz and beamed through a detection cell with a 1-cm path length and a 1-μL volume. The noise level of this detector was ~ 15 microdegrees. The MDQ for fructose was ~ 50 pmol injected onto the column, corresponding to an injected analyte concentration of 1×10^{-4} M. This detector was also used to detect optically inactive materials by indirect polarimetry. In this case, an optically active solvent composed of a 50:50 mixture of (−)-2-methyl-1-butanol and acetonitrile was used as the mobile phase. Use of this relatively exotic mobile phase is made possible by the small solvent consumption rate (20 μL/min) of the microbore column. The operating cost is less than the operating cost for ordinary acetonitrile for a conventional HPLC column. In this system, as an optically inactive analyte elutes from the column, it acts to dilute the optically active mobile phase and thus yields a shift in optical rotation. The MDQ in this case was approximately 2 nmol of dodecane, or an MDC of 4×10^{-3} M. The authors speculated that by using an optically active mobile phase with an optical rotation greater by a factor of 30 (they claim such is available), a 30-fold decrease in the detection limit should be realized. If this enhancement were achieved, the detector would be comparable in detection limits to the refractive index detectors described previously. An interesting property of indirect detection is the potential to accurately quantify an analyte without knowing its identity and, hence, without use of calibration mixtures. This application is too involved to describe in these pages but is one with obvious interest to those who must determine purity of substance without having the luxury of standards for all the impurities. Bobbitt and Yeung (34) also point out the potentially important application of this detector for determining optical purity of pharmaceuticals.

The laser-based circular dichroism detector (35) made use of the 488-nm line of an argon ion laser and involved a fairly complex optical and electronic layout. The detection cell had a 1-cm path length. Modulation to produce left- and right-handed circularly polarized light was carried out between 100 kHz and 10 MHz. These high frequencies reduced the flicker noise of the laser so that the noise approached the photon shot noise limit. The MDQ for $(+)$-Co(III)-trisethylenediamine was ~ 20 pmol, corresponding to an injected concentration of 5×10^{-5} M.

3.7. Electrochemical Detection

Electrochemical detection has seen much use in microcolumn LC, owing to the extraordinary sensitivity that may be attained with the amperometric mode of detection. Conventional "thin-layer" detection cells can also easily be modified to have low internal volume, making them suitable for use with micropacked columns of 60-μm i.d. Hirata et al. (36) used precisely this approach. They used glass columns drawn to an internal diameter of 60–80 μm with 30-μm irregular alumina particles already packed in the column before drawing. Their cell was made with a 50-μm-thick Teflon spacer gasket, resulting in a cell volume of only 0.15 μL. No detection limits were cited, but good S/N performance was demonstrated down to at least the micromolar concentration range.

Slais and Kourilova (37) described the use of columns of 500-μm i.d. with an amperometric detection cell with a volume of 0.02 μL. In their cell design, a platinum wire working electrode was positioned near the outlet of the microcolumn in order to reduce post column band spreading. The MDQ for pyrocatechol was ~ 30 fmol, corresponding to an injected concentration of 1×10^{-7} M.

Goto and co-workers (38, 39) described a thin-layer flow cell in which working electrodes were placed on opposite sides of a thin-layer channel that was 30–50 μm thick and had a cell volume of approximately 1 μL. In such a system, the potential of the two electrodes may be poised such that the analyte may be oxidized on one side of the channel at one electrode, then diffuse across the channel and be reduced at the other electrode. As this process is repeated, the signal is amplified by repeated oxidation and reduction of the same analyte molecule. Their system was used with micropacked columns of 500-μm i.d. Signal amplification factors (current relative to current in a simple one-electrode thin-layer cell) of 2- to 20-fold were obtained with this system. Detection limits were not stated in this work, but the MDQ may be estimated from chromatograms to be in the tens of femtomoles for catechols.

The majority of the work on amperometric detection with open tubular columns was described by the Jorgenson group in a serious of papers (40–48). In this work, a carbon microfiber of 5- to 10-μm diameter was inserted with the aid of a micropositioner into the outlet end of an OTLC column of 8- to 20-μm i.d. This close fit of the working electrode into the actual column outlet provided for minimal bandspreading as well as high detection efficiency. In an amperometric mode, the MDQ was $\sim$ 20 amol, corresponding to an injected concentration of 2×10^{-9} M (41). Electrode fouling was found not to be a major problem with this detector, as an electrode "cleaning" procedure, involving cycling of the electrode potential, was developed. With increased attention to shielding and electronic filtering, this has been improved to a level of 1 amole, or a concentration of 1×10^{-10} M (49). This detector has been used in a voltammetric (potential scanning) mode under computer control (42, 43). In the voltammetric mode, the detection limits were about 2 orders of magnitude poorer than in the amperometric mode. The voltammetric mode provided considerable extra resolution and information over the amperometric mode. This system was used for the analysis of the intrinsically electroactive components of single snail neurons, and the chromatovoltammograms permitted unambiguous identification and quantification of dopamine, serotonin, tyrosine, and trytophan in single neurons (44).

Oates and Jorgenson (45, 46) also showed that this system was compatible with gradient elution. In the amperometric mode, the MDQ was $\sim$ 30 amol and the MDC was 5×10^{-9} M with gradient elution of a derivatized amino acid. Detection of this derivative required use of a potential of 0.9 V (vs. Ag/AgCl reference), which is a relatively high potential and thus more prone to noise. The entire analysis system was so sensitive and required so little sample that it permitted the analysis of the free amino acid content of single cells (47). Gradient elution was required in this system, as the amino acid derivatives span a wide polarity range and could not be analyzed under isocratic contitions. This system was also used for ultramicroanalysis of the amino acid content of proteins and permitted excellent quantitative amino acid analysis of as little as 4 fmol of protein (48).

Figure 15.1 is data from a gradient elution (100% water to 60:40 water/ acetonitrile) separation of brewed tea, with detection by voltammetry. Figure 15.1a shows the chromatogram with plotted data at $+1.3$ V, Figure 15.1b, c shows the full chromatovoltammogram of the data set. It is clear that on-line voltammetry is fully compatible with gradient elution conditions. The peak labeled **T** is due to theophylline.

Goto and Shimada (50) developed a microfiber working electrode similar to that described by the Jorgenson group. They inserted 7-μm-diameter by 3-mm-long fibers into a cell made from 50-μm-i.d. fused silica tubing and attached this cell to the outlet of a 350-μm-i.d. packed capillary column. They used this system in the voltammetric mode, and an MDC of $\sim 1 \times 10^{-7}$ M was reported for dopamine.

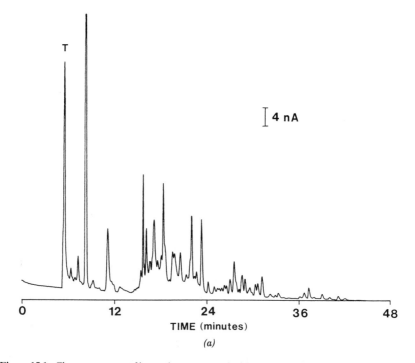

Figure 15.1. Chromatograms of brewed tea separated with gradient elution (100% water to 60:40 water/acetonitrile); (a) data at $+1.3$ V; (b, c) full chromatovoltammograms. Reprinted with permission from M. D. Oates and J. W. Jorgenson, *Anal. Chem.* **61**, 1977 (1989). Copyright 1989, American Chemical Society.

Slais and Krejci (51) described an amperometric detector that placed a 100-μm platinum wire at the outlet end of an open tubular column. The performance of this system was largely inferior to the aforementioned system, with a MDQ of roughly 1 fmol and a MDC of 2×10^{-7} M.

In a series of papers (52–54) Manz, Simon, and colleagues [see Froebe et al.

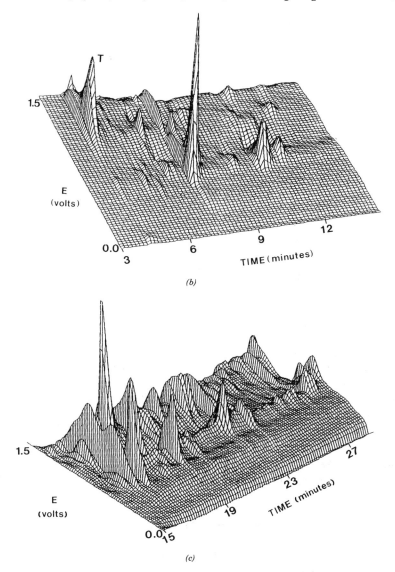

(b)

(c)

Figure 15.1*b* **and** *c*. See facing page for legend.

(52), Manz et al. (53), and Manz and Simon (54)] described a potentiometric detector based on micrometer-sized ion-selective electrodes that were inserted into the outlet end of open tubular columns. This detection mechanism gave a logarithmic (Nernstian) response vs. concentration that is unusual in a chromatographic detector. When used with 25-μm-i.d. OTLC columns the MDQ for potassium ion was determined to be 6 fmol and the MDC was 1×10^{-6} M. In a later paper these authors used this detector with 14- and 3.5-μm-i.d. OTLC columns and achieved a MDQ for iodide ion of 6 amol injected onto the columns and a 5×10^{-7} M MDC.

3.8. Photoionization Detection

Photoionization detection with 10- and 15-μm-i.d. OTLC columns was described by de Wit and Jorgenson (55). In this system, analyte was vaporized into a flowing stream of helium gas. For the less volatile analytes, vaporization could be assisted with the use of pulse heating from a 25-μm tungsten wire resistance heater placed across the outlet end of the capillary column. Once in the helium stream, analyte flowed in front of the magnesium fluoride window of a krypton discharge vacuum UV light source emitting ~ 150 mW at 123.6 nm. Light of this wavelength has sufficient energy (10.02 eV) to photoionize most analyte molecules, but it has insufficient energy to ionize water, methanol, acetonitrile, ammonia, pentane, and many other of the common mobile phase components and additives used in LC. This system showed an optimum in helium flow rate, in terms of detector response and S/N, as well as requiring a certain minimum flow rate in order to ensure that analytes did not linger in the detection volume and cause excessive peak broadening. A good helium flow rate proved to be ~ 20 mL/min, and under such conditions, the MDQ for toluene was 20 fmol injected, corresponding to an injected concentration of 5×10^{-6} M. For a less volatile analyte such as acridine, the detection limits were 80 fmol and 2×10^{-5} M. The advantages of this detector are that it is relatively general in response, not too limiting in the mobile phases with which it may be used, and relatively inexpensive to build and operate. Better detection limits might be realized if a reliable laser could be developed for use near 120 nm. Application to completely involatile analytes might be made possible by sample introduction to the gas phase ("vaporization") by an electrospray process (described below in Section 3.11 on mass spectrometry).

3.9. Flame-Based Detectors

The flame ionization detector and its varients are the workhorse detectors in gas chromatography. In LC, they could provide the general (relatively nonselective) detector that has long been sought for. This detection principle has seen limited usage in LC, owing to its incompatibility with most mobile phases, particularly the large mass load that a flow of 1 mL/min represents to a flame. In this latter regard, the low flow rates of microcolumn LC are an advantage. Krejci et al. (56) used a flame ionization detector of relatively conventional design in combination with OTLC columns with inner diameters ranging from 5 to 34 μm. The OTLC column was terminated at the end of the burner jet tip, and the capillary end could be heated from ambient temperature to 1000 K. As vaporization of effluent was required, only volatile solutes were investigated. Detector performance was not assessed in detail. Krejci et al. (57) described a similar system operated as an alkali flame ionization detector with selectivity toward halogen and phosphorous-containing compounds. Again, solutes required some volatility in order to be introduced into the flame. The MDQ was on the order of 10 fmol for tributylphosphine oxide.

McGuffin and Novotny (58, 59) described detectors for microcolumn LC based on flame photometric detection as well as thermionic detection. In the flame photometric detector (58), a 50-μm-i.d. capillary connected the column outlet to the detector. This capillary terminated 1 mm above the tip of the burner jet tip. The detector was used to monitor phosphorous compounds at 530 nm, and a MDQ of 50 pmol of phosphorous (or 50 pmol of a compound containing one phosphorous atom) was found. This detector was able to tolerate in excess of 20 μL/min of aqueous mobile phase flow without extinction of the flame, but performance was better at flow rates below 5 μL/min. Linearity of this detector was also found to be good. The authors also described a dual-flame thermionic detector (59), where the lower flame was responsible for nebulization and desolvation of the mobile phase aerosol, while the upper flame was responsible for the analytical measurement and thus was more free of matrix effects. A MDQ of ~ 5 pmol of phosphorous-containing compounds was reported. The thermionic detector was modified by Gluckman and Novotny (60) in order to provide more favorable nebulization of compounds of low volatility into the flame. The 50-μm-i.d. connecting capillary was inserted into the burner jet 1 mm below the tip of the jet and at a right angle to the flow of the flame gas. In this manner mobile phase was more effectively nebulized into the flame gas. This arrangement permitted detection of less volatile species such as thiophosphinic esters of hydroxy steroids.

Tsuda et al. (61) modified a commercial moving wire flame ionization detector in which the effluent from 500-μm-i.d. microcolumns was deposited on a wire moving directly beneath a 50-μm-i.d. tube connected to the outlet of the microcolumn. The mobile phase flow rate was from 2 to 12 μL/min of chloroform. At higher flow rates, droplets of solvent formed on the wire. The mobile phase was then evaporated and the wire passed through the detector flame, where non volatile analytes generated a signal. An MDQ of ~ 50 pmol of triglycerides was reported.

3.10. Inductively Coupled Plasma Detection

Inductively coupled plasma (ICP) emission spectrometry offers the possibility of element-selective detection for the column effluent. Jinno and Tsuchida (62) and Jinno et al. (63) reported on the use of 500-μm-i.d. packed columns interfaced to an ICP detector. Column effluent was entrained into a larger carrier flow of liquid, and these combined flows were aspirated into the base of a full-size commercial ICP spectrometer running with 1–2 kW of RF power. Organometallic compounds of copper, zinc, iron, cobalt, and chromium were investigated, and the system was used in both reversed-phase and normal-phase modes. A MDQ in the neighborhood of 1–10 ng, depending on the metal, was obtained. This corresponds to 10–100 pmol of metal, or a minimum detectable injected concentration of 1×10^{-5} to 1×10^{-4} M. The main disadvantage in this detector was the dilution introduced by the requirement for a carrier flow of liquid. UV absorption detection of the effluents provided at least 10-fold better S/N ratios for the analytes than did ICP detection, albeit without the elemental selectivity of the ICP.

3.11. Mass Spectrometric Detection

Mass spectrometry (MS) has become an important mode of detection for LC, and in particular for micro-LC and OTLC. At first glance, MS and LC do not appear compatible, owing to the mass flow used in LC (64, 65): MS deals with ions in the gas phase at low pressures, whereas LC uses liquids at atmospheric pressures or above. The low solvent flow rates used in micro-LC and OTLC create a much smaller mass-flow incompatibility problem than does conventional HPLC. A second problem stems from the fact that LC is most often utilized to separate compounds of low volatility, whereas MS can only be done on ions in the gas phase. The role of a good LC/MS interface is to bridge the two techniques, minimizing these incompatabilities, while sometimes providing for ionization of the analytes as well. The most commonly used ionization techniques with LC are electron impact ionization (EI) and chemical ionization (CI). More recently fast atom bombardment (FAB) has become an ionization technique of interest for LC, in particular for biochemical applications. The ionization technique employed is dependent on the interface.

The great interest in MS as a detection method for LC has resulted in a large number of publications on this subject. The research in this area has been focused on LC/MS interfaces, their ability to maintain chromatographic fidelity, the ionization techniques, and the kind of information obtainable with

LC/MS. Owing to the power of this analytical method, a large number of application papers have appeared in the literature as well.

A variety of interfaces for micro-LC and OTLC have been used, the most common being the direct introduction (DI) interface (of which direct liquid introduction (DLI) is an example), and a variety of in-house constructed interfaces often similar to DLI. Most interfaces were originally designed for conventional HPLC and have been modified for use with micro-LC. [In conventional HPLC only a small fraction (1%) of the effluent is introduced into the ion source, which decreases sensitivity.] These kinds of interfaces include DLI, thermospray (TSP), atmospheric pressure ionization (API), monodisperse aerosol generation interface (MAGIC), and the moving belt. Niessen addressed the impact of the LC effluent on the mass spectrometer's vacuum system (66, 67).

In the direct introduction interfaces the effluent has either passed through a laser-punctured pinhole (diaphragm) (68–70) or through a capillary tube (71–78). The approach used is a major characteristic of the interface.

In the diaphragm-based interface the LC effluent flows through a tube whose end is positioned a short distance (typically 1 mm) in front of the pinhole. The effluent fraction that does not go through the pinhole flows back around the LC effluent tube through a coaxial tube to waste. In cases where the effluent flow is not large enough to create a liquid jet, a makeup flow has to employed. Arpino et al. (69) have made theoretical projections about the formation of a liquid jet in capillary and diaphragm restrictions. Practical considerations such as plugging, freezing, and jet stability as a function of diameter and shape (for example, jet stability is adversely affected by rough edges of the orifice) have been investigated (68).

Capillary inlet interfaces have been the most widely used. Fused silica capillary tubing has been utilized extensively in the manufacturing of these interfaces (72, 73, 75–78). Capillary inlets have been used to produce liquid jets (DLI), gas jets, and nebulized jets (TSP). Mathematical predictions have been published regarding the conditions needed to form such jets (74, 75, 78). Most of these calculations were directed toward answering the question of where the vaporization of the effluent takes place. Bruins and Drenth (78) examined the major processes that govern the site of vaporization in a capillary tube: the rate of vaporization (depending on the capillary inner diameter—i.e., surface area–and the shape of the capillary) and the mass flow rate of the liquid.

In the DLI, the LC effluent, in the form of a liquid jet, is introduced directly into the ion source. Microbore LC, with flow rates of $10-50 \mu L/min$, allows the total effluent to be introduced into the ion source (78–86). Total effluent introduction yields better detection limits than do conventional HPLC systems that require splitting of the effluent. Unfortunately, at these flow rates only

CI mass spectra can be obtained. DLI with CI has been widely applied. Henion and Maylin (79) obtained full-scan chemical ionization spectra on 1–5 ng levels of drugs and their metabolites using micro-LC–DLI at 8-μL/min flow rates. The CI spectra allowed identification of these drugs.

DLI of the entire LC effluent under EI conditions can be done with flow rates encountered in packed capillary LC columns (100- to 500-μm i.d., 1–5 μL/min) (72, 73, 87, 88), or OTLC columns (2- to 20-μm i.d., < 0.1 μL/min) (74, 77, 89). Alborn and Stenhagen (72, 73) coupled packed fused columns (0.22-mm i.d.) to magnetic sector instruments. Their interface was constructed by drawing the column into a fine tip. A course packing material was used in the tip to keep the finer chromatographic bed in place. They obtained spectra similar to ordinary EI spectra, which allowed comparison with reference EI spectra, with detection limits in the low nanograms for volatile compounds.

In OTLC the flow rates are so low that in order to create a liquid jet a makeup liquid has to be added (90, 91). Niessen and Poppe (91) developed an interface in which a 4-μm diaphragm and makeup flow (at rates from 0.2 to 0.6 μL/s) were employed in order to create a liquid jet. This allowed only CI spectra to be obtained. A capillary-inlet-based interface that they developed later allowed EI spectra to be acquired (77, 92). De Wit et al. (89, 93) employed a combination of tapering and heating of the OTLC column end to create a vapor jet. The interface also allowed introduction of a CI gas so that either CI or EI spectra could be acquired. Low-picogram detection limits for herbicides were obtained (94).

The most commonly used interface for conventional HPLC, namely, TSP, has until recently not been compatible with micro-LC, since it requires flow rates above 0.5 mL/min to obtain stable ion currents. Butfering et al. (95) solved this problem by developing a dual-beam TSP interface. This interface employs two capillaries–one for the LC effluent, and one for provision of reagent ions (the electrolyte solution). The electrolyte solution flow rate was 1 mL/min, whereas the LC effluent was flowing at 30 μL/min. Normal TSP mass spectra can be obtained with this interface.

In the API interface, the mass flow problem between LC and MS is avoided completely since, as the name implies, the source operates at atmospheric pressure. Of the three approaches to API—(1) the heated pneumatic nebulizer, (2) electrospray, and (3) liquid ion evaporation (64)—the electrospray method is the only one used with micro-LC (96, 97). Bruins et al. (97) reported use of this interface with flow rates of 10 μL/min. In this method a vapor mist is created by flowing the effluent through a metal capillary at a large electrical potential with respect to the surrounding API chamber walls. The charge induced on the effluent is so large that the coulombic repulsion forces are strong enough to overcome surface tension and a fine spray is created and enters the mass spectrometer through a 100-μm-diameter sampling orifice.

Good chromatographic fidelity is maintained, and the interface can be used with a wide variety of compounds. A disadvantage of API is the limited amount of information obtained from API mass spectra due to lack of fragmentation, although MS/MS instruments having a collision cell between the two spectrometers can remedy this shortcoming.

MAGIC is the only form of DLI that allows acquisition of EI mass spectra (98, 99). Browner et al. claimed compatibility with micro-LC with flow rates at 100 μL/min and obtained detection limits of 1–10 ng for a thermally labile carbamate pesticide, under full-scan conditions (40–400 Da scan). The MAGIC interface operates by passing the LC effluent through a small orifice creating a fine liquid jet. After this jet breaks up naturally to form uniform droplets, the droplets are dispersed with a gas stream and the generated aerosol continues through an aerosol beam separator (analogous to a gas jet separator) to the ion source.

The moving belt interface, an interface that has been used with flame ionization detection, also has been used with micro-LC/MS (100–102). Hayes et al. (101) reported on a nebulizer they used for spray deposition of the effluent onto a moving belt. Flow rates were typically around 100 μL/min, and detection limits for phenols of 6 ng were obtained under full-scan EI conditions. This technique did not introduce significant extra column variance under the stated conditions.

Most of the interfaces discussed so far used either CI or EI ionization, which limits the range of applicability (64, 103). Highly nonvolatile or polar compounds, typically encountered in biochemical and biomedical applications, were either very difficult to analyze with these systems or could not be done at all. The combination of LC with FABMS would solve this problem. FABMS allows analysis of most polar and nonvolatile compounds.

Several approaches to interfacing LC and FABMS have been made. The moving belt interface allows the use of FAB ionization (104), but no work with micro-LC and moving belt FABMS has been reported. Research has recently been focused on continuous-flow FAB (CF–FAB). In continuous flow FAB the effluent flows through a specially designed probe very similar in appearance to a normal FAB probe (103, 105–110). A number of advantages of CF–FAB over static (conventional) FAB exist. The matrix background encountered in CF–FAB is significantly lower than that encountered in conventional FAB, which results in less complicated mass spectra and higher sensitivities due to improved S/N ratios.

Ito and co-workers (105), and Caprioli and Fan (108) were among the first to connect a microbore HPLC column directly to a FAB source. They used a specially designed stainless steel frit on the tip of the CF–FAB probe. The micro-LC effluent at a flow rate of $\sim 0.5\ \mu$L/min was passed through 40-μm-i.d. fused silica tubing to the fritted tip. Owing to the large surface area of the tip, immediate vaporization of the mobile phase resulted. Therefore, the glycerol and the analyte were concentrated on the surface of the tip and exposed to the FAB beam. In their work, glycerol matrix was added to the chromatographic mobile phase, and the concentration of the glycerol in the mobile phase affected both the MS sensitivity and the retention time of the solute. A typical mobile phase composition for their work was glycerol:water:acetonitrile (10:27:63).

Caprioli and co-workers (103, 107, 108) have introduced a CF–FAB probe similar to the conventional FAB probe. Their use of flow rates of $5 \mu L/min$ caused the copper probe tip to remain wet when stable operating conditions were obtained. Good sensitivity was obtained for the peptide Substance P (molecular weight = 1347), for which a detection limit of ~ 200 fmol was reported. This demonstrated significant improvements in sensitivity over that of conventional FAB.

De Wit et al. (110) investigated the combination of OTLC and CF–FAB using two approaches. In the first the matrix was to be introduced in the mobile phase and in the sample solutions. The second approach employed a coaxial tubing system in which the matrix flowed around the OTLC column and the two flows mixed at the FAB probe tip. Figure 15.2a is a schematic diagram of the "Tee" which permits coaxial insertion of the separation column inside the sheath column. The FAB matrix solution then flows in the annular region between the separation column and the sheath column. Figure 15.2b is a schematic diagram showing the arrangement at the FAB probe tip. The separation column is positioned ~ 1 mm inside the end of the sheath column, so that the two column flows mix just prior to delivery to the high vacuum of the ion source. In this latter approach the conditions for the OTLC and the matrix can be optimized separately. Detection limits of ~ 1 fmol have been obtained for selected ion monitoring of the peptide L-methionyl-L-leucyl-L-phenylalanine (Met-Leu-Phe). Figure 15.3 shows selected ion monitoring of this peptide at 54 (first three injections) and 5.4 fmol injected (last three injections).

3.12.   Infrared and Fourier Transform Infrared Detectors

The infrared (IR) region exhibits sharp absorption bands that are highly specific and characteristic of the analytes. It can thus offer structural information and in some cases, through the use of reference spectra, positive spectral identification. For these reasons IR and Fourier transform infrared (FTIR) detectors have become more common in chromatographic detection, first in GC, and later in HPLC and micro-LC. The application of IR detection to LC is complicated owing to mobile phase absorptions. Common mobile phases in reversed phase chromatography, such as alcohol and water, absorb

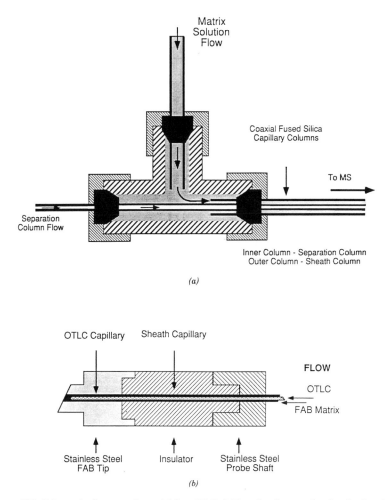

Figure 15.2 Schematic diagram of coaxial flow FAB–MS probe. See text for details. Reprinted with permission from M. A. Moseley, Ph.D. thesis, University of North Carolina at Chapel Hill, December 1990.

strongly over wide ranges in the IR. Normal phase LC however, employs a number of solvents that are relatively transparent in the mid-IR region.

There have been two approaches in using IR detection for LC: the first is to remove the solvent by depositing the effluent onto a collection device and allowing it to evaporate before detection; the second is the use of a flow cell, allowing on-line detection.

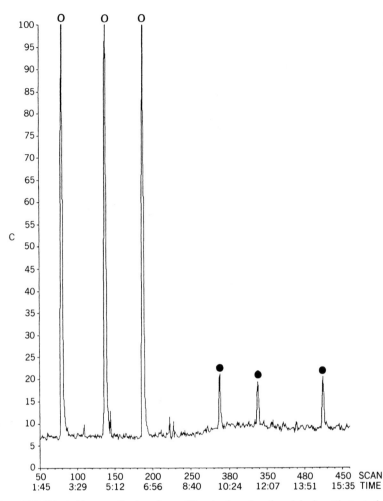

Figure 15.3 Selected ion monitoring of L-methionyl-L-leucyl-L-phenylalanine. First set of peaks obtained from 54 fmol injected; second set from 5.4 fmol injected. Reprinted with permission from M. A. Moseley, Ph.D. thesis, University of North Carolina at Chapel Hill, December 1990.

FTIR has the advantage of high-scan speed and sensitivity. This sensitivity can be further enhanced by coaddition of scans across a chromatographic peak. Johnson and Taylor (111) showed that a maximum (S/N) can be obtained by coaddition of scans $1\frac{1}{2}$ standard deviations either side of the peak maximum. Beyond this point the S/N goes down owing to the low signal at the edges of the peak. Another advantage of FTIR is the ability to obtain Gram–Schmidt reconstructed chromatograms.

The removal of solvent after deposition of the LC effluent has been tried in a variety of ways. Conroy et al. (112) used a series of small cups filled with KBr powder in which the micro-LC effluent was deposited. After solvent evaporation diffuse reflectance (DR) FTIR was utilized. This system does not continuously monitor the chromatographic effluent, and thus peaks can be missed or several peaks can be collected in one cup. The use of KBr also precludes the use of aqueous solvents.

Jinno and coworkers (113–115) combined micro-LC with FTIR using the KBr buffer memory technique. The effluent was deposited on a moving rectangular KBr crystal (8.8 mm × 35 mm × 5 mm) as a continuous narrow band (~ 1.5 mm wide). The mobile phase was evaporated with the help of a heated stream of nitrogen. The speed of collection (motion of the KBr plate) was controlled according to LC conditions. In one application the mobile phase flow rate was 5 μL/min and the collection rate was 2.4 mm/min. After collection of a chromatogram the KBr crystals were placed automatically in the beam of an IR spectrometer and IR transmission spectra were obtained. This technique can also be used with an FTIR instrument. This technique is relatively sensitive: separations of low-microgram quantities of 2,4-dinitro-phenylhydrazones, with n-hexane/dichloromethane as the mobile phase on a 0.5-mm i.d. by 15-cm-long column packed with 5-μm porous silica were reported.

Gagel and Biemann (116) introduced a device in which the chromatographic effluent was continuously nebulized with pressurized nitrogen onto a rotating disk that had a reflective surface. Typical deposition speed of the 1- to 2-mm-wide track was ~ 4 mm/min. The reflectance–absorbance FTIR spectra were measured off-line by rotating the disk again in the FTIR spectrometer. A good IR spectrum could be obtained for 125 ng of anthracene, which was an unfavorable test compound since it is relatively volatile and a poor IR absorber. This device can be used with aqueous solvents since the disk is inert to all commonly used solvents.

Many approaches for flow cells have been tried for conventional HPLC, but they do not provide enough sensitivity for micro-LC. The use of an attenuated total reflectance (ATR) cell, in which the effluent flows along the outside of the cell, requires milligram-level injections to obtain a signal (117). ATR cells allow the use of highly absorbing solvents owing to the inherently short path lengths. Micro-LC often has the advantage over conventional HPLC of increased sample concentration as a function of the amount injected, as well as higher efficiencies, which thus imporve sensitivity. Also the low solvent consumption permits the use of expensive solvents, such as $CDCl_3$, which are used in IR spectroscopy. In general most FTIR flow cell applications are done with normal phase chromatography and size exclusion chromatography (in which the solvent plays no role in the separation).

Teramae and Tanaka (118) used a homemade 0.025-mm path length NaCl flow cell (0.5-μL volume), and a 0.2-mm path length KBr flow (1.2-μL volume) with a commercially available FTIR spectrometer and micro-LC. The flow cells were simply inserted into the spectrometer's sample holder. A separation of 28 μg diethylphthalate (DEP) and 29 μg p-nitrotoluene on a 0.5-mm-i.d. styrene–divinylbenzene porous polymer (micro gel permeation) column and 4 μL/min CCl$_4$ mobile phase was performed using the 0.2-mm path length KBr flow cell detector. Calibration curves for DEP for both flow cells showed linearity over a range from 1 to 20 μg/μL (measured at 1730 cm^{-1} for the —C—O stretch). The detection limit for DEP using the 0.2-mm path was 1 μg, whereas for the 0.025-mm path it was 2 μg. The better than expected detection limit for the 0.025-mm flow cell was attributed to lessened peak broadening in the lower volume cell.

Brown et al. (119, 120) reported the detectability of selected compound classes by flow cell microbore FTIR. The MDQ for phenols and amines were reported in the low-microgram range. Ester and amide derivatives were prepared which had stronger IR absorptions and better MDQs by about a factor of 10, demonstrating that derivatization can help to improve detection limits. With a homemade FTIR flow cell (3.2-μL volume, 0.2-mm path length), detection was performed on a series of nine amines (3 μg each) separated on 50-cm × 1-mm-i.d. column packed with 10-μm silica particles and using CDCl$_3$ at 20 μL/min as the mobile phase. The Gram–Schmidt reconstructed (GSR) chromatogram showed better resolution with FTIR detection than with UV detection. The IR spectrum of indole, the last eluting compound, and the IR spectrum of indole dissolved in CDCl$_3$ were similar. The MDQ for 2,6-di-$tert$-butylphenol was determined at 4 wavelengths in the IR (3640 cm^{-1} for —OH stretch, 2960 cm^{-1} for asymmetric —CH stretch, 1426 cm^{-1} for the asymmetric —CH bend, and 1232 cm^{-1} for the —C—O stretch), with the lowest MDQ (at 1426 cm^{-1}) being 0.63 μg. Noise levels were also presented at each of these frequencies.

Amateis and Taylor (121, 122) used micro-LC with a 0.2-mm path flow cell FTIR for the analysis of coal-derived nitrogen-containing compounds. By calculating the MDQ for several compounds, they established that the IR absorbance is inversely proportional to flow rate. For quinoline a MDQ of 0.71 μg was obtained at 30 μL/min, but at 60 μL/min a MDQ of 0.80 μg was obtained.

Johnson and Taylor (111) developed a zero-dead-volume (ZDV) FTIR flow cell, which had a circular shape that provided better flow characteristics. This cell was manufactured by drilling a 0.75-mm-diameter hole in a block of either CaF$_2$ or KBr, approximately 10 mm × 10 mm × 6 mm. The micro-LC column was connected directly to the crystal in such a manner that no excessive dead volumes were introduced. The IR beam, with a focal diameter of 3 mm, was—with the help of a beam condenser—reduced to approximately the same diameter as the flow cell (0.075 mm), which it intersected at a 90° angle. The performance of this cell with respect to chromatographic fidelity and spectrometric performance (detection limit) was closely examined and compared with other FTIR flow cells. Four FTIR flow cells were evaluated (two parallel plate designs and two ZDV designs) in terms of path, volume, path/volume, and peak asymmetry caused by the flow cell. The asymmetry factor of the ZDV cell was minimal, but the parallel plate designs had regions in which dead (unswept) volumes could arise and tailing peaks were observed. The pathlength/volume ratio was also far more favorable for the ZDV designs (1.36 vs. 0.062 for the parallel plate designs). The detection limit for 2,6-di-*tert*-butylphenol was 37 ng, which is far better than previously reported (at 3641 cm^{-1}).

REFERENCES

1. M. Novotny, *HRC & CC, J. High Resolut. Chromatogr. Chromatogr. Commun.* **10**, 248 (1987).

2. N. H. C. Cooke, K. Olsen, and B. G. Archer, *LC Mag.* **2**, 514 (1984).

3. B. Ishii, K. Asai, K. Hibi, T. Jonokuchi, and M. Nagaya, *J. Chromatogr.* **144**, 157 (1977).

4. C. I. Baram, M. A. Grachev, N. I. Komarova, M. P. Perelroyzen, Yu. A. Bolvanov, S. V. Kuzmin, V. V. Kargaltsev, and E. A. Kuper, *J. Chromatogr.* **264**, 69 (1983).

5. S. Rokushika, Z. Y. Qiu, Z. L. Sun, and H. Hatano, *J. Chromatogr.* **280**, 69 (1983).

6. Y. Hirata and M. Novotny, *J. Chromatogr.* **186**, 521 (1979).

7. B. Ishii and T. Takeuchi, *J. Chromatogr. Sci.* **18**, 462 (1980).

8. M. Krejci, K. Tesarik, and J. Pajurek, *J. Chromatogr.* **191**, 17 (1980).

9. P. Kucera and G. Guiochon, *J. Chromatogr.* **283**, 1 (1984).

10. J. H. Knox and M. T. Gilbert, *J. Chromatogr.* **186**, 405 (1979).

11. F. J. Yang, *HRC & CC, J. High Resolut. Chromatogr. Chromatogr. Commun.* **4**, 83 (1983).

12. F. J. Yang, *HRC & CC, J. High Resolut. Chromatogr. Chromatogr. Commun.* **3**, 589 (1980).

13. E. J. Guthrie and J. W. Jorgenson, *Anal. Chem.* **56**, 483 (1984).

14. T. Takeuchi and E. S. Yeung, *J. Chromatogr.* **389**, 3 (1987).

15. J. C. Gluckman and M. Novotny, *HRC & CC, J. High Resolut. Chromatogr. Chromatogr. Commun.* **8**, 672 (1985).

16. J. C. Gluckman, D. C. Shelly, and M. V. Movotny, *Anal. Chem.* **57**, 1546 (1985).

17. S. Folestad, L. Johnson, B. Josefsson, and B. Galle. *Proc. Int. Symp. Capillary Chromatogr., 1981 4th*, p. 405 (1981).

18. E. J. Guthrie, J. W. Jorgenson, and P. R. Dluzneski, *J. Chromatogr. Sci.* **22**, 171 (1984).

19. P. R. Dluzneski and J. W. Jorgenson, *HRC & CC, J. High Resolut. Chromatogr. Chromatogr. Commun.* **11**, 332 (1988).

20. H. P. M. van Vliet and H. Poppe, *J. Chromatogr.* **346**, 149 (1985).

21. H. P. M. van Vliet, G. J. M. Bruin, J. C. Kraak, and H. Poppe, *J. Chromatogr.* **363**, 187 (1986).

22. K. Miyaguchi, K. Honda, and K. Imai, *J. Chromatogr.* **316**, 501 (1984).

23. K. Miyaguchi, K. Honda, T. Toyoloka, and K. Imai, *J. Chromatogr.* **352**, 255 (1986).

24. K. Honda, K. Miyaguchi, and K. Imai, *Anal. Chim. Acta* **177**, 111 (1985).

25. C. J. de Jong, N. Lammers, F. J. Spruit, C. Dewaele, and M. Verzele, *Anal Chem.* **59**, 1458 (1987).

26. A. J. Weber and M. L. Grayeski, *Anal. Chem.* **59**, 1452 (1987).

27. P. R. Dluzneski, Ph.D. thesis, University of North Carolina at Chapel Hill, (1987).

28. C. Fujimoto, T. Morita, and K. Jinno, *Chromatographia* **22**, 91 (1986).

29. D. J. Bornhop, T. G. Nolan, and N. J. Dovichi, *J. Chromatogr.* **384**, 181 (1987).

30. B. Lukacs, Ph.D. thesis, University of North Carolina at Chapel Hill (1983).

31. M. J. Sepaniak, J. D. Vargo, C. N. Kettler, and M. P. Maskarinec, *Anal. Chem.* **56**, 1252 (1984).

32. C. N. Kettler and M. J. Sepaniak, *Anal. Chem.* **59**, 1733 (1987).

33. T. G. Nolan, D. J. Bornhop, and N. J. Dovichi, *J. Chromatogr.* **384**, 189 (1987).

34. B. R. Bobbitt and E. S. Yeung, *Anal. Chem.* **56**, 1577 (1984).

35. R. E. Synovec and E. S. Yeung, *Anal. Chem.* **57**, 2606 (1985).

36. Y. Hirata, P. T. Lin, M. Novotny, and R. M. Wightman, *J. Chromatogr.* **181**, 287 (1980).

37. K. Slais and D. Kourilova, *J. Chromatogr.* **258**, 57 (1983).

38. M. Goto, G. Zou, and D. Ishii, *J. Chromatogr.* **268**, 157 (1983).

39. M. Goto, E. Sakurai, and D. Ishii, *J. Liq. Chromatogr.* **6**, 1907 (1983).

40. L. A. Knecht, E. J. Guthrie, and J. W. Jorgenson, *Anal. Chem.* **56**, 479 (1984).

41. R. L. St. Claire, III and J. W. Jorgenson, *J. Chromatogr. Sci.* **23**, 186 (1985).

42. J. G. White, R. L. St. Claire, III, and J. W. Jorgenson, *Anal. Chem.* **58**, 293 (1986).

43. J. G. White and J. W. Jorgenson, *Anal. Chem.* **58**, 2992 (1986).

44. R. T. Kennedy and J. W. Jorgenson, *Anal. Chem.* **61**, 436 (1989).

45. M. D. Oates and J. W. Jorgenson, *Anal. Chem.* **61**, 432 (1989).

46. M. D. Oates and J. W. Jorgenson, *Anal. Chem.* **61**, 1977 (1989).

47. M. D. Oates, B. R. Cooper, and J. W. Jorgenson, *Anal. Chem.* **62**, 1573 (1990).

48. M. D. Oates and J. W. Jorgenson, *Anal. Chem.* **62**, 1577 (1990).

49. R. T. Kennedy, Ph.D. thesis, University of North Carolina at Chapel Hill (1988).

50. M. Goto and K. Shimada, *Chromatographia* **21**, 631 (1986).

51. K. Slais and M. Krejci, *J. Chromatogr.* **235**, 21 (1982).

52. Z. Froebe, K. Richon, and W. Simon, *Chromatographia* **17**, 467 (1983).

53. A. Manz, Z. Froebe, and W. Simon, *J. Chromatogr. Libr.* **30**, 297 (1985).

54. A. Manz and W. Simon, *Anal. Chem.* **59**, 74 (1987).

55. J. S. M. de Wit, and J. W. Jorgenson, *J. Chromatogr.* **411**, 201 (1987).

56. M. Krejci, K. Tesarik, M. Rusek, and J. Pajurek, *J. Chromatogr.* **218**, 167 (1981).

57. M. Krejci, M. Rusek, and J. Houdkova, *Collect. Czech. Chem. Commun.* **48**, 2343 (1983).

58. V. L. McGuffin and M. Novotny, *Anal. Chem.* **53**, 946 (1981).

59. V. L. McGuffin and M. Novotny, *Anal. Chem.* **55**, 2296 (1983).

60. J. C. Gluckman and M. Novotny, *J. Chromatogr.* **314**, 103 (1984).

61. T. Tsuda, A. Nago, G. Nakagawa, and M. Masek, *HRC & CC, J. High Resolut. Chromatogr. Chromatogr. Commun.* **6**, 694 (1983).

62. K. Jinno and H. Tsuchida, *Anal. Lett.* **15**, 427 (1982).

63. K. Jinno, H. Tsuchida, S. Nakanishi, Y. Hirata, and C. Fujimoto, *Appl. Spectrosc.*, **37**, 258 (1983).

64. T. R. Covey, E. D. Lee, A. P. Bruins, and J. D. Henion, *Anal. Chem.* **58**, 1451A (1986).

65. M. L. Vestal in *The Importance of Chemical 'Speciation' in Environmental Processes* M. Bernhard, F. E. Brinckman, and P. J. Sadler, eds., p. 613 Springer-Verlag, Berlin, 1986.

66. W. M. A. Niessen, *Chromatographia* **21**, 277 (1986).

67. W. M. A. Niessen, *Chromatographia* **21**, 342 (1986).

68. J. D. Henion and T. Wachs, *Anal. Chem.* **53**, 1963 (1981).

69. P. J. Arpino, P. Krien, S. Vajta, and G. Devant, *J. Chromatogr.* **203**, 117 (1981).

70. R. Tiebach, W. Blaas, and M. Kellert, *J. Chromatogr.* **323**, 121 (1985).

71. A. P. Bruins and B. F. H. Drenth, *J. Chromatogr.* **271**, 71 (1983).

72. H. Alborn and G. Stenhagen, *J. Chromatogr.* **323**, 47 (1985).

73. H. Alborn and G. Stenhagen, *J. Chromatogr.* **394**, 35 (1987).

74. R. Tijssen, J. P. A. Bleumer, A. L. C. Smit, and M. E. van Kreveld, *J. Chromatogr.* **218**, 137 (1981).

75. J. Arpino and C. Beaugrand, *Int. J. Mass Spectrom. Ion Proc.* **64**, 275 (1985).

76. A. P. Bruins and B. F. H. Drenth, *Int. J. Mass Spectrom. Ion Phys.* **46**, 213 (1983).

77. W. M. A. Niessen and H. Poppe, *J. Chromatogr.* **385**, 1 (1987).

78. A. P. Bruins and B. F. H. Drenth, *J. Chromatogr.* **271**, 71 (1983).

79. J. D. Henion and G. A. Maylin, *Biomed. Mass Spectrom.* **7**, 115 (1980).

80. K. H. Schafer and K. Levsen, *J. Chromatogr.* **206**, 245 (1981).

81. J. D. Henion and T. Wachs, *Anal. Chem.* **53**, 1963 (1981).

82. J. D. Henion, *J. Chromatogr. Sci.* **19**, 57 (1981).

83. P. Krien, G. Devant, and M. Hardy, *J. Chromatogr.* **251**, 129 (1982).

84. F. R. Sugnaux, D. S. Skrabalak, and J. D. Henion, *J. Chromatogr.* **264**, 357 (1983).

85. P. Hirther, H. J. Walther, and P. Datwyler, *J. Chromatogr.* **323**, 89 (1985).

86. E. D. Lee and J. D. Henion, *J. Chromatogr. Sci.* **23**, 253 (1985).

87. M. Novotny, A. Hirose, and D. Wiesler, *Anal. Chem.* **56**, 1243 (1984).

88. T. Tsuda, G. Keller, and H. Stan, *Anal. Chem.* **57**, 2280 (1985).

89. J. S. M. de Wit, C. E. Parker, K. B. Tomer, and J. W. Jorgenson, *Anal. Chem.* **59**, 2400 (1987).

90. T. Tsuda, G. Keller, and H. J. Stan, *Anal. Chem.* **57**, 2280 (1985).

91. W. M. A. Niessen and H. Poppe, *J. Chromatogr.* **323**, 37 (1985).

92. W. M. A. Niessen and H. Poppe, *J. Chromatogr.* **394**, 21 (1987).

93. J. S. M. de Wit, K. B. Tomer, and J. W. Jorgenson, *J. Chromatogr.* **462**, 365 (1989).

94. J. S. M. de Wit, C. E. Parker, K. B. Tomer, and J. W. Jorgenson, *Biomed. Environ. Mass Spectrom.* **17**, 47 (1988).

95. L. Butfering, G. Schmelzeisen-Redeker, and F. W. Rollgen, *J. Chromatogr.* **394**, 109 (1987).

96. C. M. Whitehouse, R. N. Dreyer, M. Yamashita, and J. B. Fenn, *Anal. Chem.* **57**, 675 (1985).

97. A. P. Bruins, T. R. Covey, and J. D. Henion, *34th Annu. Conf. Appl. Mass Spectrom. Allied Top., Cincinnati,* Paper WPB 11 (1986).

98. R. C. Willoughby and R. F. Browner, *Anal. Chem.* **56**, 2626 (1984).

99. R. F. Browner, P. C. Winkler, B. D. Perkins, and L. E. Abbey, *Microchem. J.* **34**, 15 (1986).

100. M. J. Hayes, E. P. Lankmayer, P. Vouros, and B. L. Karger, J. M. McQuire, *Anal. Chem.* **55**, 1745 (1983).

101. M. J. Hayes, H. E. Schwartz, P. Vouros, B. L. Karger, A. D. Thruston, and J. M. McQuire, *Anal. Chem.* **56**, 1229 (1984).

102. A. C. Barefoot and R. W. Reiser, *J. Chromatogr.* **398**, 217 (1987).

103. R. M. Caprioli, T. Fan, and J. S. Cottrell, *Anal. Chem.* **58**, 2949 (1986).

104. J. G. Stroh, K. L. Rinehart, J. C. Cook, T. Kihara M. Suzuki, and T. Arai, *J. Am. Chem. Soc.* **57**, 858 (1986).

105. Y. Ito, T. Takeuchi, D. Ishii, and M. Goto, *J. Chromatogr.* **346**, 161 (1985).

106. A. E. Ashcroft, J. R. Chapman, and J. S. Cottrell, *J. Chromatogr.* **394**, 15 (1987).

107. R. M. Caprioli, W. T. Moore, and T. Fan, *Rapid Commun. Mass Spectrom.* **1**, 15 (1987).

108. R. M. Caprioli and T. Fan, *Biochem. Biophys. Res. Commun.* **3**, 1058 (1986)

109. Y. Ito, T. Takeucki, D. Ishii, M. Goto, and T. Mizuno, *J. Chromatogr.* **358**, 201 (1986).

110. J. S. M. de Wit, L. J. Deterding, M. A. Moseley, K. B. Tomer, and J. W. Jorgenson, *Rapid Commun. Mass Spectrom.* **2**, 100 (1988).

111. C. C. Johnson and L. T. Taylor, *Anal. Chem.* **56**, 2642 (1984).

112. C. M. Conroy, P. R. Griffiths, and K. Jinno, *Anal. Chem.* **57**, 822 (1985).

113. K. Jinno, *Spectrosc. Lett.* **14**, 659 (1981).

114. K. Jinno, C. Fujimoto, and Y. Hirata, *Appl. Spectrosc.* **36**, 67 (1982).

115. C. Fujimoto, K. Jinno, and Y. Hirata, *J. Chromatogr.* **258**, 81 (1983).

116. J. J. Gagel and K. Beimann, *Anal. Chem.* **58**, 2184 (1986).

117. M. Sabbo, J. Gross, J. S. Wang, and I. E. Rosenberg, *Anal. Chem.* **57**, 1822 (1985).

118. N. Teramae and S. Tanaka, *Spectrosc. Lett.* **13**, 117 (1980).

119. R. S. Brown, P. G. Amateis, and L. T. Taylor, *Chromatographia*, **18**, 396 (1984).

120. R. S. Brown and L. T. Taylor, *Anal. Chem.* **55**, 1492 (1983).

121. P. G. Amateis and L. T. Taylor, *LC Mag.* **2**, 854 (1985).

122. P. G. Amateis and L. T. Taylor, *Anal. Chem.* **56**, 966 (1984).

INDEX

(*continued from front*)